Practical Drafting for the HVAC Trades

SECOND EDITION

JOHN E. TRAISTER

PRENTICE-HALL, INC., Englewood Cliffs, New Jersey 07632

Library of Congress Cataloging in Publication Data

Traister, John E.
Practical drafting for the HVAC trades.

Includes index.
1. Heating—Equipment and supplies—Drawings.
2. Ventilation—Equipment and supplies—Drawings. 3. Air conditioning—Equipment and supplies—Drawings.
4. Mechanical drawing. I. Title. II. Title: Practical drafting for the H.V.A.C. trades.
TH7331.T7 1984 697'.0022'1 83-23123
ISBN 0-13-689308-2

Printed in the United States of America

10 9 8 7 6 5 4 3

ISBN 0-13-689308-2

PRENTICE-HALL INTERNATIONAL, INC., *London*
PRENTICE-HALL OF AUSTRALIA PTY. LIMITED, *Sydney*
EDITORA PRENTICE-HALL DO BRASIL, LTDA., *Rio de Janeiro*
PRENTICE-HALL CANADA INC., *Toronto*
PRENTICE-HALL OF INDIA PRIVATE LIMITED, *New Delhi*
PRENTICE-HALL OF JAPAN, INC., *Tokyo*
PRENTICE-HALL OF SOUTHEAST ASIA PTE. LTD., *Singapore*
WHITEHALL BOOKS LIMITED, *Wellington, New Zealand*

Preface

This book is the first complete practical guide to the drafting of heating, ventilating, and air conditioning systems. It contains much of the knowledge and experience gained over a period of many years by the author, other consulting engineers, draftsmen, and many manufacturers in the field. It is tailored to the specific needs of the person responsible for transmitting technical information through the preparation and interpretation of HVAC working drawings; therefore, the first consideration has been the requirements of the draftsmen employed by consulting engineers and mechanical contractors.

If any one word best describes this work, it is the word *practical*.

- It is arranged in a logical order of learning difficulty.
- It is designed so that each student can work on his own job and may proceed with a minimum of assistance according to his own ability.
- It provides a well-rounded course in basic drafting with emphasis on the HVAC trades and is ready for use without further preparation on the part of the instructor.
- It is usable at many educational levels.
- It bridges the gap between theory and problems encountered in actual employment at consulting engineering or mechanical contracting firms.
- It provides guidance in simplified form.

In summary, this book is a quick reference for those actively engaged in the drafting of HVAC systems, a learning method for those studying HVAC drafting and design, and a refresher for those draftsmen with wide experience in the field.

I would like to thank the many manufacturers who generously contributed data and illustrations for this book, and to those draftsmen, engineers, contractors, and my own office personnel who so willingly and enthusiastically contributed their time and experience to this project.

JOHN E. TRAISTER

Preface

[illegible]

Contents

SECTION I—BASIC DRAFTING

SECTION II—DUCT DRAWING

SECTION III—ISOMETRIC DUCT DRAWING

SECTION IV—SECTIONAL DRAWING AND DETAILS

SECTION V—APPENDICES

SECTION I

Basic Drafting

UNIT 1

Introduction to HVAC Drafting

> OBJECTIVES: To define HVAC drafting
>
> To describe construction documents
>
> To describe the HVAC draftsman's job

GENERAL

Engineers, designers, and draftsmen are constantly working on plans for heating, ventilating, and air-conditioning systems in new buildings as well as plans for the renovation of existing buildings. Much of the design of such systems is accomplished by the use of mathematics, good judgment, and drafting. Mathematical formulas are used in calculating the heat loss and heat gain of various structures, the required volume of air, the sizes of ducts needed to carry this air, and similar details. Good judgment is gained from experience and is used to determine the type of fuel used for a particular building and to select the best HVAC (*H*eating, *V*entilating, and *A*ir *C*onditioning) system and equipment. Drafting is the means of using lines, symbols, dimensions, and

Fig. 1-1. Two HVAC student draftsmen.

notations to convey the engineer's calculations and design to the workmen on the job.

Drafting is really a language—an instrument of communication. Therefore, a potential draftsman must master this language so that his drawings will show a complete description of the finished product and so that the workmen can understand what is expected of them.

PREREQUISITES FOR BEING A DRAFTSMAN

Draftsmen are made, not born, but there are a few personal characteristics that, by the very nature of the work, a person should have to be a successful draftsman regardless of his previous training or background. These characteristics include:

1. An orderly mind
2. Imagination and ability to visualize
3. Ability to concentrate
4. Patience
5. Neatness
6. Reasonable amount of mathematical ability

Any potential draftsman who lacks the preceeding background should take steps to improve himself. Two HVAC student draftsmen are shown in Fig. 1-1.

BUILDING CONSTRUCTION

Much of the material in this unit deals with the relationship of the HVAC draftsman to the building construction industry. Familiarity with this relationship is considered necessary to give the draftsman a proper background for approaching his work more intelligently.

Relation of the Architect to the Job

Whenever the construction of a new building is contemplated, an architectural firm is usually commissioned to prepare complete working drawings and specifications for the building. These construction documents usually include the architectural drawings which show the design and building construction details—the floor-plan layouts, vertical elevations of all building exteriors, various cross sections of the building, and other details of construction. While the number of drawings will vary depending on the size and complexity of the job, the drawings will almost always fall into the following five general groups:

1. *Site Work*—The site or plot plan usually will include the location of the building on the property as well as the outside utilities which will serve the building. Topography lines are sometimes included on the site plan also, especially when the building site is on a slope.
2. *Architectural*—The architectural drawings usually will include: elevations of all exterior faces of the building; floor plans showing walls and partitions for each floor; and sufficient cross sections to indicate clearly the various floor levels and details of the foundation, walls, floors, ceilings, and roof construction. Large-scale detail drawings may also be included.
3. *Structural*—Structural drawings are usually included for long-span wood-truss construction, and all drawings of reinforced-concrete and structural-steel construction are prepared by structural consulting engineers.
4. *Electrical*—The electrical drawings will cover the complete design and layout of electrical wiring systems for light, power, and communication.
5. *Mechanical*—The mechanical drawings will cover the complete design and layout of the plumbing, piping, heating, ventilating, and air-conditioning systems and related mechanical construction. Electrical-control wiring diagrams for the heating and cooling systems are very often included on the mechanical drawings also.

In general, the HVAC drawings will indicate the location of all HVAC equipment; the supply, makeup, and return air ducts; the fans for ventilation; the size and location of grilles, registers, and diffusers; and any other items falling under the HVAC specifications. The drawings will also include equipment schedules and large-scale construction details, when they are required. On very small jobs the architect may merely show a one-line diagram of the ductwork on the architectural floor plans with the required volume of air and British thermal units (Btu) to each area. The ductwork is usually sized, in this case, by the mechanical contractor installing the job.

Fig. 1-2 shows an illustration of a good HVAC drawing, while Fig. 1-3 shows an HVAC drawing which is very sketchy, leaving much doubt as to what exactly is wanted.

Relation of the Consulting Engineer to the Job

When electrical and mechanical systems in buildings became more extensive and complex, architects began hiring consulting engineers to design and lay out these systems. The consulting engineer usually will act as liasion between the architect and the electrical and mechanical contractors; the consulting engineer will handle the details of the electrical and mechanical construction from the time the design and layout of the work is started, through the bidding and construction sequences, to the final approval and acceptance of the finished job.

Although consulting firms vary in both size and types of services offered, most have an electrical, a plumbing, and a heating, ventilating, and air-conditioning design department. It is in the latter that most HVAC draftsmen are employed.

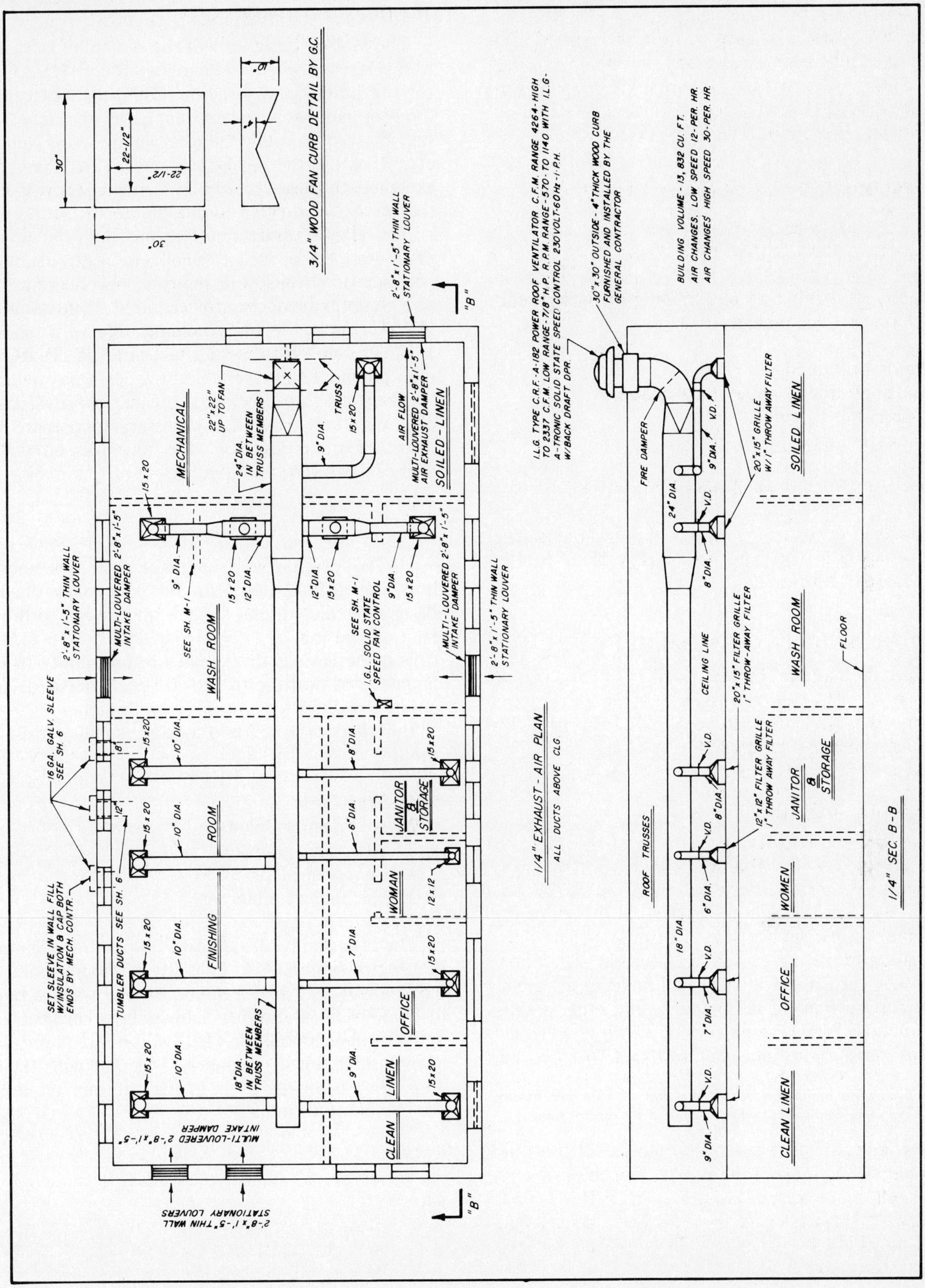

Fig. 1-2. Illustration of a good HVAC drawing, clearly showing all details of construction.

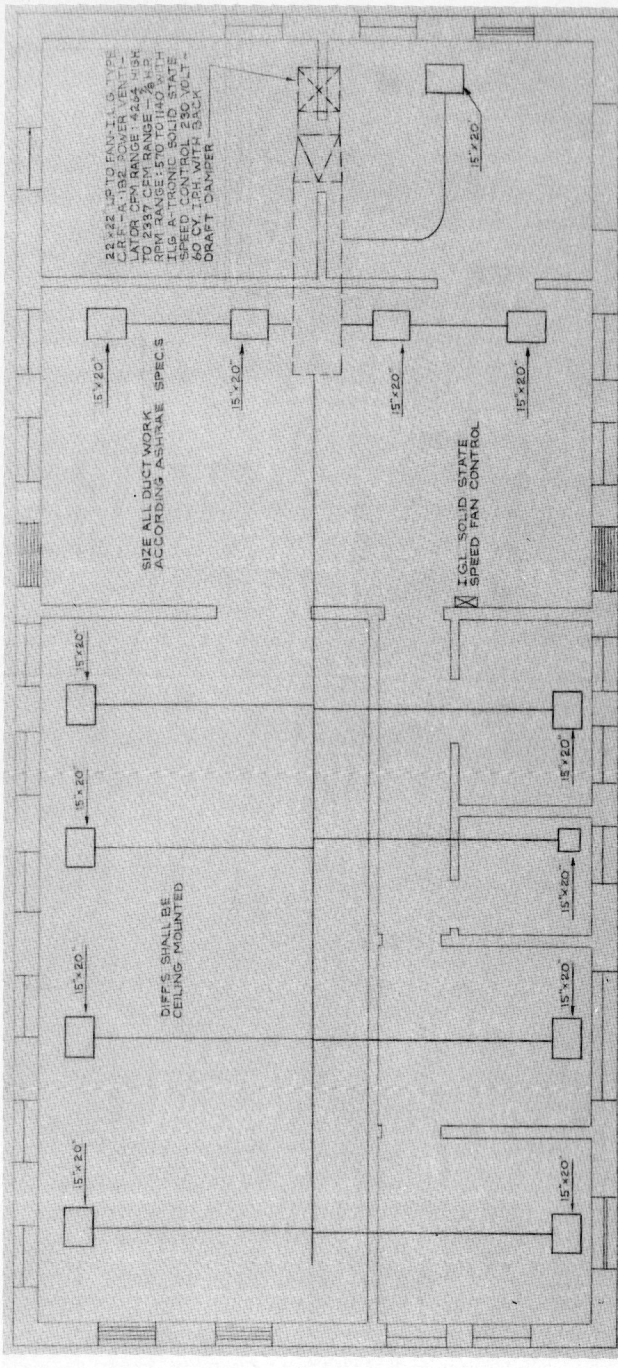

Fig. 1-3. Illustration of another HVAC drawing of only fair quality, leaving some doubt about what is required for construction.

The HVAC Draftsman

The HVAC designer will use the architect's drawings as a reference when designing a suitable HVAC system for the building in question. This involves calculating the heat gain for air conditioning, the heat loss for heating, the required air changes per hour for ventilation, etc. Usually the designer's layout is then roughly sketched as shown in Fig. 1-4 and, in turn, is given to the HVAC draftsman to complete.

The HVAC draftsman's responsibility is to provide drawings of the design which are neat, detailed, and accurate and which will indicate, beyond any question of a doubt, exactly what is required to install a correct HVAC system in the building. Fig. 1-5 shows the engineer's sketch from Fig. 1-4, after an HVAC draftsman completed his work.

The units to follow are designed to provide the student with the knowledge necessary to prepare HVAC drawings of high quality. More examples of such drawings can be found in Appendix D of this book.

SUMMARY

HVAC drafting is a language that uses lines, symbols, dimensions, and notations to convey the HVAC designer's calculations, layout, and design to the workmen on the job.

It is the HVAC draftsman's responsibility to use the language of drafting to provide neat, detailed, and accurate drawings for workmen on the job.

Building construction drawings usually consist of five groups:

1. Site plans
2. Architectural plans
3. Structural plans
4. Electrical plans
5. Mechanical plans

The HVAC layout is shown on the drawings through the use of symbols and lines which should indicate, beyond any question of a doubt, exactly what is required.

The specifications for a building or project are the written description of what will be required by the owner or architect/engineer. Together with the drawings, they form the basis of the contract requirements for the construction.

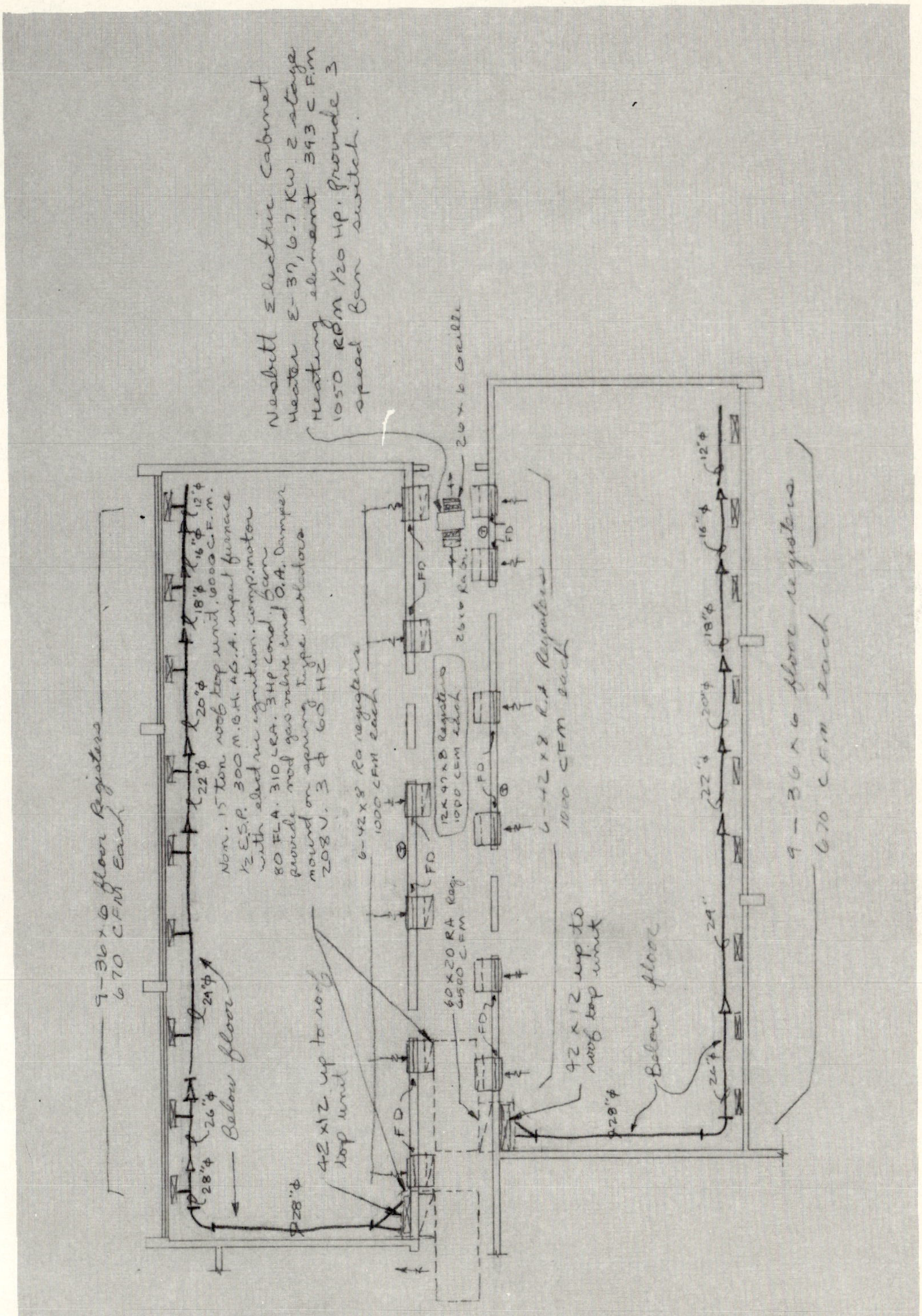

Fig. 1-4. Illustration of an engineer's rough sketch before drafting.

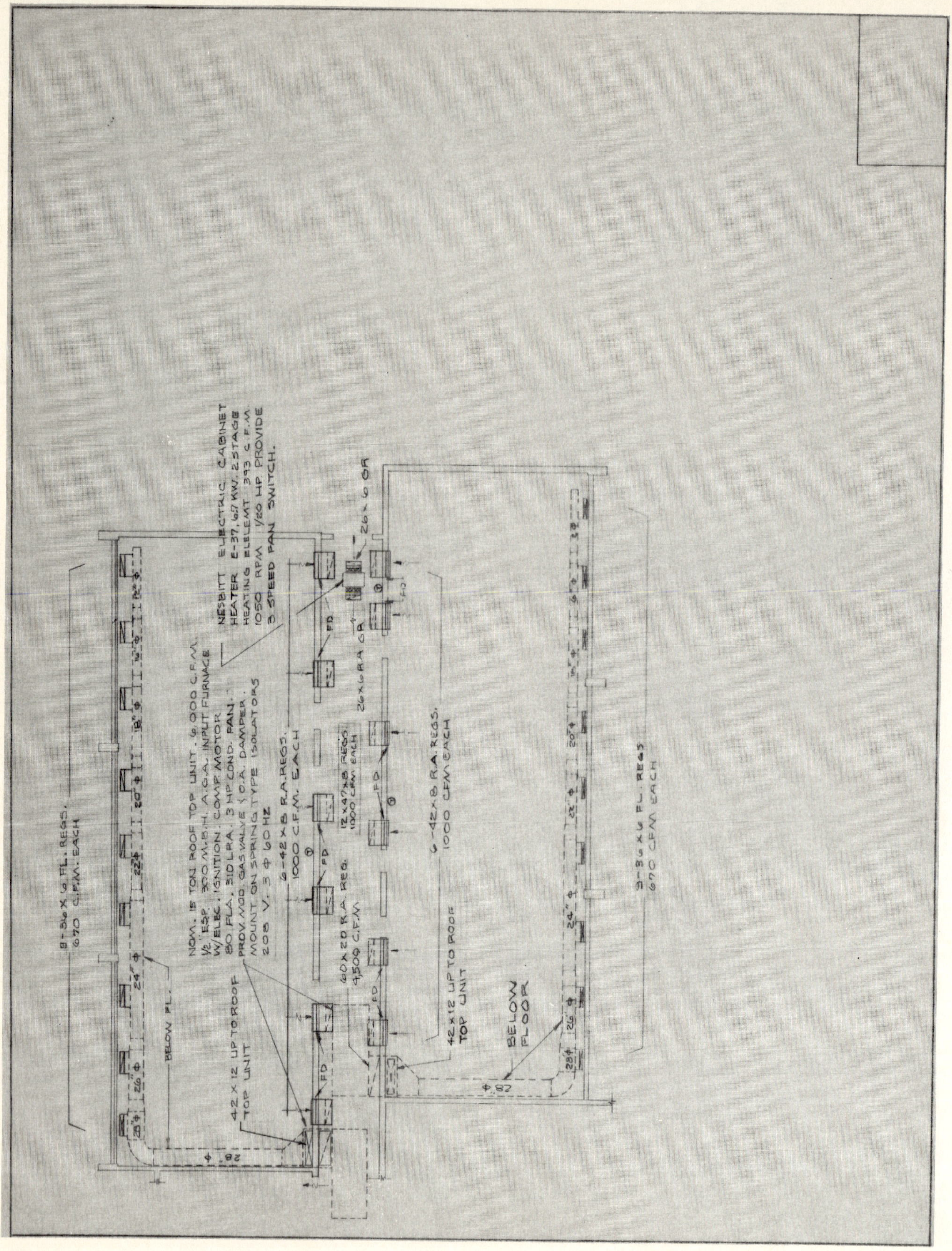

Fig. 1-5. The same engineer's sketch after it has been neatly drawn by an HVAC draftsman.

UNIT 1 — ASSIGNMENT

Read Unit 1 before completing this assignment. Check the correct answer for the following True or False questions. If any part of the statement is false, the statement should be marked "False." If the answer is not known to any particular question, it is recommended that you read the entire unit through again rather than hunting for the answer in the text. Then, after the second reading, you may search through the paragraphs for the correct answer if you are still in doubt.

	True	False
1. The term *HVAC* is an abbreviation for "Heating, Ventilation, and Air Currents."	_____	_____
2. One of the personal characteristics a person should have to be a successful draftsman is the ability to visualize.	_____	_____
3. On most construction jobs involving a new building, an architectural firm is commissioned to prepare complete working drawings and specifications.	_____	_____
4. HVAC drawings fall under the mechanical portion of building plans and specifications.	_____	_____
5. Drawings prepared by HVAC draftsmen should be neat and detailed, but accuracy is of little importance.	_____	_____
6. Information concerning the HVAC system in a building is shown on working drawings by means of lines, symbols, and written notes.	_____	_____
7. In most cases, working drawings are required even on simple buildings.	_____	_____
8. According to Unit 1, building construction drawings fall into five general categories.	_____	_____
9. Services most often offered by consulting engineering firms include electrical, HVAC, plumbing, and structural engineering.	_____	_____
10. A reasonable amount of mathematical ability may be considered an attribute for any HVAC draftsman.	_____	_____

UNIT 2

Tools of the HVAC Draftsman

OBJECTIVES: To learn the use of the tools needed for HVAC drafting

To familiarize the draftsman with proper line work for HVAC drawings

GENERAL

High-quality work in any trade or profession can be accomplished only by the correct use of high-quality tools. The drafting profession is no exception. Neatness and accuracy are the two most important qualities of any drawing, and these qualities can only be secured by the use of good-quality tools in first-class condition.

The equipment commonly used by the HVAC draftsman is listed in the next paragraph. This equipment list covers the minimum essential tools needed to do good work on the usual kind of HVAC drawings. The equipment should be of the best quality that the draftsman can afford.

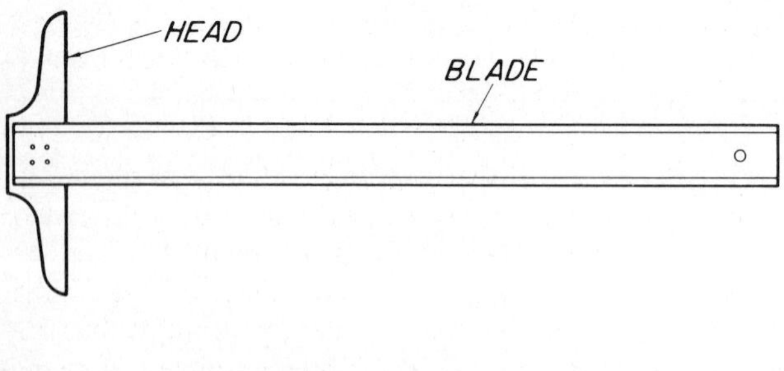

Fig. 2-1. T-square.

EQUIPMENT FOR HVAC DRAFTING WORK

Drawing Board	Assorted Lead
T-square	Lead Pointer
Drafting Machine	Drafting Tape
Architect's Scale	Eraser
Engineer's Scale	Erasing Shield
30-60° Triangle	Lettering Guide
45° Triangle	Various Templates
Protractor	Dust Brush
Set of Instruments	Tracing Paper
Irregular Curves	Dry-Clean Pad
Lead Holders	

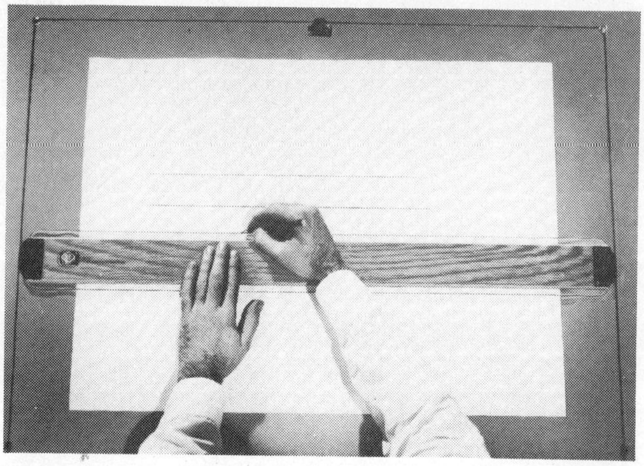

Courtesy Keuffel & Esser Co.

Fig. 2-2. Parallel straightedge.

Drawing Board—Drawing boards vary greatly in size, from very small boards to large vertical boards 6 or 7 feet high and several feet long. Regardless of size, the drawing board should have a smooth flat surface on which the drawing paper is placed. It should be made of soft pine or bass wood and should be at least 18 inches wide by 24 inches long for the student. For

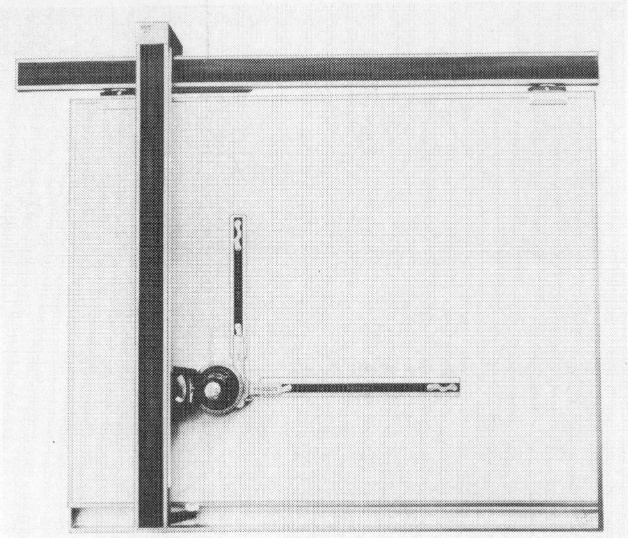

Courtesy Keuffel & Esser Co.

Fig. 2-3. Auto Flow Mark II drafting machine.

professional application, a board 31 inches wide by 42 inches long should be considered as minimum since most construction drawings are made on 24-by-36-inch tracing paper.

The left edge of the drawing board (or the right, if the draftsman is left-handed) must be perfectly straight so that it can serve as an accurate guide for the T-square. A cover for the drawing board is also recommended.

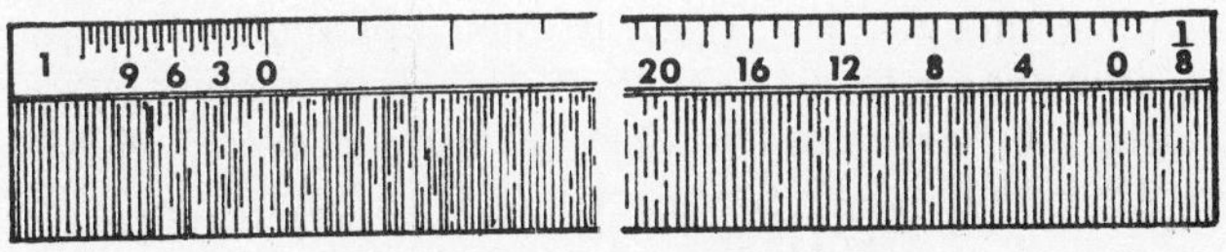

(A) 1″ = 1′ – 0″. (B) ⅛″ = 1′ – 0″.

Fig. 2-4. Typical graduations on an architect's scale.

T-Square—The common T-square (Fig. 2-1) is shaped like the letter **T** and consists of two parts called the blade and the head. The head of the T-square is always held against the left working edge of the board by right-handed draftsmen, and against the right edge of the board by left-handed draftsmen. Horizontal lines are drawn along the upper edge of the blade from left to right (right to left for left-handed draftsmen). The length of the T-square should be enough to cover the entire length of the drawing board.

In most professional drafting rooms, a parallel edge, such as the one in Fig. 2-2, is used instead of a T-square. The parallel edge is used just like the T-square except that the ends of the parallel edge are permanently attached to the board by a wire-cable arrangement. These wires, passing through pulleys within the parallel edge,

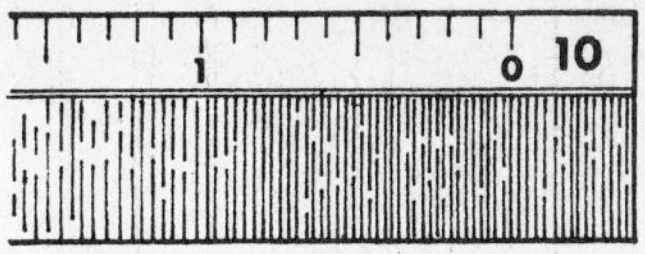

Fig. 2-5. Engineer's scale with ten graduations per inch.

keep the edge parallel at all times as it is moved up and down the board.

Drafting Machine—The drafting machine (Fig. 2-3) is also used in some offices to take the place of the T-square/parallel edge, triangles, and scales. While the draftsman may save much time with this machine, he must learn to use the basic tools first, then decide for himself if he desires to use the drafting machine—many do not.

Architect's Scale—The architect's scale (Fig. 2-4) is used for measuring. The basic unit at the end of the scale represents one foot and is subdivided into 12 parts to represent inches. On the larger scales, such as 1 inch = 1 foot, the inch division is further subdivided so that the smallest subdivision may represent ¼ of an inch. On the smaller scales, such as ⅛ inch = 1 foot, the basic unit is not divided into as many divisions. The smallest division on the ⅛-inch scale represents 2 inches. The use of the architect's scale will be fully explained in Unit 4.

Engineer's Scale—The engineer's scale (Fig. 2-5) is similar to the architect's scale except that an inch is divided into 10, 20, 30, 40, 50, and 60 parts. This scale is useful for map drawings or building-site plans. The engineer's scale marked 10 is read 1 inch = 100

(A) 45° triangle. (B) 30-60° triangle.

Fig. 2-6. Triangles.

feet or some other integral power of 10, such as 1 inch = 1000 feet, etc. The use of the engineer's scale will be fully explained in Unit 21.

Triangles—Triangles are the instruments used by the draftsman in connection with the T-square or parallel edge to draw both vertical lines and sloping lines. One has angles of 90, 60, and 30 degrees and is called the 30-60 degree triangle. The minimum size of this triangle should be 10 inches. The 45-degree triangle has one 90-degree angle and two 45-degree angles. The 45-degree triangle should be at least 6 inches long. Both

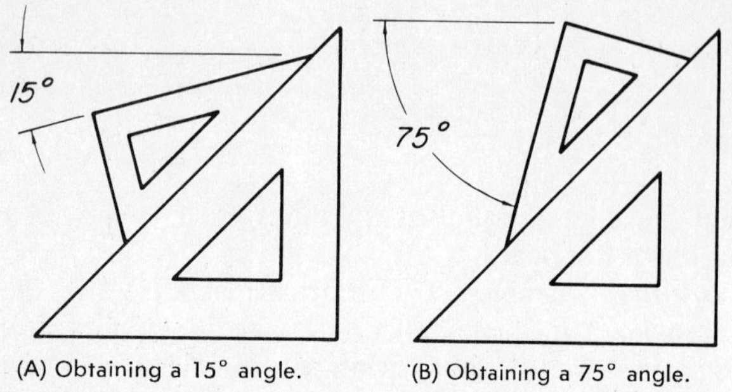

(A) Obtaining a 15° angle. (B) Obtaining a 75° angle.

Fig. 2-7. Combining two triangles to obtain other angles.

types of triangles (Fig. 2-6) are held against the upper edge of the T-square or parallel edge, and lines are drawn from the bottom toward the top.

By using both triangles in combination, angles of other degrees can be drawn. Two such combinations are illustrated in Fig. 2-7.

The adjustable triangle, illustrated in Fig. 2-8, is becoming more popular and can save the draftsman much time. However, the student should become familiar with the use of conventional drafting tools before attempting to use the various "shortcut" devices.

Fig. 2-9 illustrates the simultaneous use of the drawing board, T-square, and triangles. Note how the drawing paper is taped to the board.

Protractor—The protractor is a device for laying off

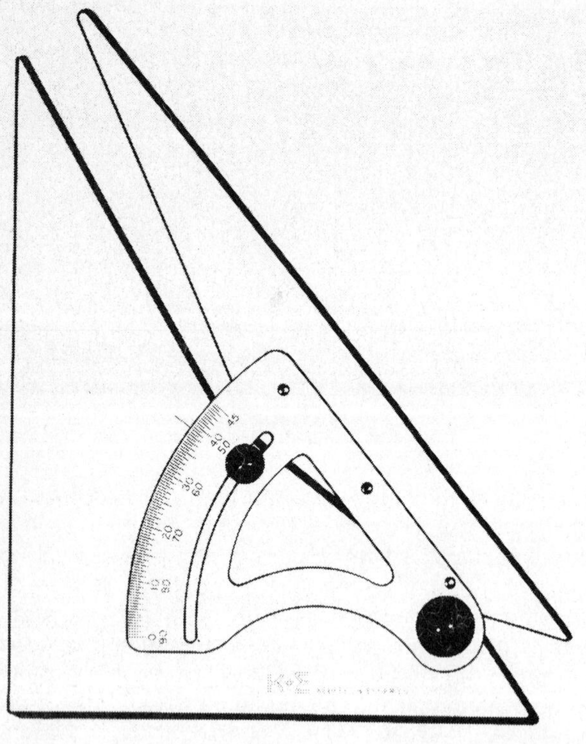

Courtesy Keuffel & Esser Co.

Fig. 2-8. Adjustable triangle.

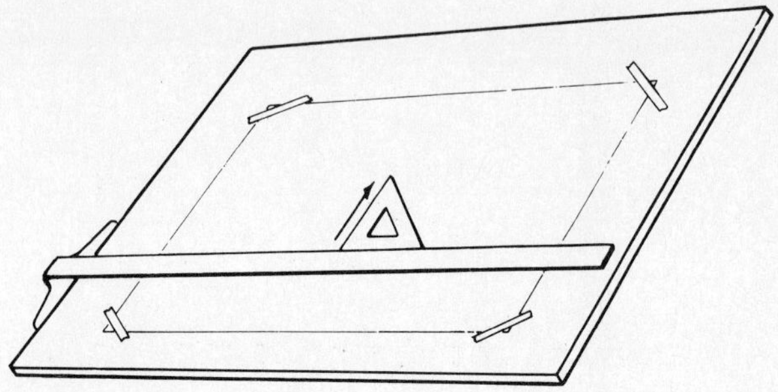

Fig. 2-9. Simultaneous use of the drawing board, T-square, and triangles.

and measuring angles on drawings. The best form for HVAC drafting consists of a full circle of plastic graduated in degrees or fractions of a degree.

Fig. 2-10 illustrates the most common form of circular celluloid protractor. To lay off an angle, the center (0) of the protractor is placed at the vertex of the angle, with the edge of the bar coinciding with the line to which the angle is referred. A light mark is then made on the drawing at the proper graduation of the protractor arc. After the protractor is removed, a line is drawn, joining this mark with the vertex. Angles can be measured in a similar manner.

Set of Instruments—Many draftsmen possess a set of drawing instruments, although plastic templates are rapidly taking their place. A typical set consists of a compass used for drawing arcs and circles, a divider used for laying out or measuring distances, and a beam compass used for the precise drawing of arcs or circles having radii greater than 6 inches. It is recommended that the HVAC draftsman not purchase a set of drawing instruments until the situation dictates.

Fig. 2-11 illustrates a set of good drawing instruments which are inexpensive. This set will handle 99 percent of all HVAC drafting requiring the use of tools to draw arcs and circles.

Irregular Curves—Irregular curves will not be used too frequently in HVAC drafting, but the draftsman should have a knowledge of their use. Irregular curves are used for ruling curved lines other than circle arcs and are the most difficult drafting instruments to use skillfully. Fig. 2-12 shows two types of irregular curves which are useful to the HVAC draftsman.

Lead Holders—Mechanical lead holders (Fig. 2-13) have all but replaced wooden pencils in most drafting rooms. They are supplied in various shapes and sizes and are made from metal, plastic, or various combinations of the two. Most lead holders are operated by pressing a button at the top of the holder which instantly projects the lead, permitting a quick adjustment to any desired length. Upon release of the button, the holder automatically grips the lead firmly and securely in any desired position. The lead is then pointed with a mechanical lead pointer or with a sandpaper board.

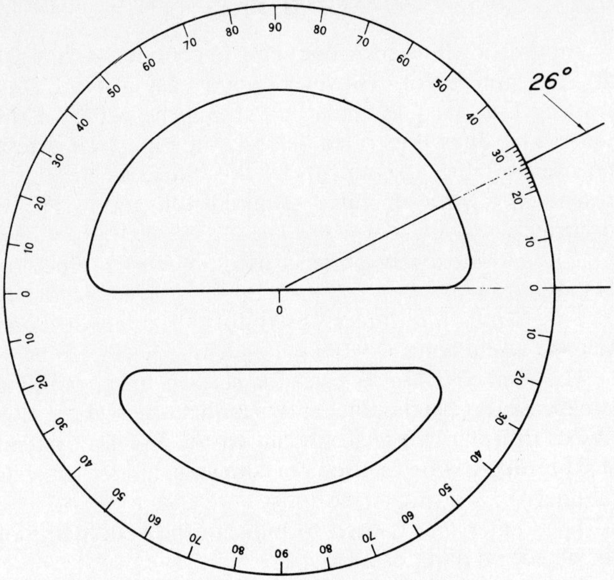

Fig. 2-10. A circular celluloid protractor.

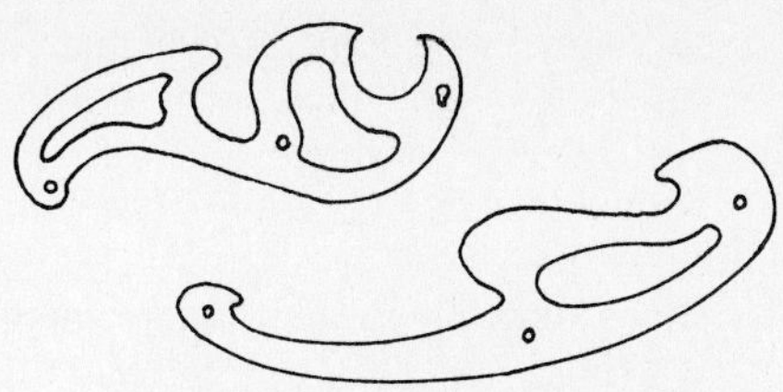

Fig. 2-12. Two types of irregular curves.

Fig. 2-13. A typical lead holder.

Fig. 2-14. Close-up of lead in lead holder, showing conical-shaped point.

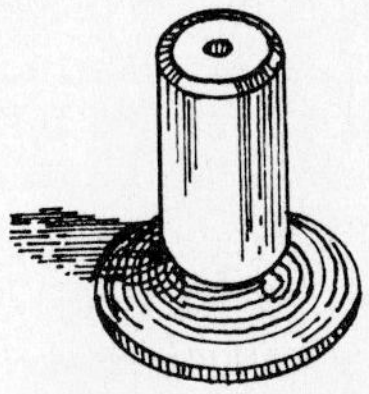

Fig. 2-15. Portable lead pointer.

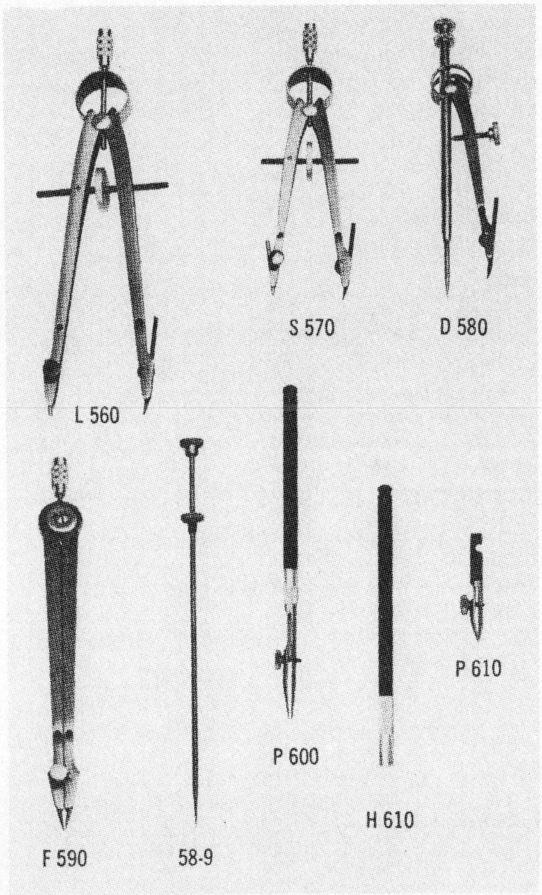

Fig. 2-11. A set of drawing instruments.

Fig. 2-16. A roll of drafting tape.

The HVAC draftsman should obtain a minimum of two lead holders.

Lead—Drawing lead is graded according to its degree of hardness, which ranges from 7B, the softest and blackest, to 9H, the hardest. Although practically all manufacturers use the same system of marking their lead, wide variations in hardness will be found between different manufacturers. It is for this reason that most draftsmen usually prefer to use only one brand of lead. The following hardnesses are recommended for HVAC drafting.

6H—For guidelines and light sketch lines.
4H—For medium-weight lines such as building outlines for HVAC drawings.
2H—For drawing HVAC equipment, ducts, and similar items.
H—For sharp dark lines, such as border lines, and for lettering.

A conical-shaped point is preferred for all HVAC drawing and lettering (Fig. 2-14).

Lead Pointer—A lead pointer is used to point refill drawing lead used in mechanical lead holders. Three types are in general use: the portable type, the desktop type, and the clamp-on type. The portable type, as illustrated in Fig. 2-15, is ideal for the HVAC draftsman.

Drafting Tape—Thumbtacks and staples have been used for many years to hold the drawing paper in place on the board. Both damage the board as well as the drawing paper, and they obstruct the movement of the T-square also.

Masking or drafting tape is the preferred method of securing drawing paper to the board. Merely stick small pieces of tape across each corner of the paper. Drafting tape (Fig. 2-16) will not damage either the board or the paper and will offer little or no obstruction to the T-square or parallel edge.

Eraser—Since we all make mistakes, the eraser needs little explanation. Use soft or medium-soft pencil erasers with beveled edges. (Fig. 2-17).

Erasing Shield—When there are many lines close together, and only one line needs to be erased, the others can be protected by using an erasing shield, such as the one illustrated in Fig. 2-18. Erasing shields are thin metal plates with various-shaped openings in them. The shield is placed on the drawing so that the line to be erased appears in one of the openings. Only

Fig. 2-17. Medium-soft pencil eraser with beveled edges.

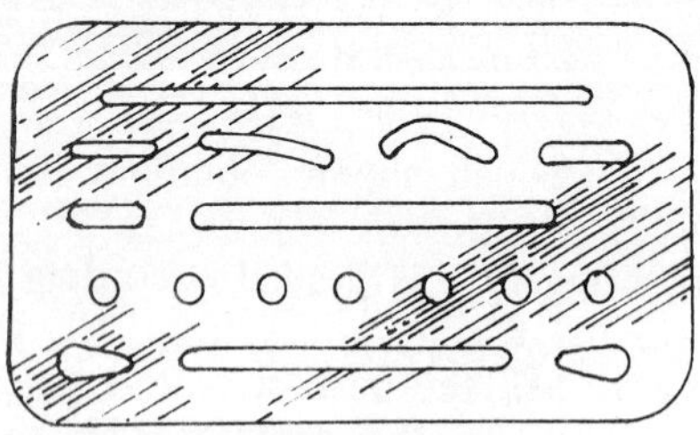

Fig. 2-18. Erasing shield.

the line in this opening will be erased because the thin metal shield will protect any lines near the one removed.

DRAFTING LINES

All drafting lines have one thing in common—they are all the same color. However, good easy-to-read contrasting lines can be made by varying the width of the lines. The lines illustrated full size in Fig. 2-19 are to be used for the assignments in the book, as they represent the types of lines required for good HVAC drafting.

The *extraheavy full line* is used for border lines on HVAC drawings.

The *heavy full line* is used for all ductwork and HVAC equipment.

The *light full line* is used for section lines, building backgrounds (outlines), and similar uses where the object to be drawn is secondary to the HVAC system.

The *dotted line* is used for showing hidden objects or lines.

The *center line* is used to indicate the center line of an object or parts of an object.

The *cutting plane line* is used to show the position of the cutting plane.

The *light break line,* which is drawn partly with a straightedge and partly with freehand zigzags, is used for long breaks.

The *heavy break line*—drawn freehand—is used for short breaks.

Further explanations on the use of these lines will be given with the assignment in each unit.

SUMMARY

High-quality tools and the ability to use them correctly are necessary for any type of high-quality work.

The student draftsman should master the use of conventional drafting tools before attempting to use the various "shortcut" devices.

All drafting lines are the same color; that is, all are black. Therefore, contrast can be obtained only by

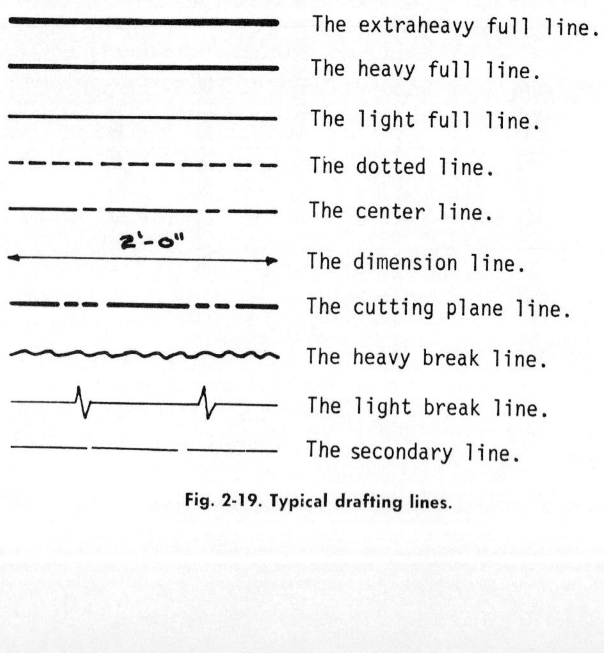

Fig. 2-19. Typical drafting lines.

varying the width of the line and by using lead of different degrees of hardness.

The draftsman's tools should be of the best quality he can afford.

A conical-shaped point on the drawing lead is preferred for all HVAC drawing and lettering.

UNIT 2 — ASSIGNMENT

1. Clean the dust from your drafting board with your dust brush. Then place a sheet of 11-by-17-inch tracing paper on the board. Adjust the paper so that its edge is parallel to the upper edge of the T-square. Holding the paper in this position, fasten the corners of the paper to the board with small strips of drafting tape. Your board should now look similar to the one shown in Fig. 2–22.

   Measure ¼ inch in from all sides of the paper, making a very light dot to indicate these measurements. Now draw a border line completely around the paper, as shown in Fig. 2–23. Use the extra-heavy full line for this border.
2. When you have completed the steps in No. 1 above, look at the various lines in Fig. 2–19. Beginning one inch down from the top border line of your drawing sheet, begin drawing the various lines shown in Fig. 2–19—in the order shown; that is, begin with the extra-heavy full line, then drop below that about ¼ inch and draw the heavy full line, drop ¼ inch below this line and draw the light full line, etc., until all of the lines in Fig. 2–19 have been drawn. You can start the line about one inch from the left-hand border line and end about one inch from the right-hand border line.

   Once one complete set of these lines has been drawn, drop down from the last (the secondary line) and begin a new series as before. Continue drawing the lines until you are about one inch up from the bottom border line. Then stop.

   In drawing the above lines, the following suggestions will help you to make quality lines:

   a. Use 2-H lead (H lead may be used for the extra-heavy line).
   b. Keep the lead sharp at all times.
   c. Tilt the pencil about 60 degrees from the paper in the same direction in which the line is being drawn (Fig. 2–20). Also, tilt the pencil about 15 degrees away from the edge of the T-square or triangle, as shown in Fig. 2–21.
   d. Rotate the pencil slowly in your hand as the line is being drawn. This will keep the lines uniform in width. Failure to rotate the pencil will cause the lead to become flat on one side, resulting in a nonuniform and "fuzzy" line.
   e. Use from three to six strokes for each visible line.

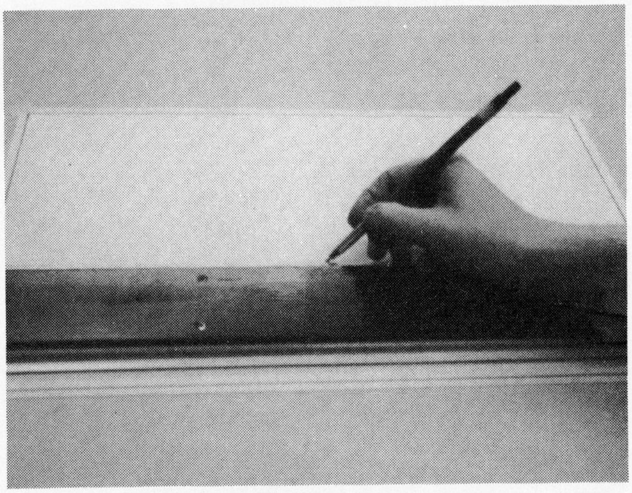

Fig. 2-20. Tilt pencil about 60° from paper.

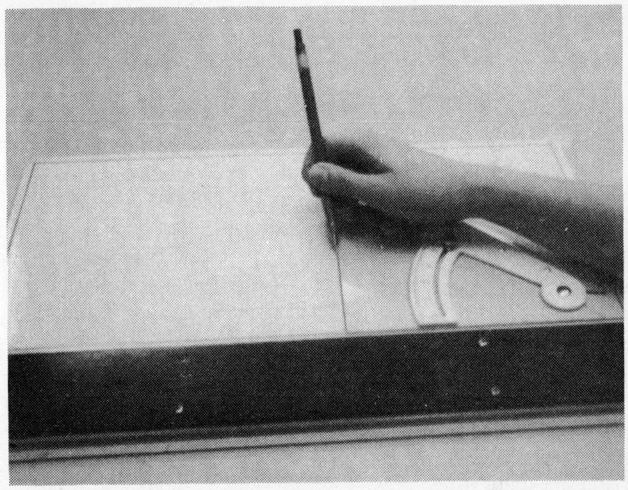

Fig. 2-21. Tilt pencil about 15° from triangle.

Fig. 2-22. Attaching paper to drawing board.

Fig. 2-23. Paper in Fig. 2-22 with border line added.

3. Remove the first drawing sheet and secure another 11-by-17-inch sheet of tracing paper on the board, after thoroughly cleaning it. Draw border lines around this sheet, as you did previously, using the following suggestions:
   a. Use 2-H lead (H is also used) and draw a sharp glossy jet-black line by making at least 6 strokes back and forth.
   b. Rotate your pencil at all times.
   c. Move your straightedge slightly away from the line which was just drawn—not enough to make two separate lines, but just enough to make the first line wider.
   d. Widen the first line by drawing another line, using six more strokes. If done properly, you will have a neat, extra-heavy solid line.

   Remove the sheet from the board, being careful not to fold or damage the sheet as you will need it for future assignments.
4. Take another sheet of 11-by-17-inch tracing paper and repeat No. 3 of the assignment. This is necessary because you will need these for the assignment in the next Unit.

UNIT 3

Lettering

OBJECTIVES: To teach the shape of the different letters and numerals

To show the proper sequence and direction of the strokes

To learn the use of the lettering guide

GENERAL

The appearance of any drawing depends to a great extent upon the appearance of the lettering on the drawing. Figures, dimensions, and explanatory notes are used extensively on practically all HVAC drawings, and it is essential that they be lettered on the drawing in a neat manner.

THIS IS AN EXAMPLE OF VERTICAL LETTERING.

(A) Vertical lettering.

THIS IS AN EXAMPLE OF INCLINED LETTERING.

(B) Inclined lettering.

Fig. 3-1. Single-stroke Gothic letters.

Lettering is used instead of script (writing) on drawings because it is more legible; therefore, script writing is not permissible.

There are several styles of lettering in use, but for general use on HVAC drawings, the single-stroke Gothic letters are the most popular. This type of lettering may be either vertical or inclined (Fig. 3-1). However, this book will give instructions for forming the vertical single-stroke Gothic letters only.

CHARACTERISTICS OF GOOD LETTERING

In order that the lettering on a drawing may have a neat appearance, the letters must be well formed, must have good relative widths, and must be spaced so that the distances between them appear to be uniform. Methods of forming all letters and numerals are listed in Fig. 3-2.

Single-stroke lettering does not mean that an entire letter is formed with a single stroke; rather, it means that a line of the required thickness is made with one stroke. The direction of the strokes are indicated by the small arrows in Fig. 3-2, and the order in which the various strokes should be made is shown by the numbers placed near the arrows. Where there is no number, as in the case of the letter *I*, the entire letter is formed in one stroke.

GUIDELINES

Before you begin practicing the art of lettering, you should learn to make guidelines and use them whenever you are lettering on drawings. Such guidelines enable draftsmen to form neat and professional-looking letters and numbers at all times.

AMES LETTERING GUIDE

The lettering guide illustrated in Fig. 3-3 is a drafting aid that enables the draftsman to draw evenly spaced parallel guidelines for lettering and numbering require-

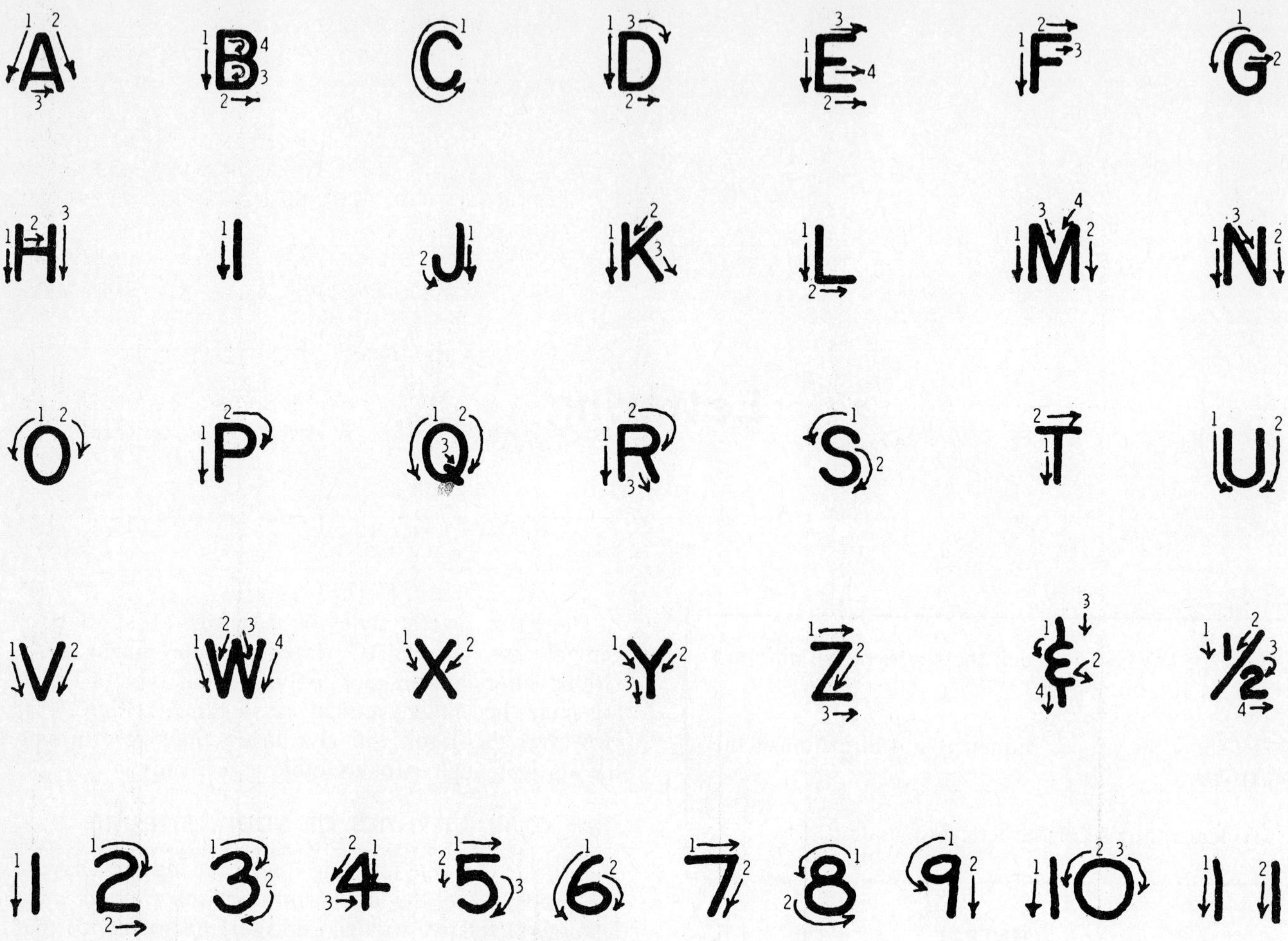

Fig. 3-2. Method of forming letters and numerals.

ments on working drawings. This type of guide is referred to as an Ames lettering guide. It is used by sliding it along the T-square with the point of a pencil placed in the proper hole.

With the Ames lettering guide it is possible to draw guidelines and slope lines for letters and numbers from $\frac{1}{16}$ inch to 2 inches in height.

The numbers from 10 through 2 on the disc denote the height of letters in thirty-seconds of an inch. Assume you want letters ¼ inch high. Rotate the disc so that the 8 is at the frame index, as shown in Fig. 3-4.

To turn the disc clockwise, hold the frame at the upper left-hand corner with the left hand. Rotate the disc with the right hand. To turn the disc counterclockwise, use the reverse procedure.

Use 6H or harder drawing lead for all guidelines. To use the guide, position it with the readable side up on the straightedge, as shown in Fig. 3-4. Insert a conical-shaped lead into the proper hole, and with the lead holder perpendicular to the drawing paper, slide the lettering guide across the drawing paper at the location desired for guidelines. Make certain at all times that the guide frame rests flat on the straightedge. The pencil lead is then moved from hole to hole with the column to produce a series of guidelines for lettering.

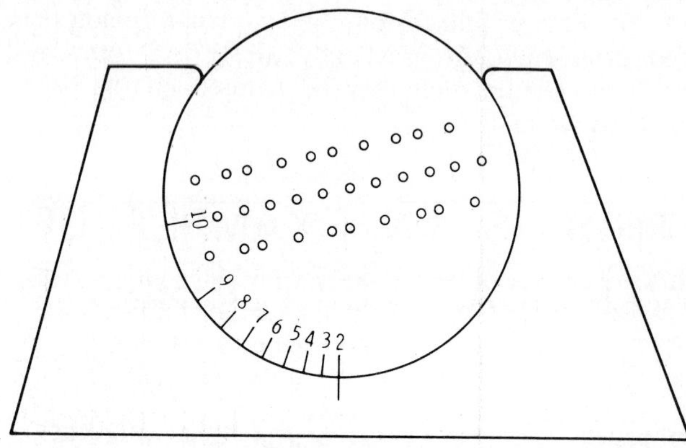

Fig. 3-3. Ames lettering guide.

If the number 10 is placed on the center mark, the holes in the disc form three vertical columns. The center column is the one used to make guidelines for all lettering requirements in this book. This column may

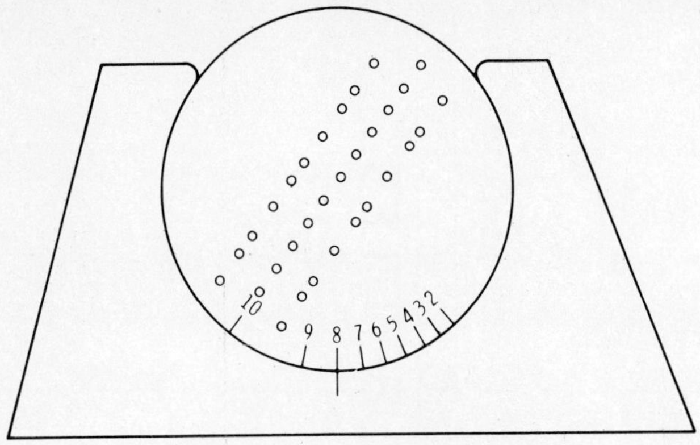

Fig. 3-4. Ames lettering guide with disc set at No. 8.

be called the ½-ratio column because all of the lines are equally spaced. When using this column for guidelines, start with the top hole and then use every second hole as shown in Fig. 3-5. For multiple lines, this gives the same distance between lines as the height of the letter. However, in most HVAC drafting the spaces between the letters are made half the height of the letter, as shown in Fig. 3-6.

Figs. 3-7 through 3-11 show the recommended adjustments of the lettering guide for various applications.

USE OF TRIANGLE IN LETTERING

The use of a triangle in lettering will help the beginning student to form the vertical strokes uniformly.

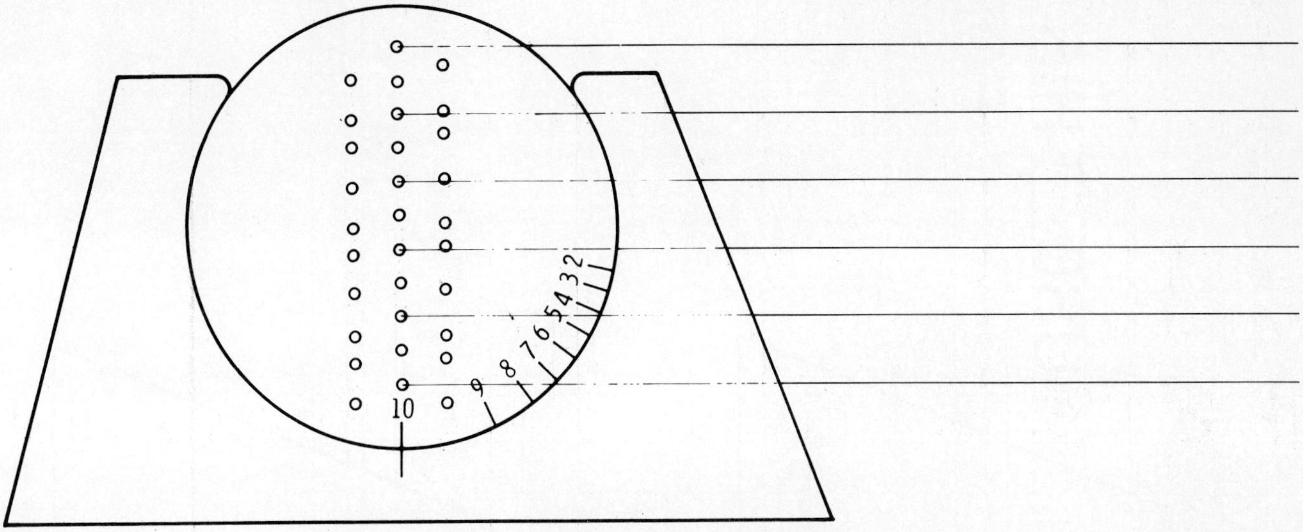

Fig. 3-5. Lettering guide set at No. 16 showing method of drawing guidelines; equal spaced.

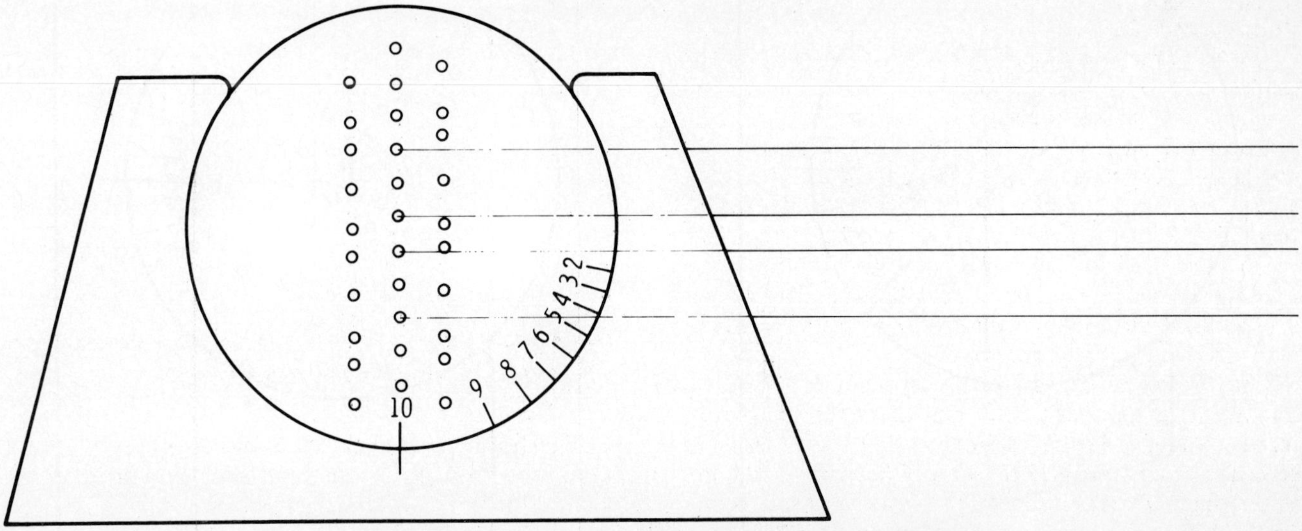

Fig. 3-6. Lettering guide set at No. 10 showing method of drawing guidelines when spaces between lines are to be half the height of the letters.

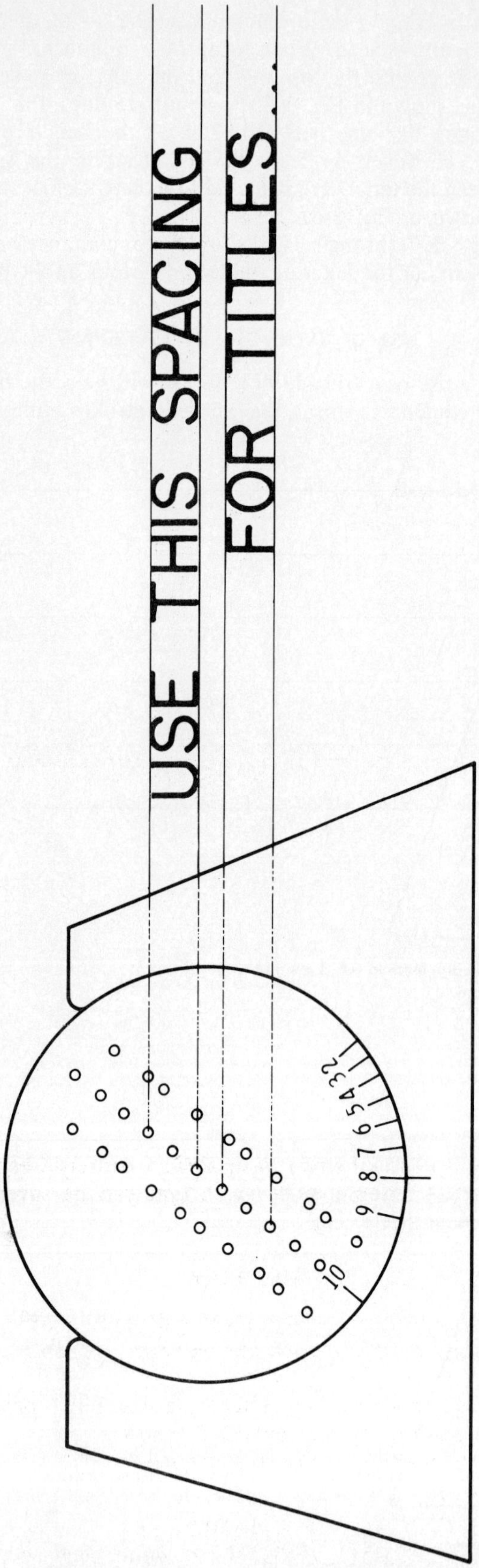

Fig. 3-7. Lettering guide adjustment for sheet titles.

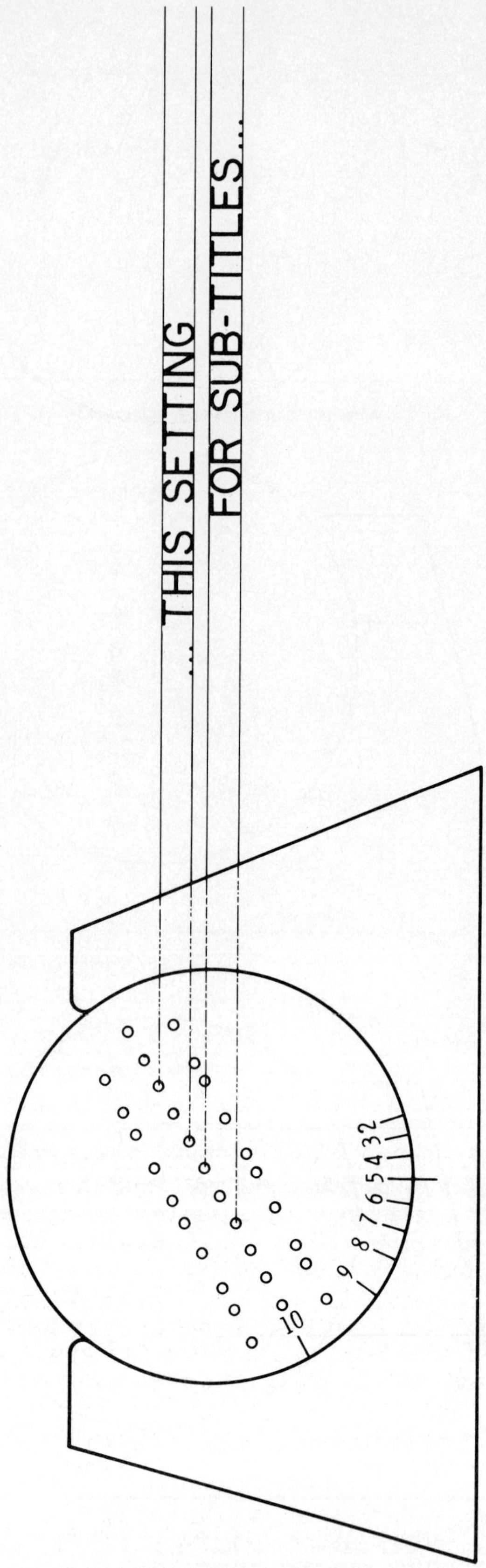

Fig. 3-8. Lettering guide adjustment for subtitles.

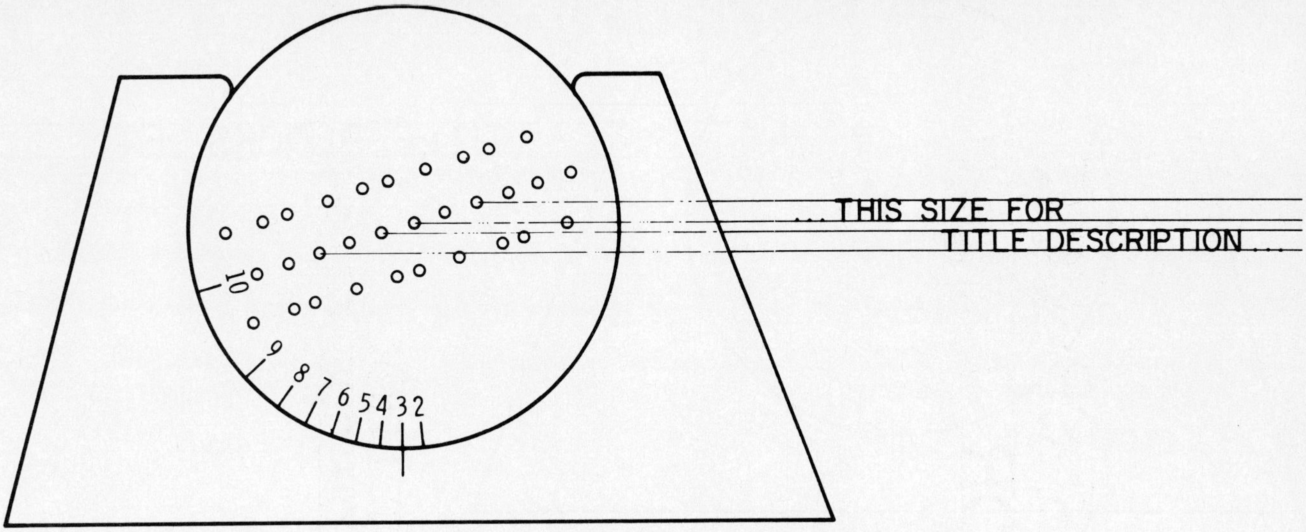

Fig. 3-9. Lettering guide adjustment for title descriptions.

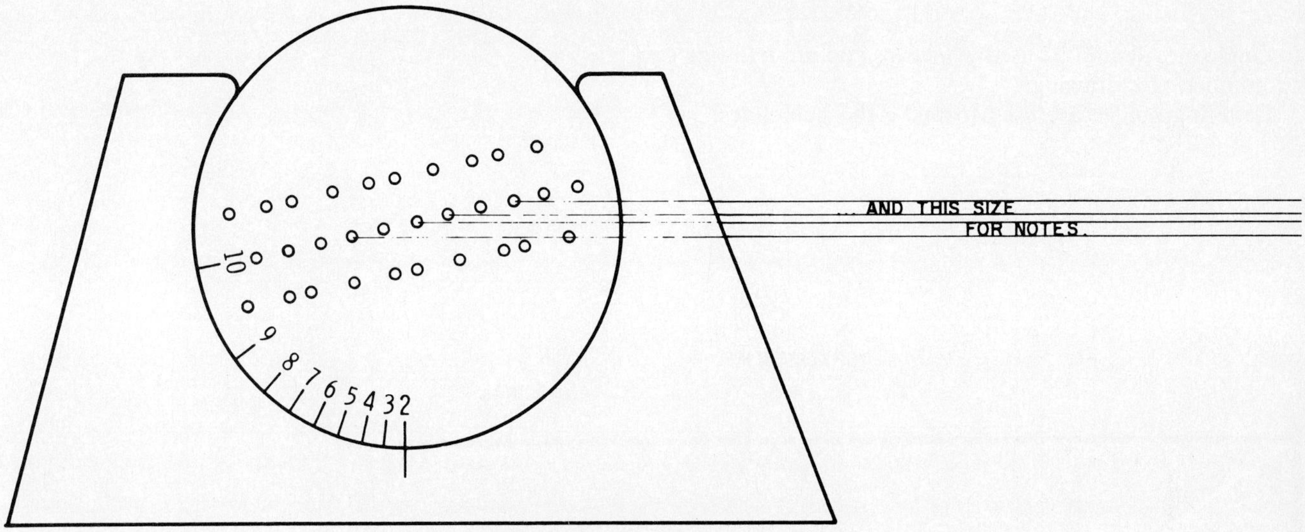

Fig. 3-10. Lettering guide adjustment for notes or equipment identification.

For example, if the word "LETTER" is to be lettered, first draw the guidelines with a 6H lead. Then with the use of a straightedge and a 90-degree triangle, make the vertical stroke of the letter *L;* then draw freehand the horizontal stroke of the letter *L.* Again form the vertical stroke of the letter *E* with the straightedge and triangle; then form the three remaining strokes of the letter freehand. The two vertical strokes of the two *Ts* are made with the 90-degree triangle, while the horizontal tops of the *Ts* are made freehand. The next letter *E* is made as before. The letter *R* is formed by using the triangle to form the vertical stroke and then drawing freehand the remaining strokes.

The lettering method described previously is a great help in making uniform vertical strokes which, in turn, help to maintain a neat appearance. However, the draftsman should strive to master an entirely freehand method of lettering since much time can be saved by all freehand lettering.

SUMMARY

The appearance of any drawing depends a great deal upon the appearance of the lettering and numbering on the sheet.

Single-stroke Gothic letters are the most popular style in use today for HVAC drawings.

Neat lettering requires that the letters

1. Be well formed.
2. Have good relative widths.
3. Be spaced so that the distance between them appears to be uniform.

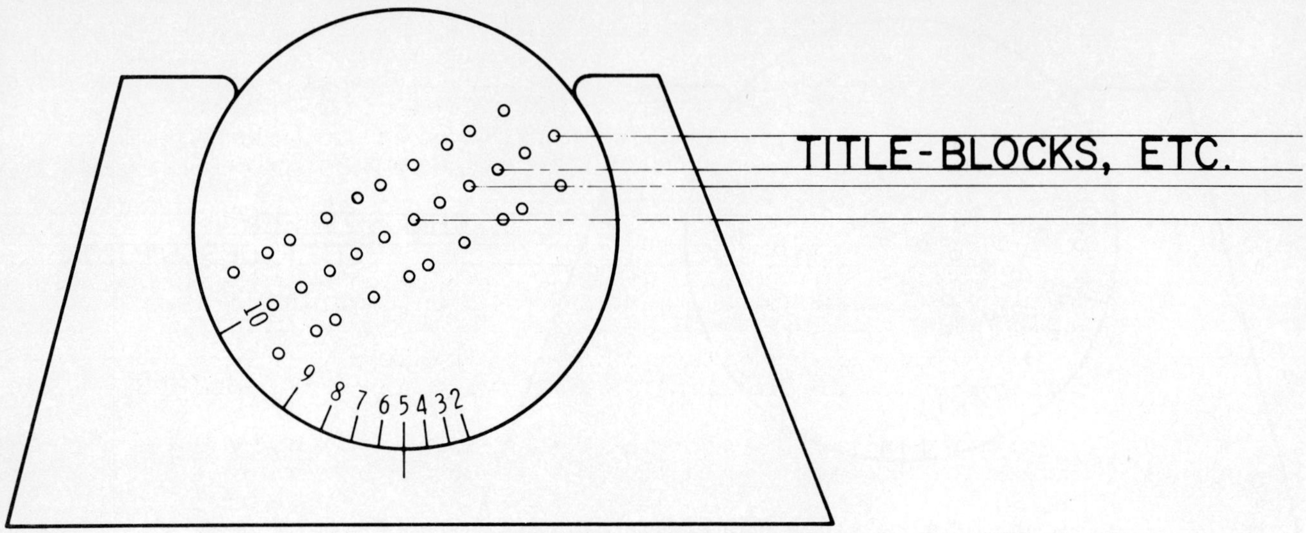

Fig. 3-11. Lettering guide adjustment for title blocks, etc.

Guidelines should be used whenever you are lettering or numbering a drawing.

Lettering guides are used to make the guidelines.

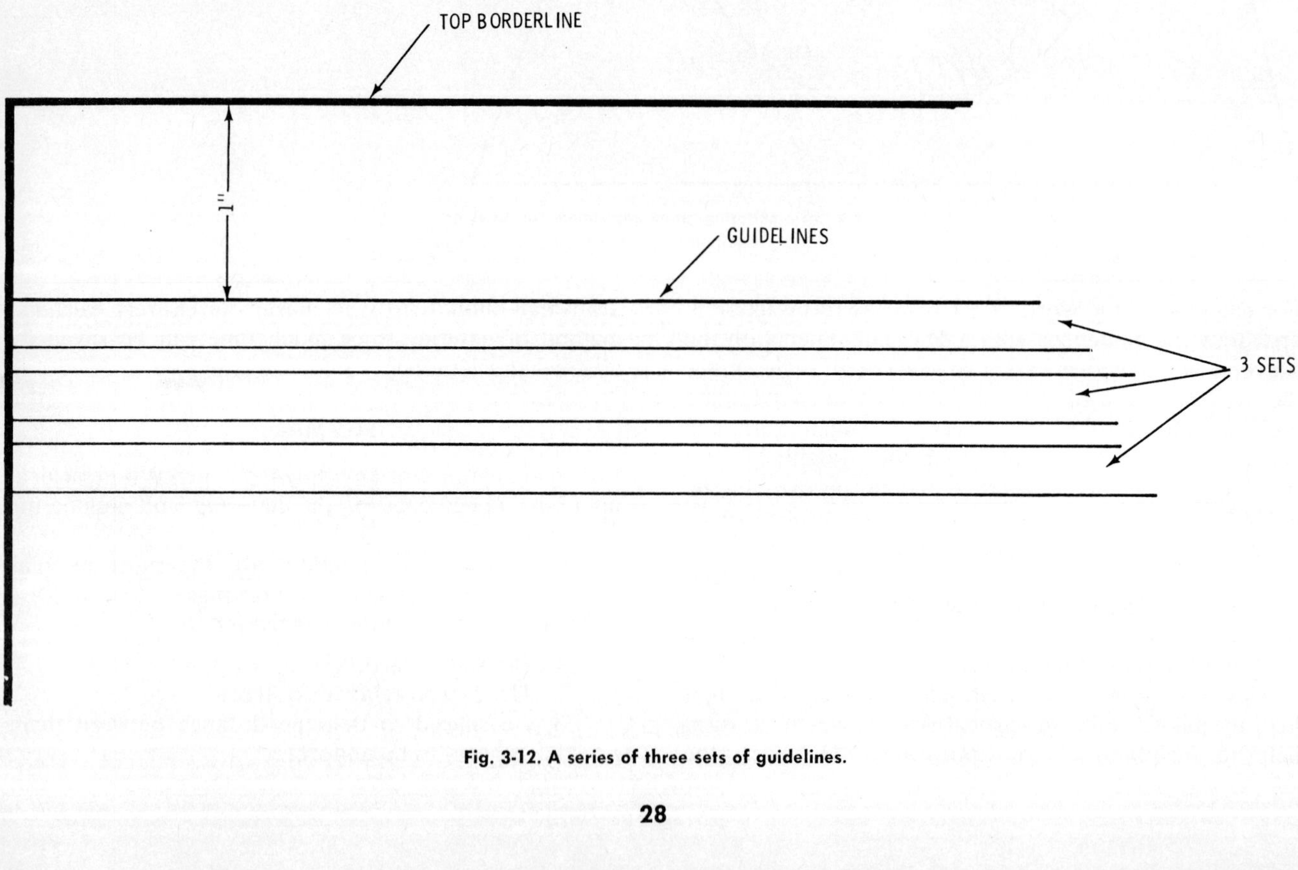

Fig. 3-12. A series of three sets of guidelines.

UNIT 3 — ASSIGNMENT

1. Secure to the drafting board one of the sheets of tracing paper (with border lines) that you prepared in your last assignment. Set your Ames lettering guide at the number 8 on the center mark. Beginning 1 inch below the top border line of the tracing paper, draw a series of three sets of guidelines as shown in Fig. 3–12. Use 6-H lead. Draw these guidelines from the left border to the right border.

Measure down 1 inch from the bottom guideline; make a light dot. Draw a series of three sets of guidelines as shown in Fig. 3–7.

Measure down 1 inch from the bottom guideline in the series just made; readjust the Ames guide to the number 5; then make another series of three sets of guidelines as shown in Fig. 3–8.

Measure down 1 inch from the bottom guideline in the series just made; readjust the Ames guide to the number 4; and make another series of three sets of guidelines as shown in Fig. 3–9.

Measure down 1 inch from the bottom guideline in the series just made; readjust the Ames guide to the number 3; make another series of three sets of guidelines as shown in Fig. 3–10.

Measure down 1 inch from the bottom guideline in the series just made; readjust the Ames guide to the number 2; make another series of three sets of guidelines as shown in Fig. 3–11.

The sheet of tracing paper should now look similar to the one shown in Fig. 3–13.

Use your triangle to make one very light vertical line ½ inch from the left border and ½ inch from the right border. Both lines should extend from the top border to the bottom border.

2. Across the top set of guidelines (the ¼-inch-high ones), letter the alphabet and numbers from 1 to 10, over and over until all the space is used up on these three sets of guidelines.

Go to the next set of guidelines and letter the alphabet and numbers from 1 to 10 until all the space is used up in this set. Continue on to each set until all guideline space is used up.

Fig. 3-13. Example of finished sheet for the assignment in this unit.

UNIT 4

Scale Drawings: Floor Plans

> OBJECTIVES: To explain the method of laying out floor plans
>
> To demonstrate the use of the architect's scale
>
> To discuss the purpose of scale drawings

THE WORKING DRAWINGS

Unit 1 discussed how the architect/engineer conveys ideas to the workmen by the use of drawings and written specifications. The written specifications give details of the type or grade of material, the class of workmanship, and a general description of the work required from the contractor and his workmen on the job. The drawings usually will include floor plans, elevation views, sections, and details. Together these documents must give all details of construction so that the workmen will know exactly what is required of them in constructing the building and the various systems within the building.

ORTHOGRAPHIC PROJECTION OF A BUILDING

An orthographic-projection drawing of a building would consist of the views illustrated in Fig. 4-1.

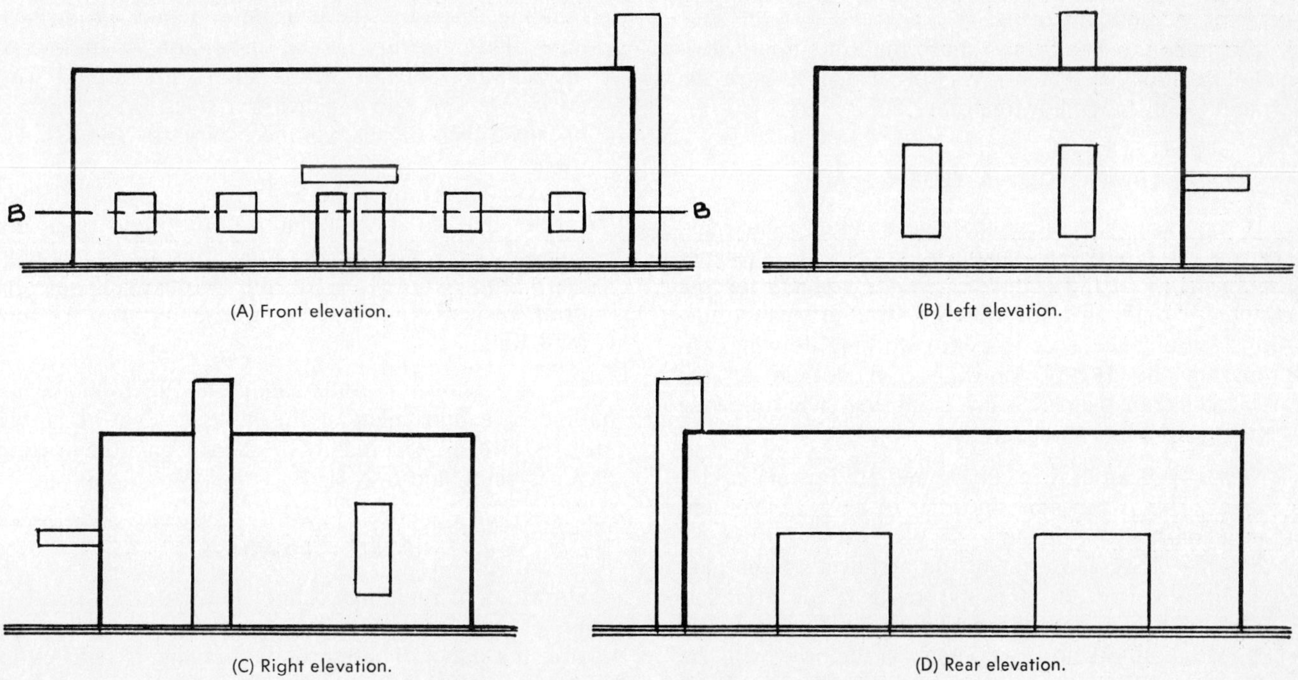

(A) Front elevation. (B) Left elevation. (C) Right elevation. (D) Rear elevation.

Fig. 4-1. Orthographic projection of a building.

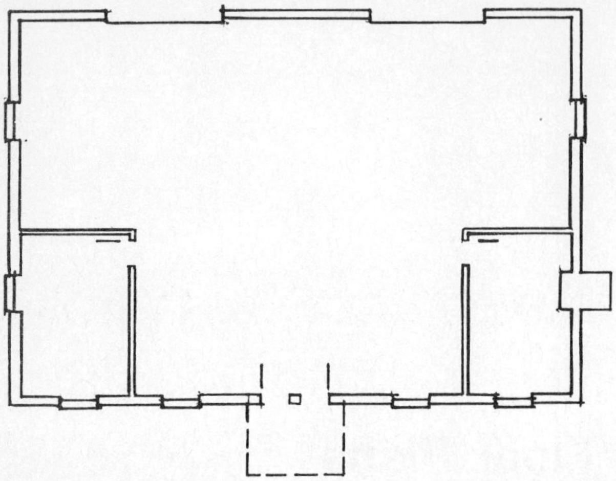

Fig. 4-2. Floor plan of building shown in Fig. 4-1.

Fig. 4-3. Draftsman preparing to trace a blueprint.

One drawing shows the building as though the observer were looking straight at the front; one as though the observer were looking straight at the left side; one as though the observer were looking straight at the right side; and one as though the observer were looking straight at the rear of the building. These drawings are called building elevations.

Referring again to Fig. 4-1, assume that a horizontal cut is made through the building along a line already indicated as B-B on the front elevation. Then imagine that the top part is removed and a drawing is then made looking straight down at the remaining part. This drawing is known as a floor plan. It indicates the outside wall lines, interior partitions, windows, doors, and similar details. See Fig. 4-2 for a drawing of the floor plan for the building in Fig. 4-1.

If a building should have more than one floor, horizontal cuts may be taken at varying distances from the ground in order to show the other floors.

LAYING OUT A FLOOR PLAN

It should now be clear that the architect plans and designs buildings; the engineer or HVAC designer usually lays out the complete HVAC systems for the architect's buildings; and the HVAC draftsman transforms the engineer's designs into working drawings. In preparing the HVAC drawings, an outline of the architect's floor plan(s) is layed out first. The following steps describe this procedure:

Step 1—Clean drawing board and all drawing equipment. This is necessary in order to have a clean and neat finished drawing.

Step 2—Take the print of the architect's floor plan and place it on the drawing board. If the print has been folded, rub out the creases as much as possible. Now line up the horizontal lines on the print with the T-square. When the print and the T-square are lined up perfectly, secure the print to the board with small pieces of drafting tape.

Step 3—Place a sheet of tracing paper over the print, line them up, and secure with drafting tape.

Step 4—Insert 4H lead into the lead holder. The light solid line will be used throughout for the floor plan. On architectural drawings, the outline of the floor plan usually will be drawn with the heavy solid line, but in HVAC drawings the HVAC equipment and ductwork are the most important objects. Therefore, the floor-plan outline is secondary and is drawn with a light solid line.

Step 5—With the use of the T-square, trace all necessary horizontal lines on the floor plan of the building. Now use the triangle and trace all vertical lines. This drawing should show just an outline of the building and all the partitions, doors, and windows. Do not show dimension lines, wall finishes, or any other details not necessary for the HVAC work.

Step 6—Check over the tracing very carefully; make certain that no doors, windows, partitions, etc., are omitted. Now remove the tape securing the tracing paper and remove the tracing. Before removing the print, recheck your tracing against the print for any omissions.

Fig. 4-3 shows a draftsman preparing to make an outline of a floor plan of the building. Note that the print is already secured to the board and the tracing paper is being laid over it.

SCALE DRAWINGS

Drawings of buildings cannot be made full size because of the size of buildings as compared to the size of drawing paper. Therefore, the drawing is reduced in size so that all the distances on the drawing are drawn

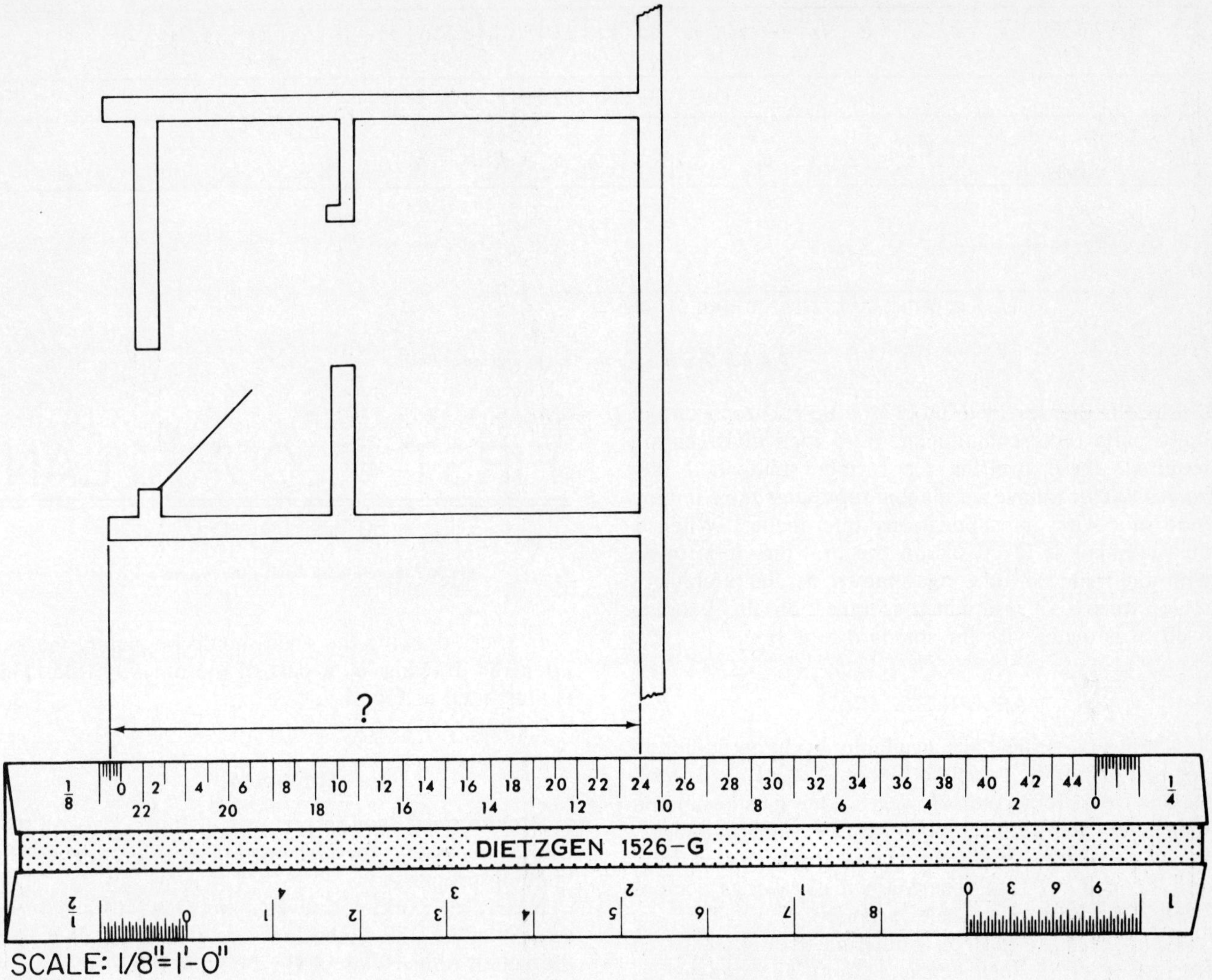

Fig. 4-4. Using the 1/8" architect's scale.

smaller than the actual dimensions of the building, all dimensions being reduced in the same proportion. The ratio, or relation, between the size of the drawing and the size of the building is indicated on the drawing (1/4 inch = 1 foot, for example), and the dimensions are the actual dimensions of the building, not the distance that is measured on the drawing.

The most common method of reducing all the dimensions in the same proportion is to choose a certain distance and let that distance represent 1 foot. This

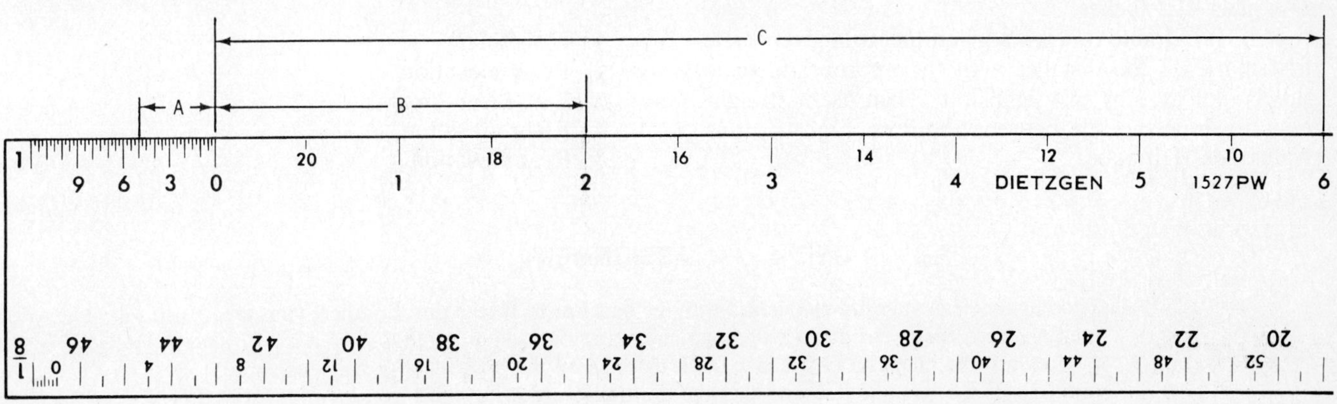

Fig. 4-5. Using the 1" architect's scale.

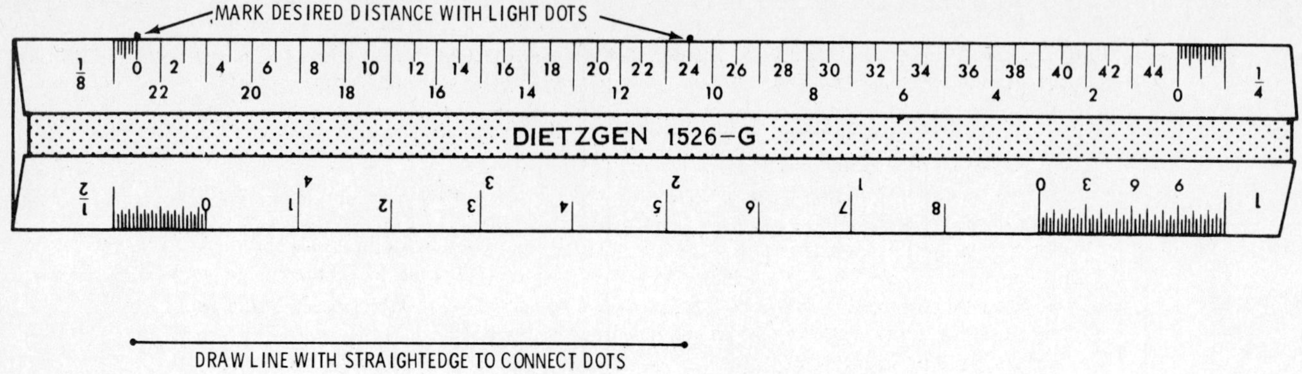

Fig. 4-6. Drawing a given line to a given scale.

distance is then divided into 12 parts, and each one of these parts represents 1 inch. If ½-inch divisions are required, these twelfths are further subdivided into halves. We now have a scale that represents the common foot rule with its subdivisions into inches. When a measurement is layed off on the drawing, it is made with the reduced foot rule known as the architect's scale; when a measurement is taken on the building itself, it is made with the standard foot rule.

ARCHITECT'S SCALE

Architect's scales are available in many different degrees. That is, 1 inch = 1 foot, ¼ inch = 1 foot, ⅛ inch = 1 foot, etc. However, the most common scales for architectural plans are $\frac{1}{16}$ inch = 1 foot, ⅛ inch = 1 foot, ¼ inch = 1 foot, ½ inch = 1 foot, and ¾ inch = 1 foot. The following drawings demonstrate how the various scales are used.

Fig. 4-4 shows part of a building floor plan drawn to a scale of ⅛ inch equals 1 foot. The dimensions in question are found by placing the ⅛-inch architect's scale on the drawing and reading the figures. It can be seen that the dimensions read 24 feet 6 inches.

Fig. 4-5 shows a portion of an architect's scale with the 1 inch = 1 foot side turned up. The dimensions of the various lines are as follows: *A* equals 5 inches; *B* equals 2 feet 4 inches; *C* equals 6 feet 9 inches; and *D* equals 4½ inches.

If it is desirable to draw a given line to a given scale, first mark off the distance with the appropriate scale; this is indicated by two light dots. Then use a straightedge to draw the line between the dots. Fig. 4-6 illustrates this technique.

FIRST FLOOR PLAN

SCALE: 1/4" = 1'-0"

Fig. 4-7. Example of way to show drawing scale.

In every drawing the scale should be plainly marked upon the drawing, as a part of the drawing title. This is illustrated in Fig. 4-7.

SUMMARY

Architectural working drawings must give all the details necessary for the workmen to construct the building properly.

Nearly all working drawings are drawn to a reduced scale; that is, all dimensions are drawn smaller than the actual dimensions of the building.

When a measurement is laid off on the drawing, it is made with the reduced foot rule known as the *architect's scale*.

When a measurement is taken on the building itself, it is made with the stnadard foot rule.

An orthographic-projection drawing of a building consists of

1. Floor plan
2. Front elevation
3. Left elevation
4. Right elevation
5. Rear elevation

UNIT 4 — ASSIGNMENT

1. Thoroughly clean your drawing board with the dust brush. If lead dust or other stains are present, wash the board down with soap and water on a damp cloth.

Refer to the blueprint (Appendix D) in the back of the book. Cut or tear this entire print out of the book and secure it to the drafting board with drafting tape.

2. Using one of the tracing sheets you prepared for the assignment in Unit 2, place it over the print so that the left-hand side of the house is approximately one-inch to the right of the left-hand border on the tracing sheet. Align it and then secure to the board with drafting tape. The top wall of the house in the floor plan should be about ½-inch from the top border line on the tracing paper.

Now trace the building floor plan using 6-H lead and include only the walls, partitions, windows, doors, and the like. Do not include dimensions, room names, plumbing fixtures, appliances, or other items that are not part of the building structure itself.

3. Prepare another sheet of tracing paper with border lines as described in Unit 2. Then, using the same procedure just described in No. 2 of this assignment, trace the building outline of the office building in Appendix D of this book.

4. Give the missing dimensions in the illustration of the architect's scale in Fig. 4–8. *Hint:* The answer to A is 6′-4″.

A. __________ B. __________ C. __________ E. __________ F. __________

G. __________ H. __________ I. __________

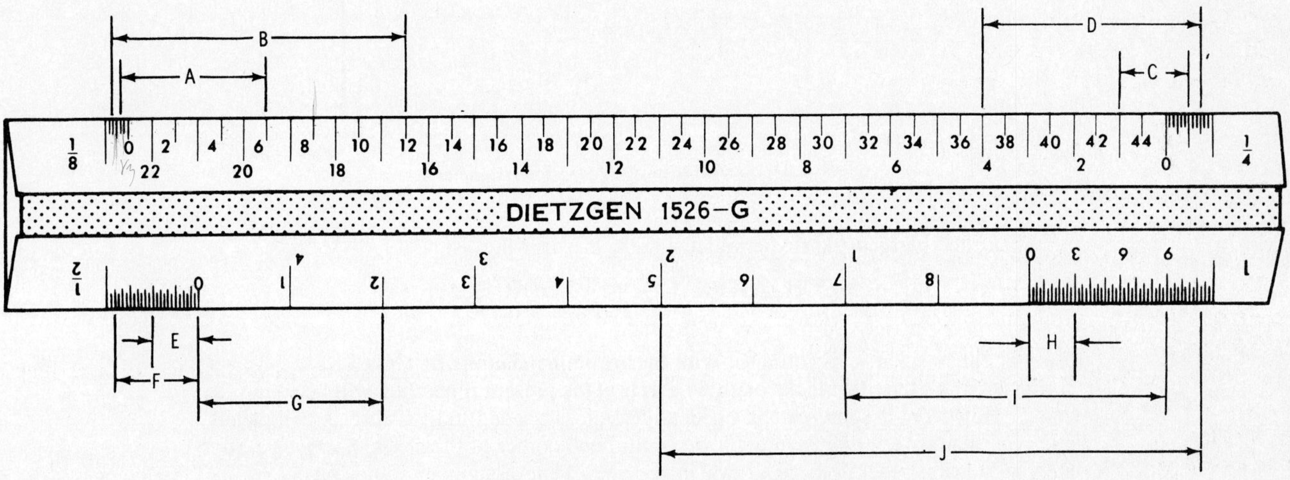

Fig. 4-8. Illustration for assignment problem.

5. Sometimes it becomes necessary to draw a part of a floor plan or other drawing to a larger scale than is originally shown on the drawing in order to show certain details more clearly. The mechanical room in the floor plan of the commercial building in Appendix D is a good example. This floor plan is shown at ⅛ inch = 1 foot. The duct connections to the air-handling unit would be much clearer if this plan were increased in size to ¼ inch = 1 foot.

Prepare another sheet of 11-by-17-inch tracing paper as indicated earlier. Then draw the mechanical-room floor plan to a scale of ¼ inch = 1 foot. Do not include any of the HVAC equipment or ductwork. *Hint:* Use the ⅛ = 1 foot architect's scale to measure the dimensions of the partitions. Then use the ¼ inch = 1 foot architect's scale to transfer these dimensions to your larger drawing. Center the drawing on your drawing sheet.

Do not transfer the entire floor plan, just the mechanical room. However, adjoining partitions should be noted on your drawing and use a break line to "cut" them as indicated in Fig. 4–9.

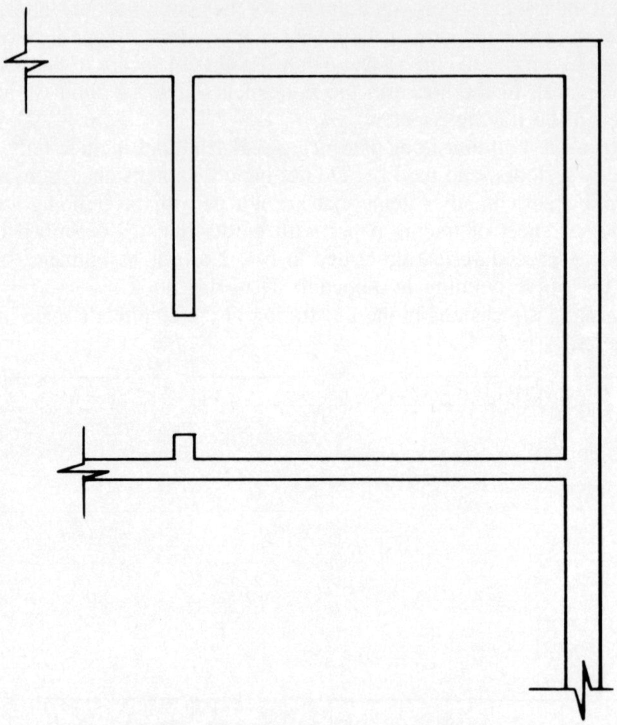

Fig. 4-9. Mechanical room floor plan for assignment.

When the enlarged plan is completed, title the drawing as:

MECHANICAL ROOM FLOOR PLAN
¼″ = 1′-0″

Use guidelines sized in accordance with the recommendations of Unit 3.

Do not place the title blocks on these sheets at the present time; they will be studied in Unit 5. However, if some means of identification is needed until Unit 5 is completed, letter your name very lightly in the lower right-hand corner of the sheet. It can be erased later, and a title block will be drawn in for one of the assignments in Unit 5.

UNIT 5

Titles and Schedules for HVAC Drawings

OBJECTIVE: To describe the purpose of title blocks and schedules and how they are applied to working drawings

GENERAL

The title block for a working drawing is usually boxed in the lower right-hand corner of the sheet; the size of the block varies with the size of the drawing and also with the information required. For 11-by-17-inch sheets, the space to be reserved for the title block should be a minimum of 3 inches long and 2 inches high.

CONTENTS OF A TITLE BLOCK

In general, the title of a HVAC drawing should contain the following:

1. Name of the project
2. Address of the project
3. Name of the owner or client
4. Name of the architectural and/or engineering firm
5. Date of completion
6. Scale(s)
7. Initials of the draftsman, checker, and designer, with dates under each
8. Job number
9. Sheet number
10. General description of the drawing

Every architectural/engineering firm has its own standard form for drawing titles, and they are often preprinted directly on the tracing paper or else printed on "stick-on" paper which is placed on the drawing. A typical architectural title block is illustrated in Fig. 5-1.

Often the consulting engineering firm will be listed too, which means that an additional title will be applied to the drawing usually next to the architectural title block. Fig. 5-2 shows completed architectural and engineering title blocks as they appeared on an actual drawing.

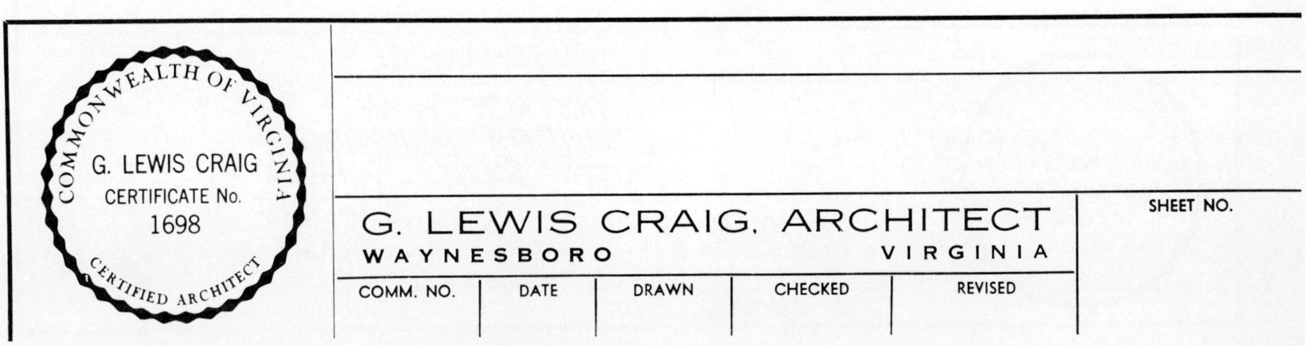

Fig. 5-1. Typical architectural title block.

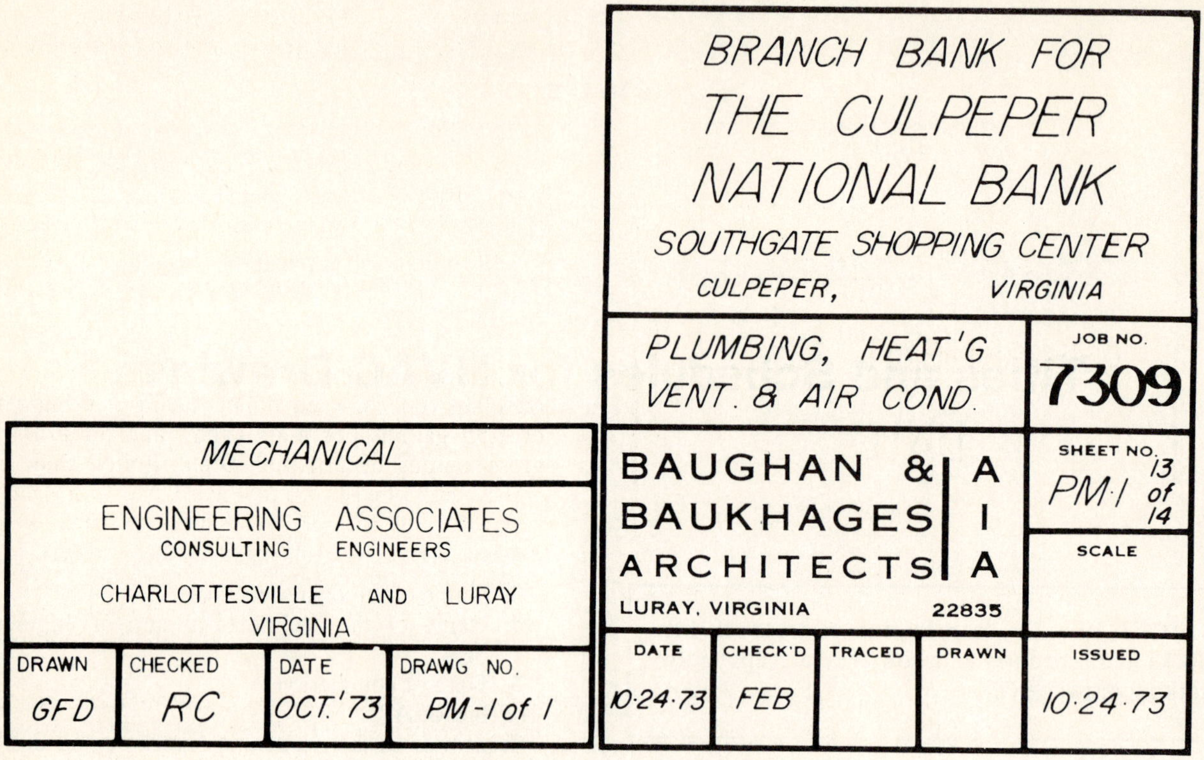

Fig. 5-2. Architectural and engineering title blocks.

Drawing the Title Block

Once the title block lines have been measured and drawn on the tracing paper in the proper form, the information should be lettered freehand in single-stroke capitals. However, the contents of the title block should first be written out, on a separate piece of paper, for two important reasons. The first reason is to make certain that all information required has been obtained, and the second reason is to give the draftsman an idea

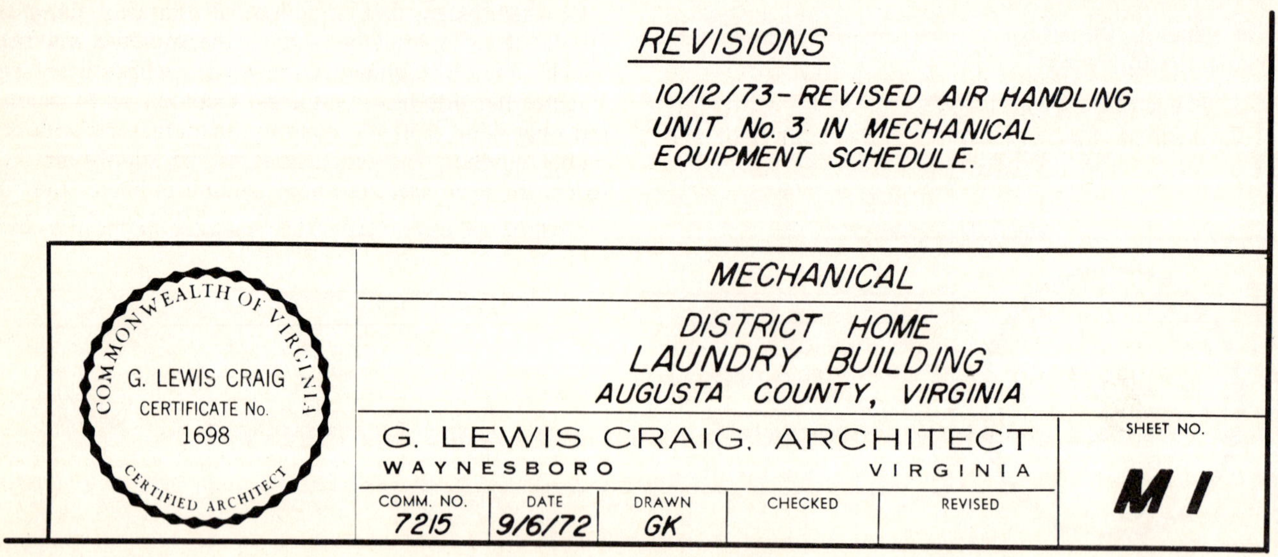

Fig. 5-3. Example of revision notes.

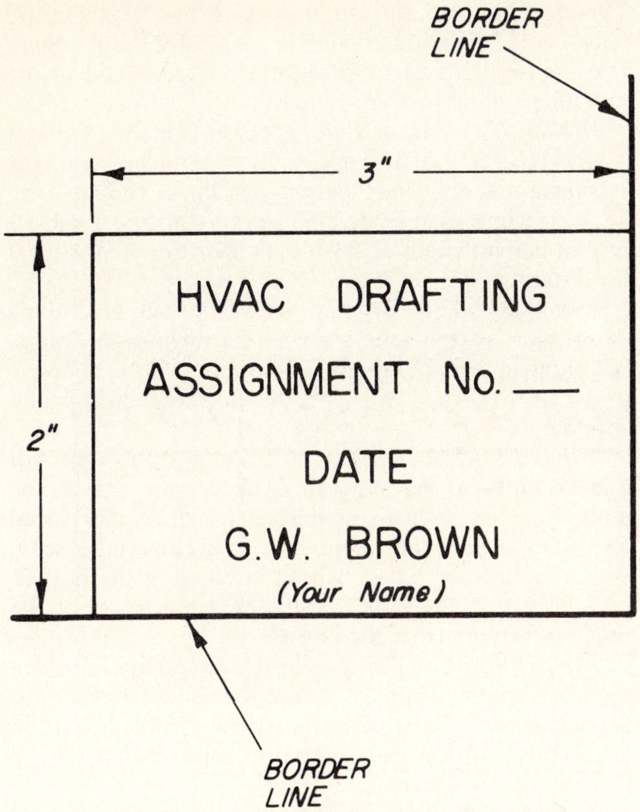

Fig. 5-4. A suggested title block for exercise problems.

of the total space required for the information. The latter enables the draftsman to obtain good spacing for a neat appearance of the title block.

Revisions

Often HVAC drawings will have to be partially redrawn or revised. Sufficient space should be left near the title block for dates and description of revisions. A space about 2 inches high and the width of the title block is usually sufficient (Fig. 5-3). A suggested title block for use on drawing exercises in this book is shown in Fig. 5-4.

SCHEDULES

A schedule is a systematic method of presenting notes or lists of equipment on a drawing in tabular form. When properly organized and thoroughly understood, schedules are not only powerful timesaving devices for the draftsman, but they can also save the specification writer and the workmen on the job much valuable time.

Grille Schedule

A grille and diffuser schedule (Fig. 5-5) shows the manufacturer and catalog number of each grille and diffuser; it also shows the dimensions of each and the volume of air in cubic feet per minute (CFM) that each will handle. A column for remarks is also included to facilitate the installation of the item.

Hot-Water-Boiler Schedule

Fig. 5-6 shows an example of a hot-water-boiler schedule used on an actual construction project.

If this schedule were not indicated on the drawings, a description of the boiler would have to appear in the written specifications and would be similar to the following:

> Contractor shall furnish and install, as indicated on plans, one oil-fired boiler having a firing rate of 2.55 gallons per hour (gph) and an I=B=R gross output of 261,000 Btuh. Boiler shall be constructed of cast iron in accordance with ASME (*A*merican *S*ociety of *M*echanical *E*ngineers) requirements for low-pressure heating boilers and bear the ASME symbol. Each section shall be factory tested at 2½ times maximum working pressure of 50 pounds for water. Boiler shall have I=B=R ratings.
>
> Boiler shall be composed of cast iron sections with vertical flues. Manufacturers will furnish boilers with an aluminized steel canopy. Water or steam trim as required. Oil burners shall include controls as herein described.
>
> Boiler shall be equipped with tankless heater having a rating of 7 gpm. If more than one heater is required to meet requirements, a factory-assembled copper manifold will be installed by the contractor as recommended by the manufacturer to connect the heater in parallel arrangement.

Diffuser Schedule

Fig. 5-7 shows another type of diffuser schedule used on the HVAC drawings for a large residence. The left-hand column indicates the room or area in which the diffusers are located; the next column—reading to the right—indicates the quantity, manufacturer's name, model number, and dimensions of the supply air diffusers for each area; the next column indicates the air volume in cubic feet per minute (CFM); and the last column is left empty for remarks.

Exhaust Fan Schedule

The exhaust fan schedule illustrated in Fig. 5-8 was used for the same residence previously discussed. The left-hand column gives the room or area designation, and the remaining column indicates the manufacturer, model number, cubic feet per minute, and the sound levels in decibels of the conventional fans. The combination heat and fan units in the laundry room and master bath have the heat output in British thermal units per hour, as well as the controls, listed.

GRILLE SCHEDULE

MARK	MFGR	MODEL	SIZE	CFM	REMARKS
1	KRUEGER	SH1·0	9 x 9	146	OBD ALUMINUM
2		SH 3	6 x 6	50	
3				60	
4				75	
5				80	
6				100	
7			9 x 9	144	
8				152	
9				170	
10				175	
11				181	
12				226	
13		SH 4	6 x 6	65	
14				100	
15			9 x 9	175	
16				222	
17		S580H	8 x 6	150	
*18			14 x 8	292	
19				300	
*20			18 x 8	354	
21			24 x 8	500	
22		S580V	34 x 12	1000	
23		SH 3	9 x 9	205	
24	KRUEGER	S580V	8 x 6	167	C.R. OBD ALUM.
25		S580V	18 x 8	362	C.R. OBD ALUM.

*INCLUDE FILTER FRAME 5FF HINGED

Fig. 5-5. Sample grille and diffuser schedule.

Radiation Schedule

Hot-water or steam baseboard radiators may be described in a schedule such as the one illustrated in Fig. 5-9. This schedule is explained as follows:

Symbol—R/1 as indicated in the schedule means that this line describes radiator No. 1, which is also identified in the same manner on the floor plans. However, the schedule describes the radiator in detail, while the floor plan gives the symbol R/1 only.

Serving—This particular schedule was used on a bank project. Note that the second column indicates that radiator No. 1 serves the Note Department, radiator No. 2 serves the vault and office area, etc.

Capacity MBH—(thousand British thermal units per hour)—This column indicates the heat output, in British thermal units, of each unit. It can be seen that radiator No. 1 (R/1) has a capacity of 18.1 MBH, or an output of 18,100 Btuh.

Enclosure Length—This column gives the physical length of each radiator listed. This information is provided for two reasons: first, to indicate the required length in case the contractor submits a substitute radiator and second, to aid the workmen on the job in "roughing-in" for the piece of equipment.

Active Length—The active length is the actual length of the radiation tubing inside the housing, and it is given for the same reasons described under Enclosure Length. Therefore, in radiator No. 1—which has an enclosure length of 24 feet and an active length of 23 feet—there is 6 inches of space (1 foot total) on each end of the radiation tubing for pipe connections.

Rating—Btu Lineal Foot—One reason this column is provided is that in case another brand of radiator is substituted, the heat output per lineal foot will be kept the same as indicated. Since radiator No. 1 (R/1) has an active length of 23 feet at 790 Btu/ft, the total head output is $23 \times 790 = 18{,}170$ Btu.

Prototype—This column gives the manufacturer's description of the unit used in the engineer's design, but another manufacturer's unit usually will be allowed, provided it is equal in capabilities to the one originally specified.

Type of Enclosure Back—A column of this type will appear only on certain jobs. The reason is that the units specified with an unfinished back will be placed against a solid wall where the back cannot be seen, while the finished backs will be installed at the bottom of a floor to ceiling glass wall where the backs of the unit can be seen from the outside.

Type of Sleeve—The type-of-sleeve column indicates the blank (nonactivated) connections between the radiator enclosures. Such sleeves are either inside or outside telescoping, as needed, and are adjustable up to 7 inches at each sleeve.

Remarks—The remaining column is used for added remarks necessary to clearly indicate what the engineer requires. In this schedule, radiator No. 4 (R/4) shall have an enclosure the full length of the wall on which it is to be installed. In some cases a "blank" extension may be necessary to acquire the desired length.

Cabinet Unit Heater Schedule

The cabinet unit heater schedule shown in Fig. 5-10 was used on HVAC working drawings to describe the nine different types of unit heaters installed in a bank building. Each of the unit heaters described in this schedule contained a hot-water coil and a fan to circulate the air over the coil. The hot-water coil received its heated water (180°F) from a centrally located hot-water boiler similar to the one described in the schedule in Fig. 5-6.

Six of the unit heaters described in this schedule can be seen on the floor plan drawing in Fig. 5-11. The unit heaters have been shaded-in on the drawing to make them stand out, and a descriptive note has been included for each. The schedule then gives a detailed description of each type and includes the following:

1. Identification Number
2. Serving (areas used)
3. CFM (air volume of each)
4. Capacity MBH (heat output in thousand British thermal units per hour)
5. GPM (gallons of hot water per minute required)

HOT WATER BOILER SCHEDULE							
BOILER NO.	FUEL	BTU/GAL.	BTUH-INPUT	OIL-GPH	IBR-NET	IBR-GROSS	FLUE
*A-504	№2 OIL	261 MBH	300,150	2.55	227,000	261,000	8"x 12"

*AMERICAN STANDARD

Fig. 5-6. Sample hot-water boiler schedule.

SUPPLY DIFFUSER SCHEDULE LOWER LEVEL			
GUEST B.R.	2-CARNES MODEL 7292-A x 4'0" LONG	157-CFM EACH	
GAME RM.	BASE BOARD ELECTRIC		
BATH #1	1-CARNES MODEL 200-12x4 REG.	NO VOLUME CONTROL 50-CFM EACH	STACK HEAD DAMPER
BAR	1-CARNES MODEL 200-12x4 REG.	NO VOLUME CONTROL 121-CFM EACH	STACK HEAD DAMPER
KITCHENETTE	1-CARNES MODEL 200-12x4 REG.	NO VOLUME CONTROL 64-CFM EACH	STACK HEAD DAMPER
GALLERY	1-CARNES MODEL C-40 FLOOR 3"x 5'0" 2-CARNES MODEL 7261-A x 3'0" LONG	200-CFM EACH 100-CFM EACH	
STUDY	1-CARNES MODEL 7292-A x 6'0" LONG	282-CFM EACH	
BED RM. #1	1-CARNES MODEL 7261-A x 6'0" LONG	188-CFM EACH	
BED RM. #2	1-CARNES MODEL 7261-A x 6'0" LONG	188-CFM EACH	
BATH #2	1-CARNES MODEL 200-12x4 REG.	NO VOLUME CONTROL 50-CFM EACH	STACK HEAD DAMPER
LAV. #1	1-CARNES MODEL 200-12x4 REG.	NO VOLUME CONTROL 50-CFM EACH	STACK HEAD DAMPER
LAV. #2	1-CARNES MODEL 200-12x4 REG.	NO VOLUME CONTROL 50-CFM EACH	STACK HEAD DAMPER
RETURN AIR DIFFUSER SCHEDULE			
GAME RM.	BASE BOARD ELECTRIC		
GUEST B.R.	CARNES MODEL 7295 x 4'0"	250-CFM EACH	
GALLERY	CARNES MODEL 7295 x 4'0"	400-CFM EACH	
BED RM. #1	CARNES MODEL 7295 x 4'0"	238-CFM EACH	
BED RM. #2	CARNES MODEL 7295 x 4'0"	238-CFM EACH	
STUDY	CARNES MODEL 7295 x 4'0"	332-CFM EACH	

Fig. 5-7. Sample of residential diffuser schedule.

6. Fan Motor (watts, voltage, phase, and hertz)
7. Type
8. Prototype (manufacturer and model number)
9. Remarks (as indicated)

NOTES

At times it is desirable to use a few general notes rather than a schedule or elaborate written specifications—especially on projects where only a few special items must be described. Fig. 5-12 shows how such notes would appear on HVAC working drawings.

The set of notes shown in Fig. 5-12 indicates the manufacturer and general type of thermostat; the manufacturer, type finish, etc., of the supply air diffusers; the manufacturer, type, finish, etc., of the wall and ceiling (clg.) grilles; and two plumbing notes.

SUMMARY

A title block for a working drawing is used to identify the drawing and the person or persons who prepared it.

EXHAUST FAN SCHEDULE	
BATH No. 1	NUTONE MODEL QT-110-110 CFM DECIBELS SOUND LEVEL 2.5
BATH No. 2	NUTONE MODEL QT-80-80 CFM DECIBELS SOUND LEVEL 1.5
LAY # 1	" " " " " " " " "
LAY # 2	" " " " " " " " "
LAUNDRY RM.	2 EA NUTONE HEAT-A-VENT # 9275-4.265 B.TUH HEAT 40 CFM EA
	WITH 2 EA ON/OFF HEAT/VENT CONTROLS
MASTER BATH	1 EA NUTONE HEAT-A-VENT # 9275-4.265 B.TUH-HEAT 40 CFM WITH
	ONE EA ON/OFF HEAT CONTROL WALL MOUNTED THERMOSTAT

Fig. 5-8. Sample of residential exhaust fan schedule.

RADIATION SCHEDULE									
SYMBOL	SERVING	CAPACITY MBH	ENCLOSURE LENGTH	ACTIVE LENGTH	RATING - BTU LIN. FT.	PROTO-TYPE	TYPE OF ENCLOSURE BACK	TYPE OF SLEEVE	REMARKS
R 1	NOTE DEPT.	18.1	24'	23'	790	NESBITT ARCH SILL LINE - STYLE GN12	UNFINISHED	INTERNAL SLEEVE AT MULLIONS	
R 2	VAULT OFFICE	11.8	15-1/2'	15'	790	NESBITT ARCH SILL LINE - STYLE GN12	FINISHED	INTERNAL SLEEVE AT MULLIONS	
R 3	LOAN DEPT.	24.5	32'	31'	790	NESBITT ARCH SILL LINE - STYLE GN12	FINISHED	INTERNAL SLEEVE AT MULLIONS	
R 4	OFFICE	6.5	15'	7'	975	NESBITT ARCH SILL LINE - STYLE FBN-14	UNFINISHED	NONE	ENCLOSURE FULL LENGTH OF WALL
R 5	TELLERS	27.6	36'	35'	790	NESBITT ARCH SILL LINE - STYLE GN12	FINISHED	INTERNAL SLEEVE AT MULLIONS	
R 6	TOILET	2.1	3'	3'	780	NESBITT ARCH SILL LINE - STYLE G-10A	UNFINISHED	NONE	

Fig. 5-9. Sample of radiation schedule.

CABINET UNIT HEATER SCHEDULE									
No.	SERVING	CFM	CAPACITY MBH	GPM	FAN MOTOR		TYPE	PROTOTYPE	REMARKS
					WATTS	V/Ø/H_2			
UH 1	BASEMENT REAR ENTRANCE	200	15.0	1.0	110	120/1/60	RECESSED HORIZONTAL CLG. MTD.	TRANE MODEL E46 FORCE FLOW	
UH 2	REAR STAIRWELL	300	20.0	1.0	160	120/1/60	VERTICAL CABINET	TRANE MODEL B42 FORCE FLOW	BUILT-IN THERMOSTAT
UH 3	BASEMENT STORAGE	200	15.0	1.0	110	120/1/60	RECESS. HOR. CLG. MTD W/O.A. CONN	TRANE MODEL E46 FORCE FLOW	PROVIDE BRICK VENT FOR O.A.
UH 4	BASEMENT VAULT	200	15.0	1.0	110	120/1/60	HORIZONTAL EXPOSED CAB CLG. MTD.	TRANE MODEL D16 FORCE FLOW	
UH 5	BASEMENT ELEVATOR LOBBY	200	15.0	1.0	110	120/1/60	VERTICAL CABINET	TRANE MODEL B42 FORCE FLOW	BUILT-IN THERMOSTAT
UH 6	MAIN STAIRWELL	400	30.0	1.5	150	120/1/60	VERTICAL CABINET	TRANE MODEL B42 FORCE FLOW	BUILT-IN THERMOSTAT
UH 7	MAIN VAULT	200	15.0	1.0	110	120/1/60	HOR. EXP. CABINET CLG. MTD.	TRANE MODEL D16 FORCE FLOW	
UH 8	ENTRANCE	400	30.0	1.5	150	120/1/60	RECESSED HORIZONTAL CLG. MTD.	TRANE MODEL E46 FORCE FLOW	
UH 9	BOOK VAULT	200	15.0	1.0	110	120/1/60	HOR. EXP. CABINET CLG. MTD.	TRANE MODEL D16 FORCE FLOW	

Fig. 5-10. Cabinet unit heater schedule.

A schedule is a systematic method of presenting notes or lists of equipment in tabular form.

Schedules are timesaving devices when properly used by draftsmen.

Schedules save the specification writer much time.

Schedules also save workmen on the job much time as well as others who are required to work with the drawings.

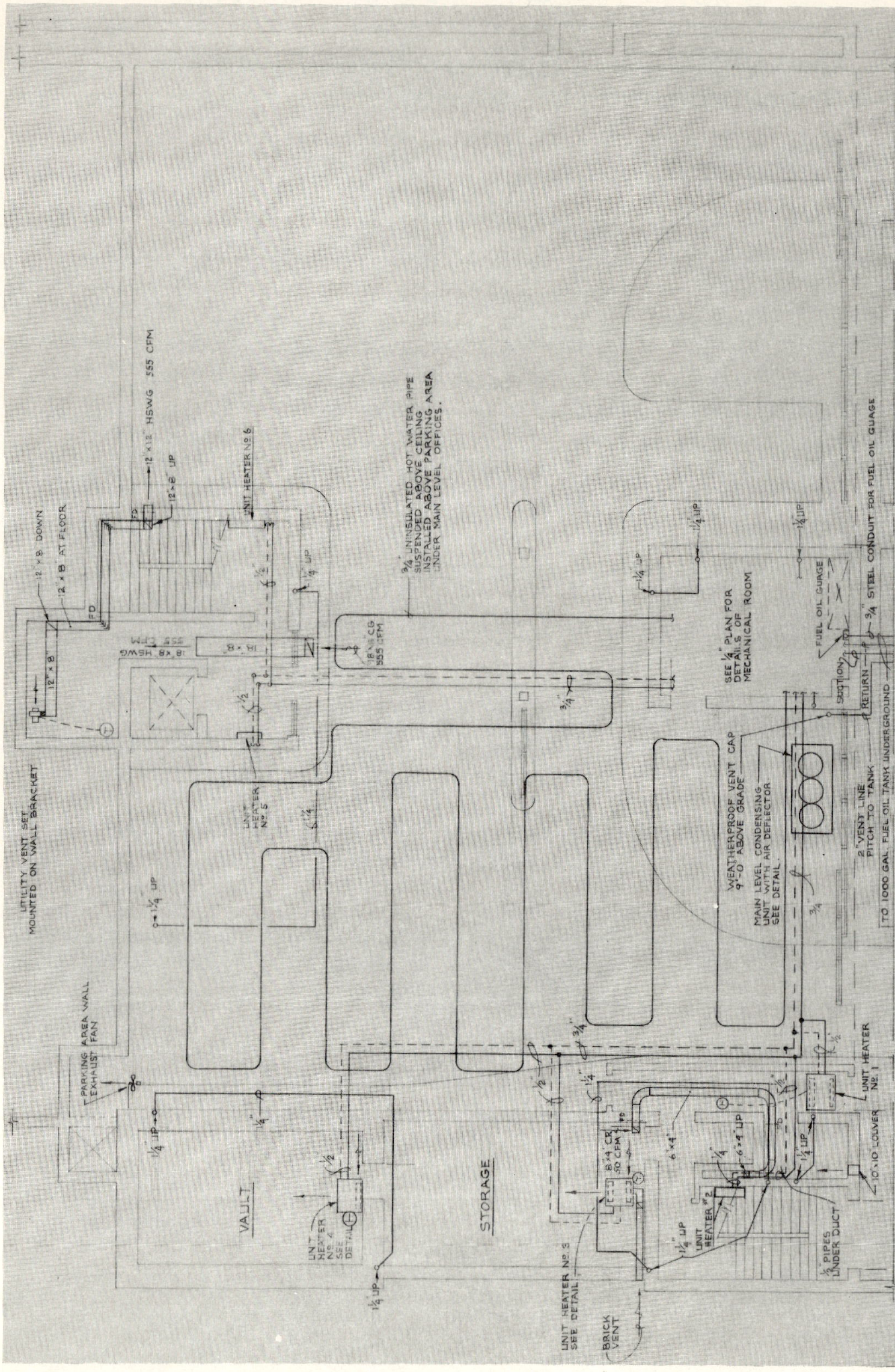

Fig. 5-11. Floor plan of bank building showing location of unit heaters described in the schedule in Fig. 5-10.

NOTES

THERMOSTAT:
TRANE (1 STAGE HEATING & 1 STAGE COOLING) W/ON, OFF,
AUTO. & HEAT-COOL SUB BASE

DIFFUSERS:
CARNES TYPE DFE W/No. 4 DAMPER & WE FINISH

WALL GRILLES:
CARNES TYPE CRE-51Q

CLG. GRILLES:
CARNES TYPE A-CT W/WHITE No. 8-377 FINISH

HOSE BIBBS:
FREEZE PROOF, REMOVABLE HANDLE, RECESSED
WITH COVER. BRASS, JOSAM

CONDENSATE DRAIN:
RUN TO NEAREST D.S.

Fig. 5-12. Sample of notes appearing on HVAC drawing in lieu of a schedule.

UNIT 5 — ASSIGNMENT

1. Thoroughly clean your drawing board with the dust brush. If lead dust or other stains are present, wash the board down with soap and water on a damp cloth.
2. Secure a blank sheet of tracing paper to your drawing board with drafting tape; then draw border lines around the sheet as described in Unit 2.
3. Leave the drawing sheet just prepared in place on your drawing board and look at the grill schedule in Fig. 5–5. Use your architect's scale to measure the grill in Fig. 5–5; that is, the height and width of the entire schedule, widths of the columns, etc. Then draw this schedule in the upper left-hand corner of the drawing sheet that you already have secured to your drawing board. The let-hand side and the top of the schedule should be about ¼ inch from the left and top borders, respectively, of the drawing sheet. Again, use the architect's scale to lay out the lines in the schedule, but draw it twice the size that it appears in the book. This means if you use the ⅛-inch scale for the book measurements, use the ¼-inch scale to transfer the dimensions to the drawing sheet.
4. Beginning about one inch below the bottom of the schedule drawn in No. 3 above, draw the schedule shown in Fig. 5–6. Use the architect's scale to take measurements, but this time draw it to the same size as the one in the book. In other words, if you use a ¼-inch scale to measure the schedule in Fig. 5–6, use the same scale for your drawing.
5. Place a title block, as described in Fig. 5–4, on this sheet and all sheets previously required on assignments. The assignment you did for Unit 3 will have lines drawn through the space where the title block is to go. Draw the title block in the lower right-hand corner of this sheet first. Then, use your erasing shield to remove all lines within the area of the title block.
6. Use an Ames lettering guide and draw in guidelines, where required on these schedules, and title blocks by using the recommendations found in Unit 3. Your instructor will designate the exact wording and arrangement of the title block for each sheet.

SECTION II

Duct Drawing

HVAC Graphic Symbols

OBJECTIVES: To teach the various graphic symbols used in drafting

To apply these symbols to HVAC drawings

GENERAL

Ductwork is used to transmit air from the air-handling equipment to the space to be conditioned. Fig. 6-1 shows a simplified diagram of how such a system operates.

Although the HVAC draftsman usually is not required to have a detailed knowledge of duct design, he should have a basic understanding of duct design in order to approach his work more intelligently. The units to follow will give the draftsman this knowledge and the ability to apply it to HVAC drawings.

HVAC DRAWINGS AND RELATED SYMBOLS

The purpose of an HVAC drawing is to show the location of air-handling and other HVAC equipment; to show the sizes and routes of the various ductwork (both supply and return air); to show the air outlets, including the volume and velocity of air from each; and other necessary information required in a duct

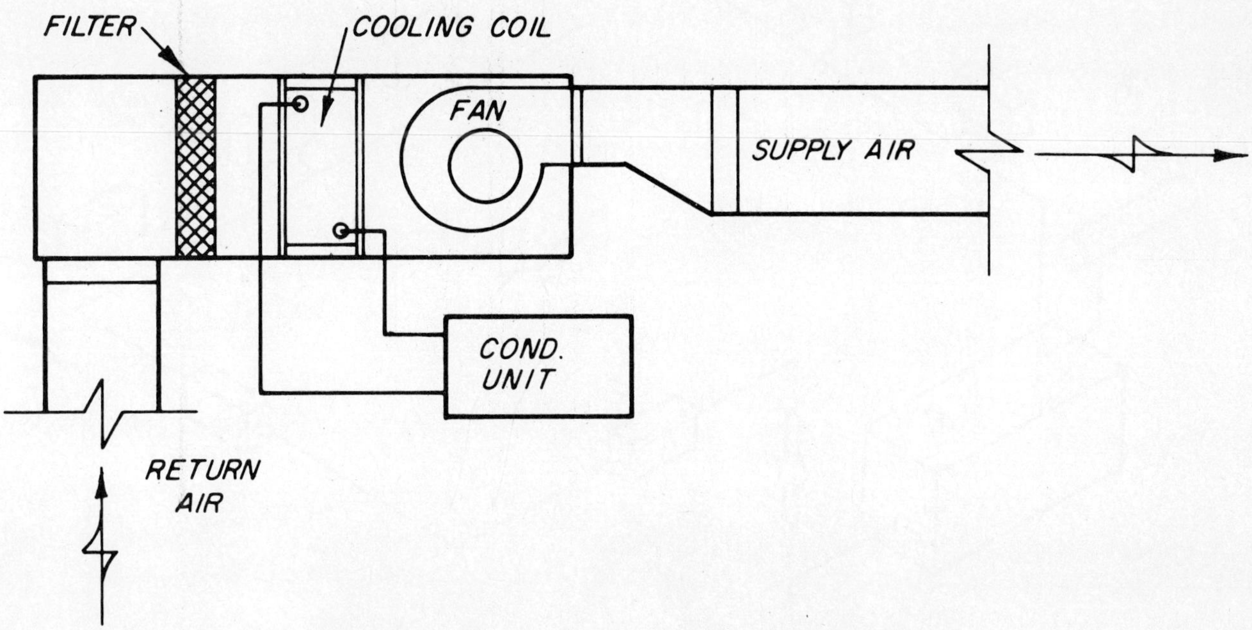

Fig. 6-1. Simplified diagram of how an air-conditioning system operates.

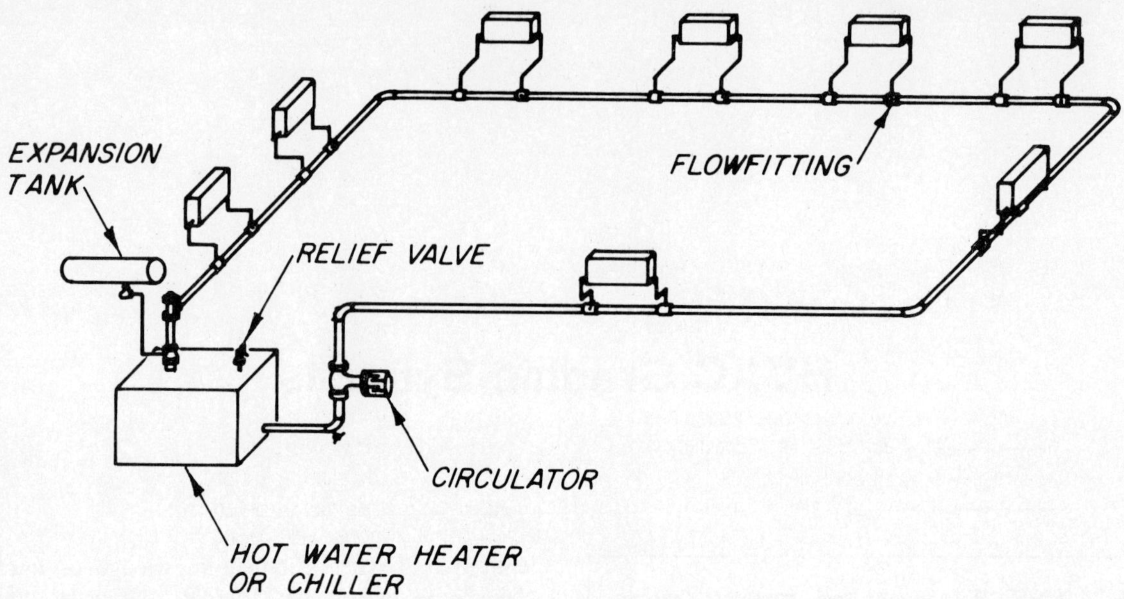

Fig. 6-2. Pictorial drawing of heating system.

system. In the preparation of such drawings, symbols are used to simplify the draftman's work and to save time. Figs. 6-2 and 6-3 illustrate this.

The pictorial drawing in Fig. 6-2 clearly indicates how the system is to be installed. However, this drawing would require considerably more time to draw than the drawing shown in Fig. 6-3, and the drawing in Fig. 6-3 is exactly the same as Fig. 6-2 except symbols are used instead of pictorial drawings, of the objects.

In drawing HVAC plans, most engineers and designers use the symbols adopted by the ANSI (*A*merican *N*ational *S*tandards). However, consulting engineering firms, as well as contracting-design firms frequently

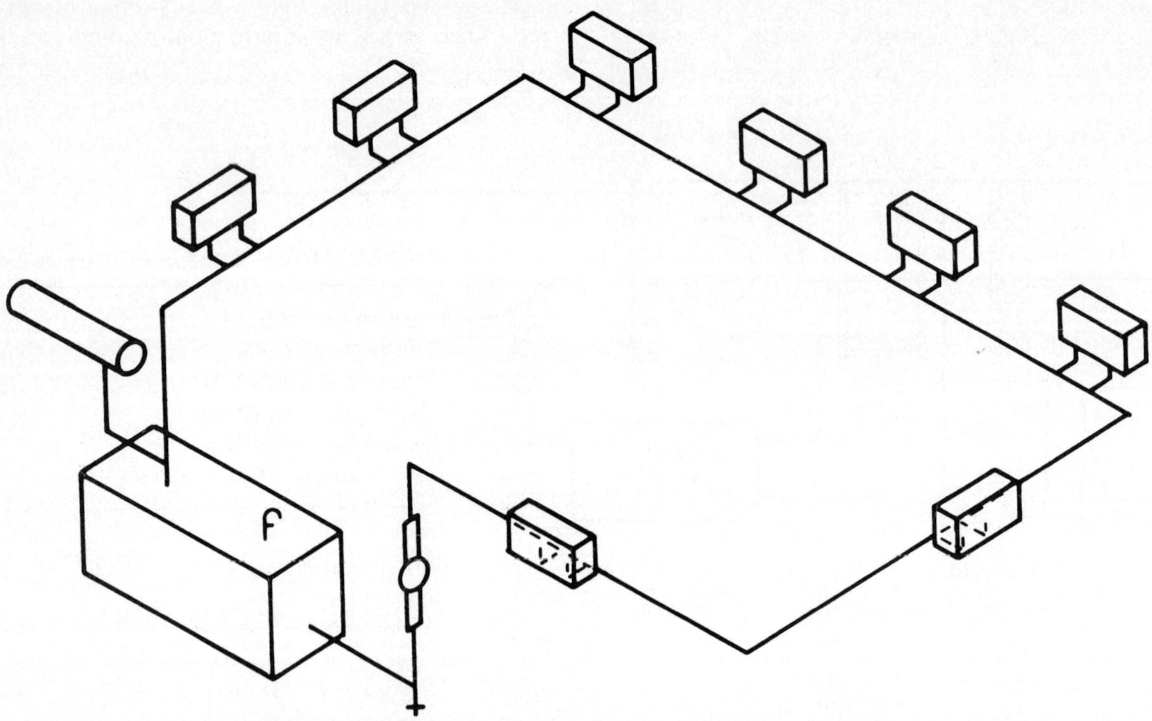

Fig. 6-3. Pictorial drawing of heating system using symbols and simplified lines.

MECHANICAL DRAWING SYMBOLS

NOTE: THESE ARE STANDARD SYMBOLS AND MAY NOT ALL APPEAR ON THE PROJECT DRAWINGS; HOWEVER, WHEREVER THE SYMBOL ON THE PROJECT DRAWINGS OCCURS, THE ITEM SHALL BE PROVIDED AND INSTALLED.

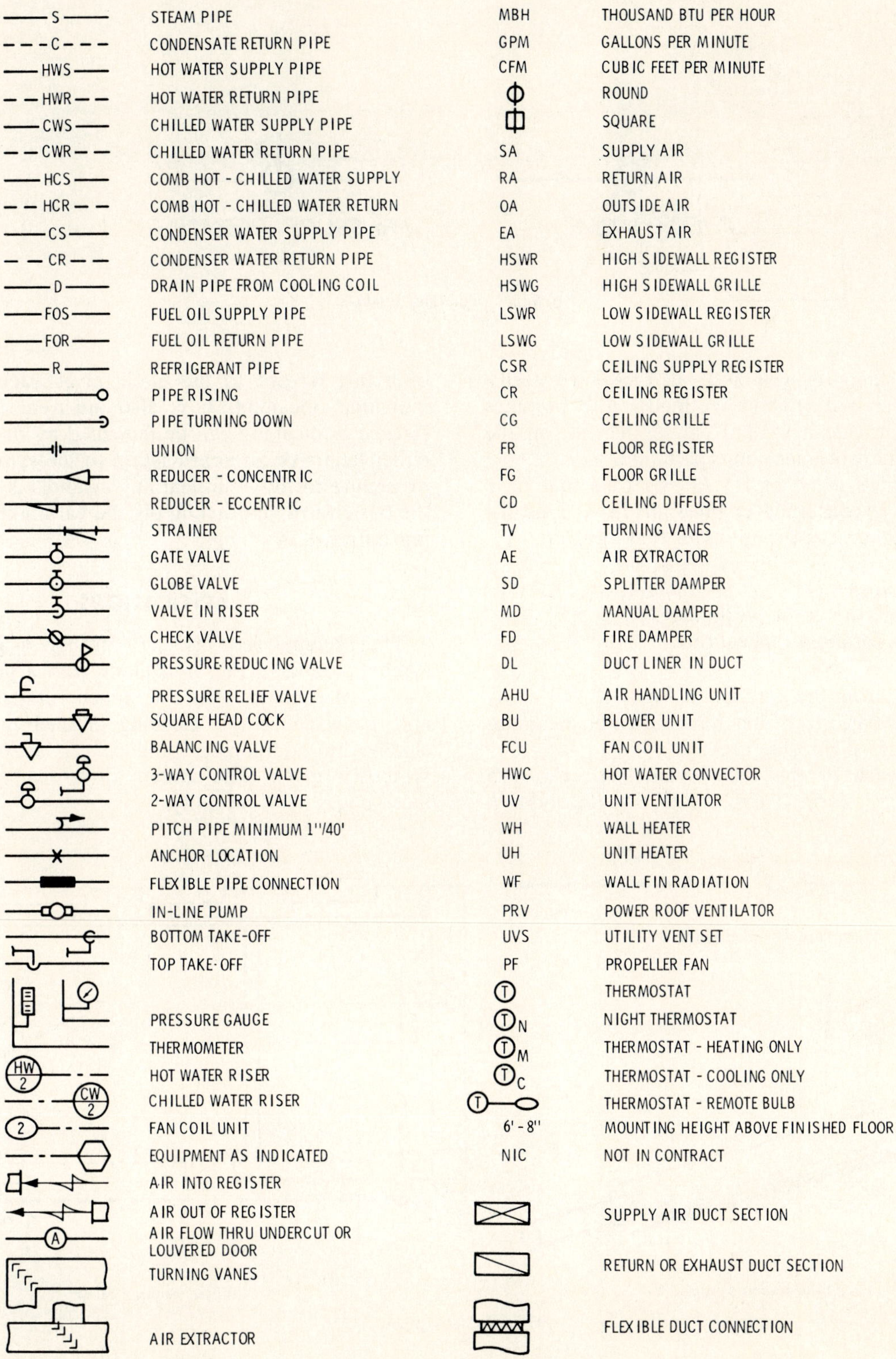

Fig. 6-4. Mechanical drawing symbols.

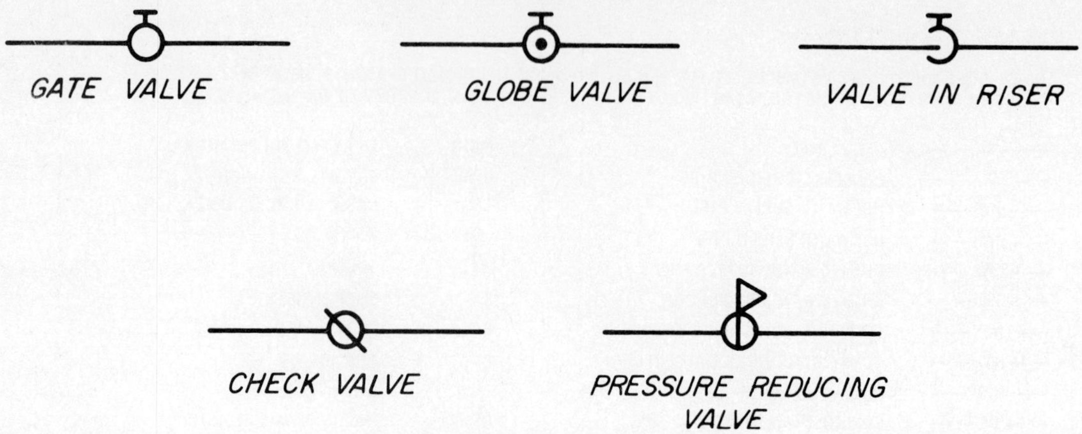

Fig. 6-5. Various valve symbols.

modify these standard symbols to fit their own needs. When these standard symbols are modified, it becomes necessary to provide a symbol list, or legend, on the drawings for the workmen's interpretation.

Fig. 6-4 shows a list of HVAC symbols that have been used by a consulting engineering firm. This list represents a good set of symbols in that they are

1. Easy to draw
2. Easily interpreted by workmen
3. Sufficient for most applications

It is evident from the symbols in Fig. 6-4 that many have the same basic form but have different meanings because of some slight difference.

Note that each of the valve symbols in Fig. 6-5 has the same basic form (a circle), but that the addition of a line or dot to the circle gives each symbol an individual meaning. It is also apparent that the difference in meaning can be indicated by the addition of either letters or an abbreviation to the symbol. A good procedure to follow in learning symbols is to first learn the basic form and then apply the variations for obtaining different meanings.

APPLICATIONS

The sketches in Figs. 6-6 through 6-58 show the symbols found in Fig. 6-4, along with a pictorial illustration of each, so that the student draftsman will be able to visualize the meaning of each symbol more clearly. In addition, a practical application of each symbol is given. The student draftsman should study these pages very carefully.

(A) Recommended symbol.

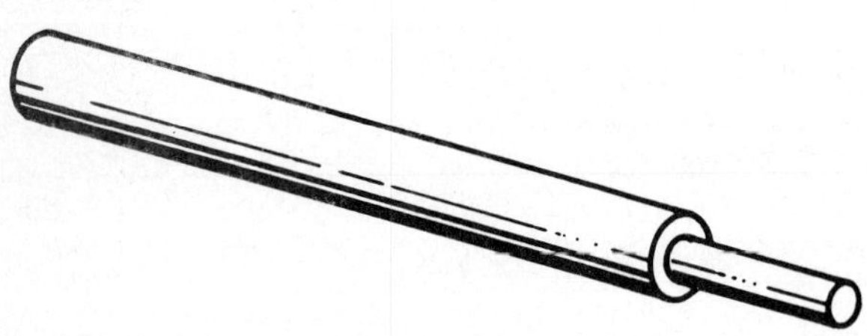

(B) Illustration of pipe.

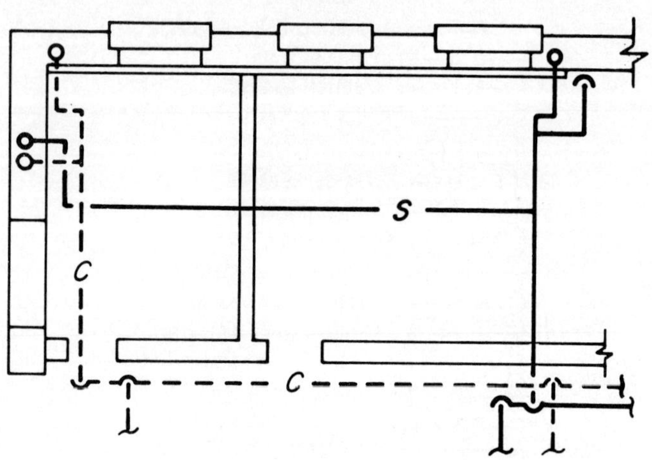

(C) Practical application.

Fig. 6-6. Steam pipes.

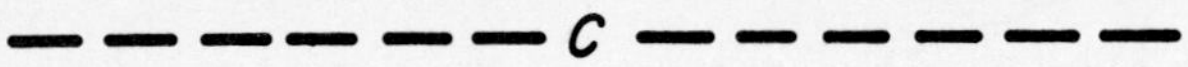

(A) Recommended symbol.

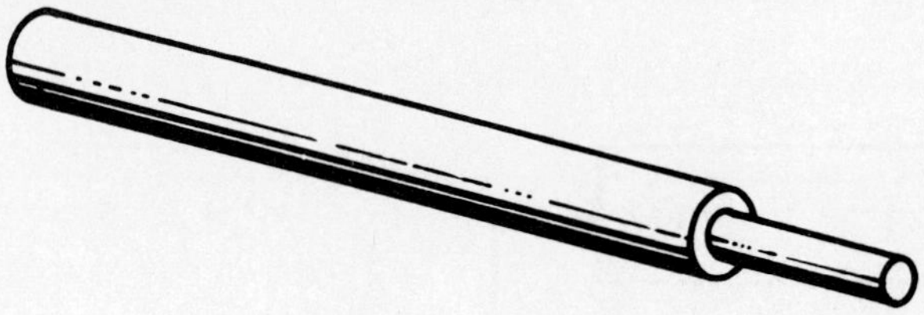

(B) Illustration of pipe.

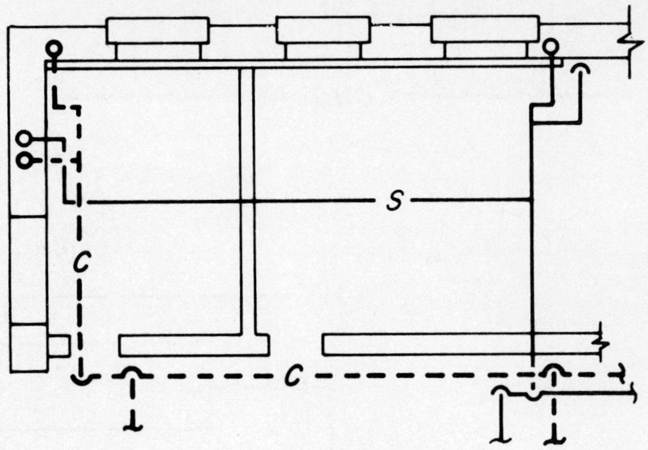

(C) Practical application.

Fig. 6-7. Condensate return pipe.

(A) Recommended symbol.

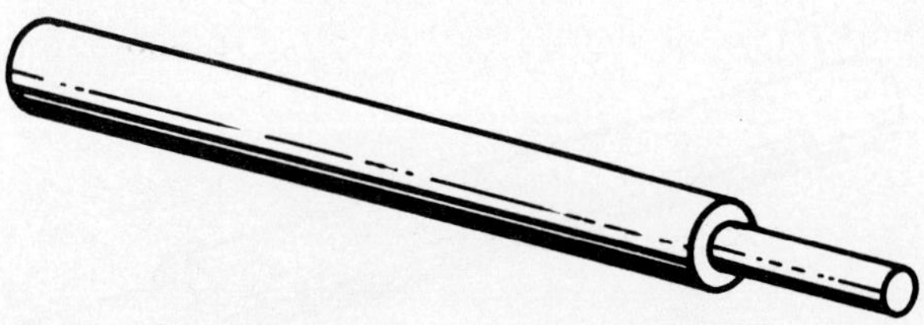

(B) Illustration of pipe.

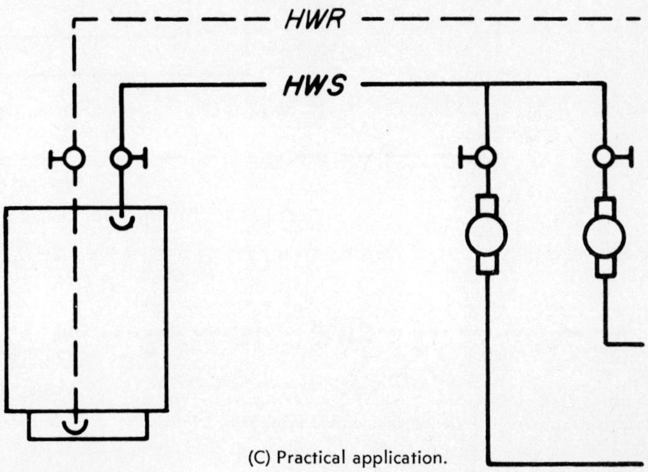

(C) Practical application.

Fig. 6-8. Hot-water supply pipe.

(A) Recommended symbol.

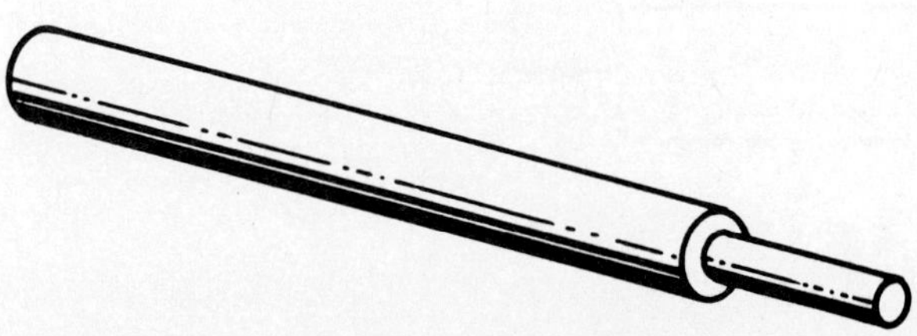

(B) Illustration of pipe.

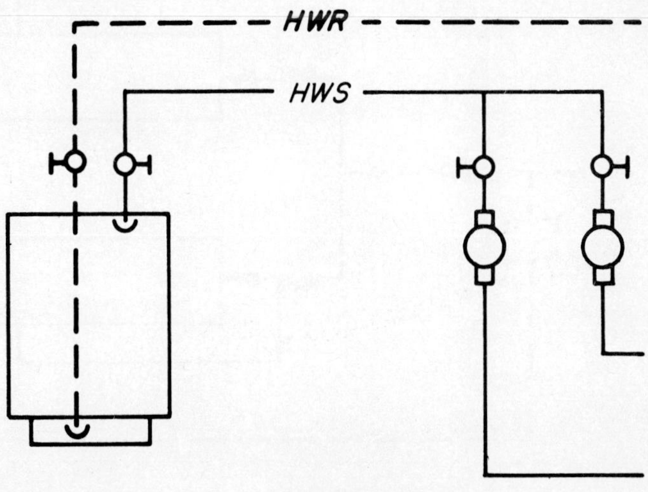

(C) Practical application.

Fig. 6-9. Hot-water return pipe.

(A) Recommended symbol.

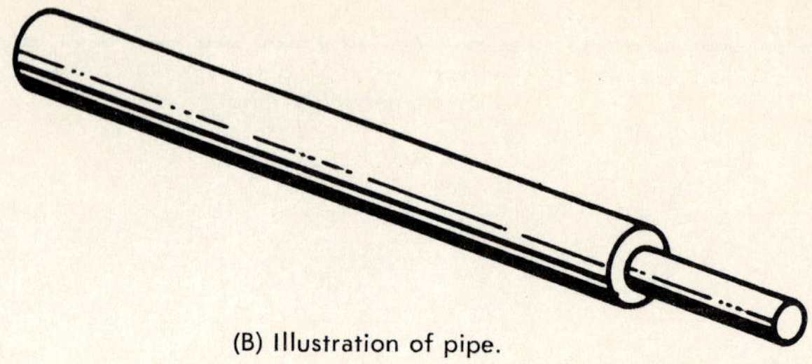

(B) Illustration of pipe.

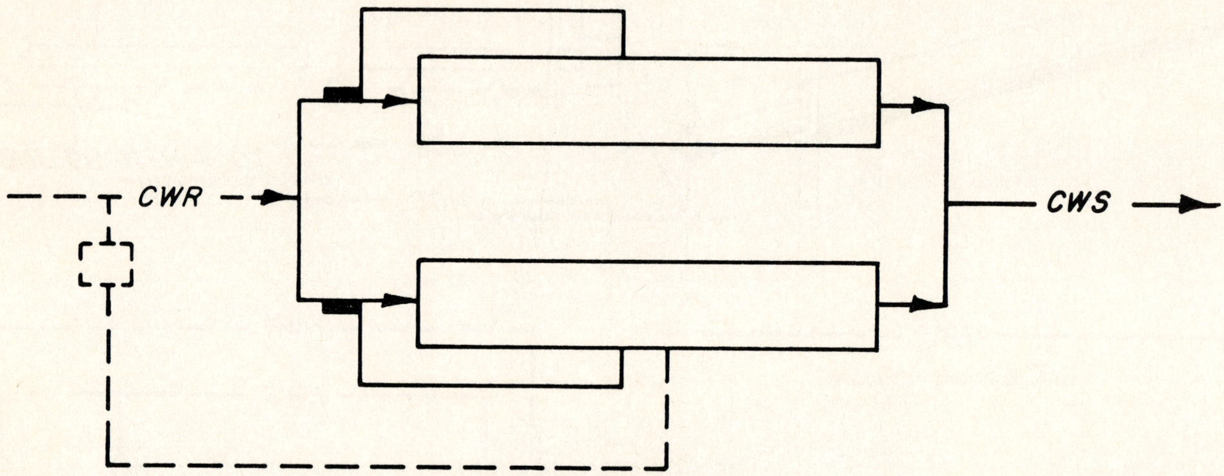

(C) Practical application.

Fig. 6-10. Chilled-water supply pipe.

(A) Recommended symbol.

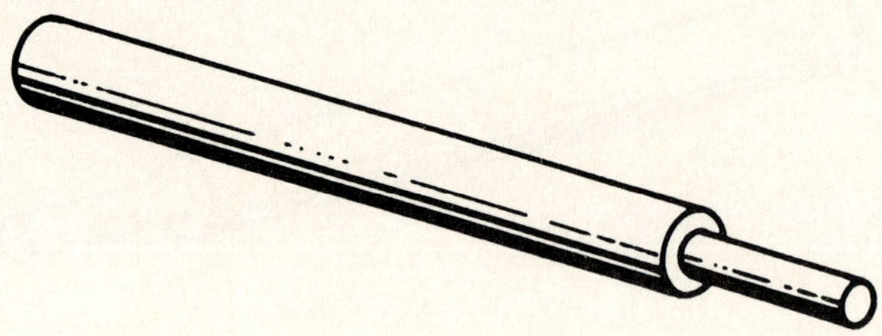

(B) Illustration of pipe.

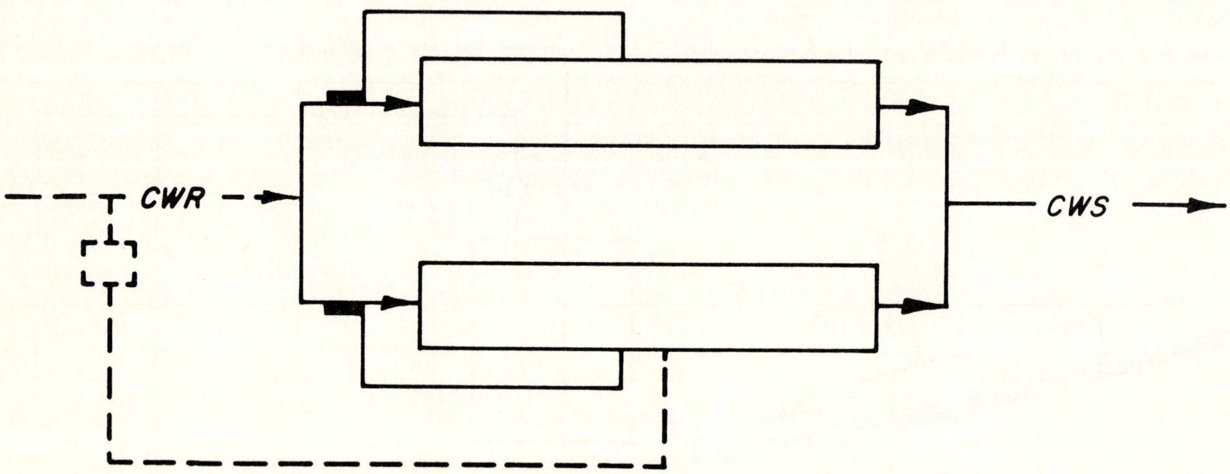

(C) Practical application.

Fig. 6-11. Chilled-water return pipe.

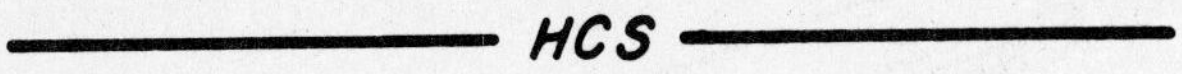

(A) Recommended symbol.

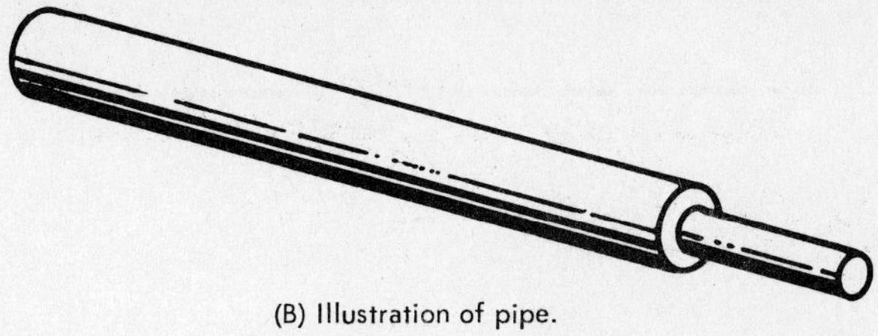

(B) Illustration of pipe.

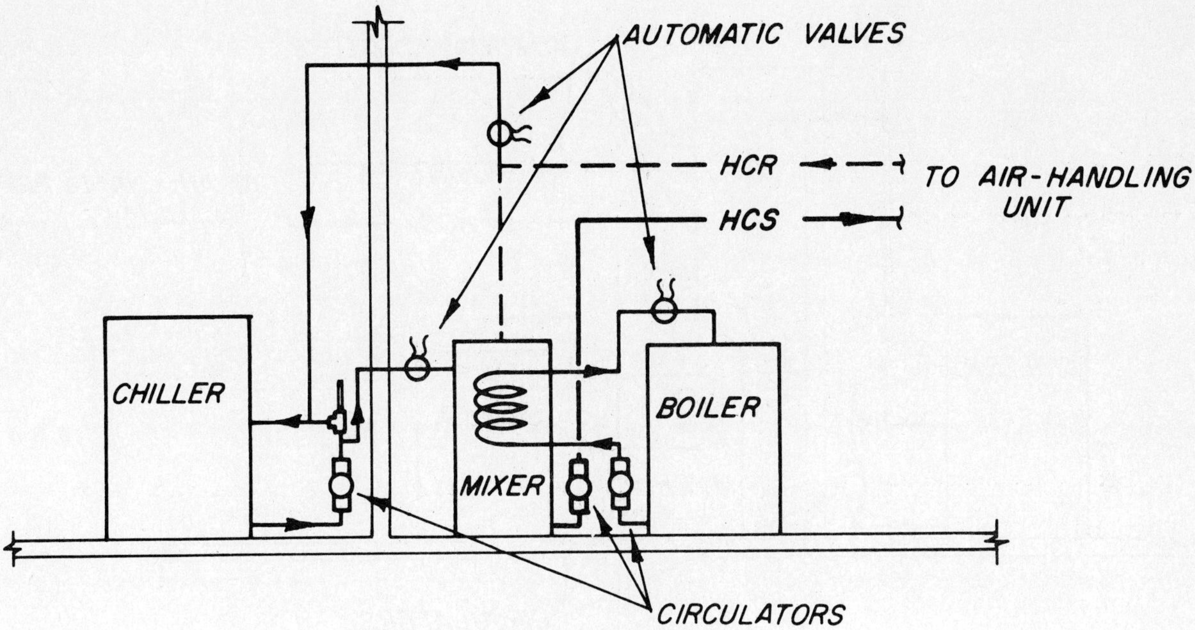

(C) Practical application.

Fig. 6-12. Combination hot- and chilled-water supply pipe.

(A) Recommended symbol.

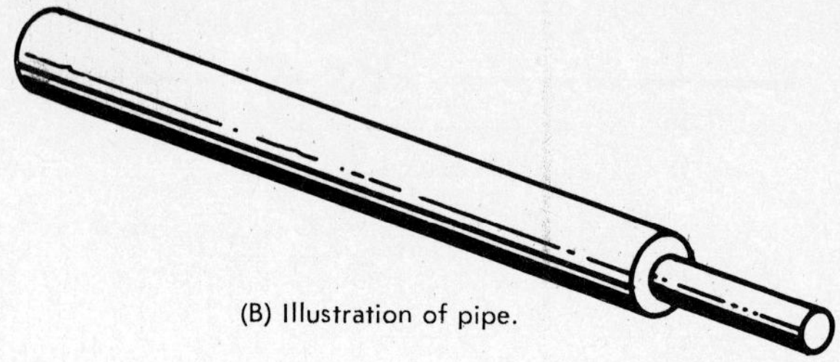

(B) Illustration of pipe.

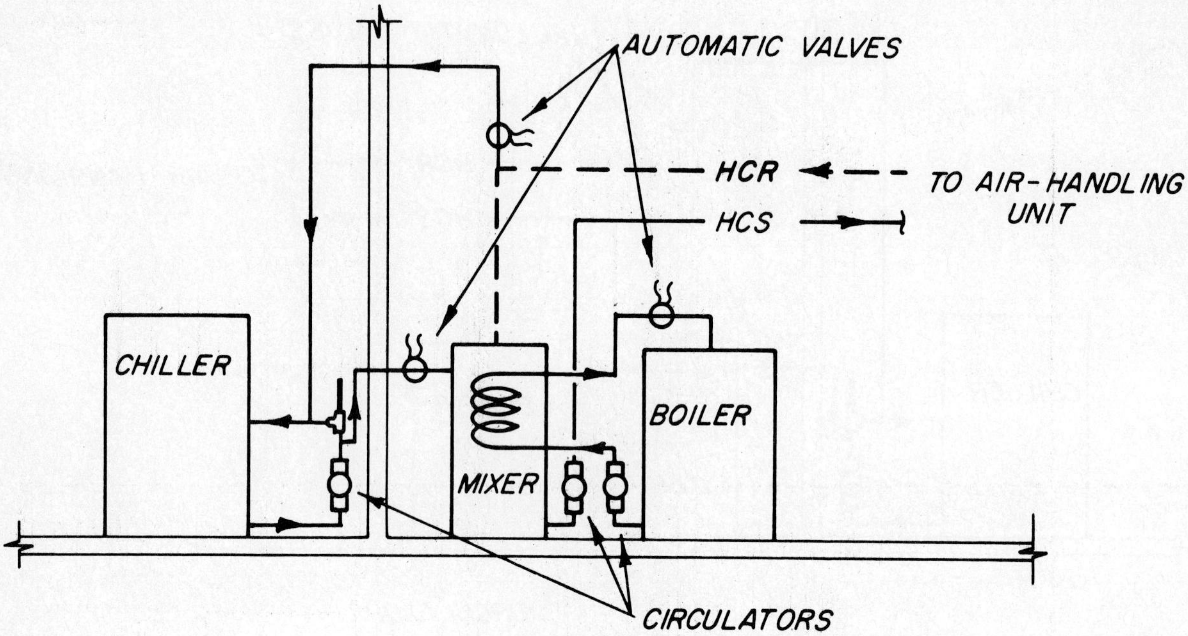

(C) Practical application.

Fig. 6-13. Combination hot- and chilled-water return pipe.

(A) Recommended symbol.

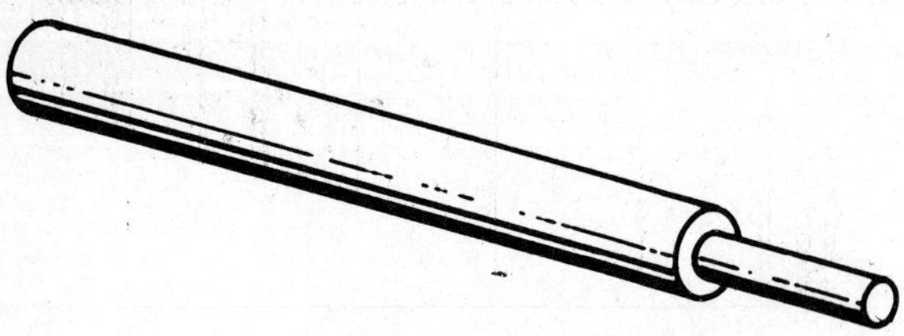

(B) Illustration of pipe.

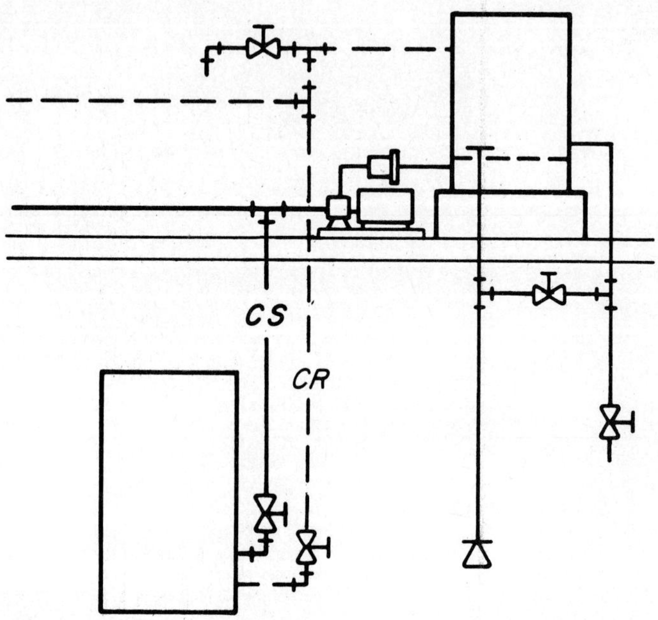

(C) Practical application.

Fig. 6-14. Condenser water supply pipe.

(A) Recommended symbol.

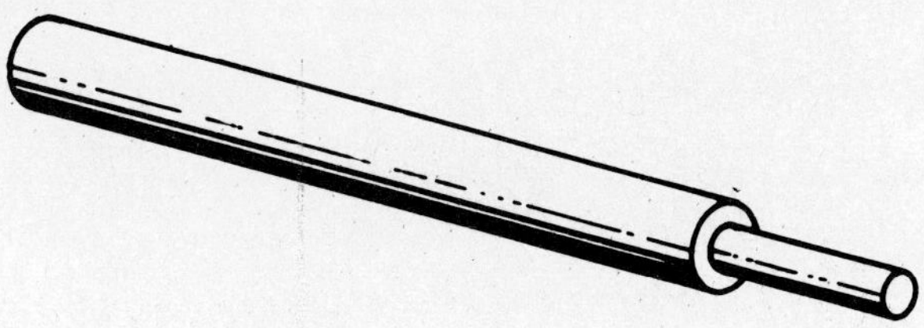

(B) Illustration of pipe.

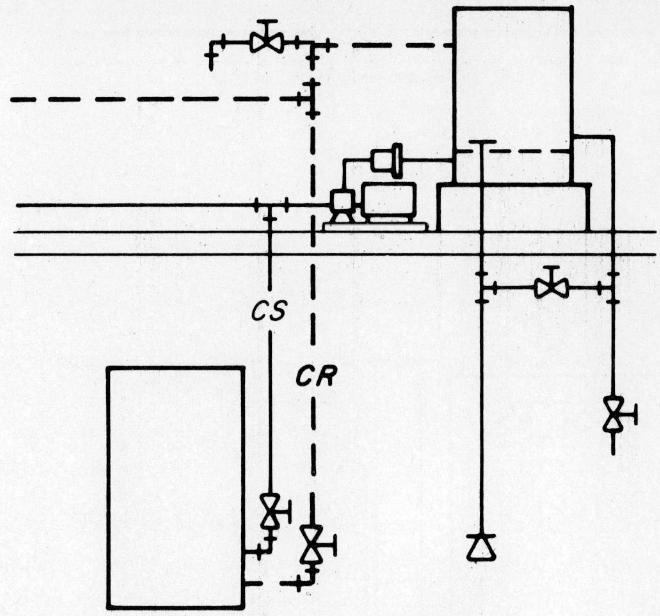

(C) Practical application.

Fig. 6-15. Condenser water return pipe.

(A) Recommended symbol.

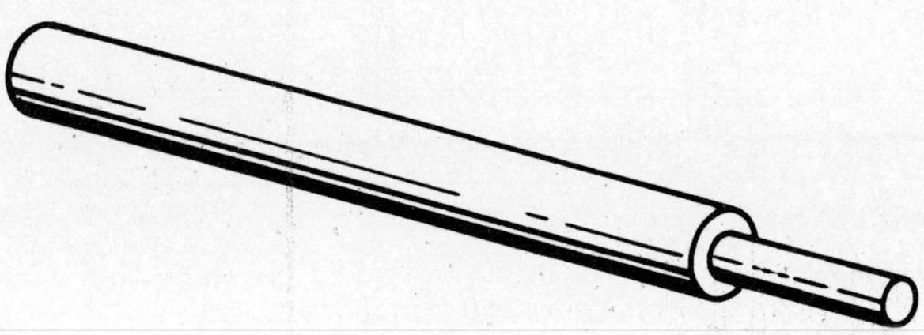

(B) Illustration of pipe.

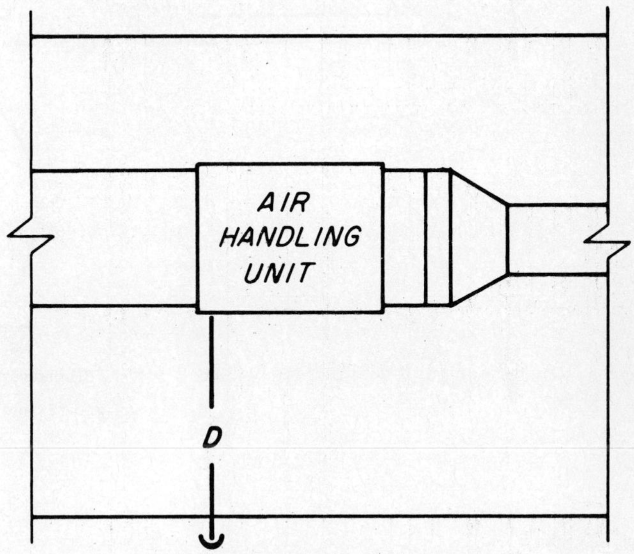

(C) Practical application.

Fig. 6-16. Drain pipe from cooling coil.

(A) Recommended symbol.

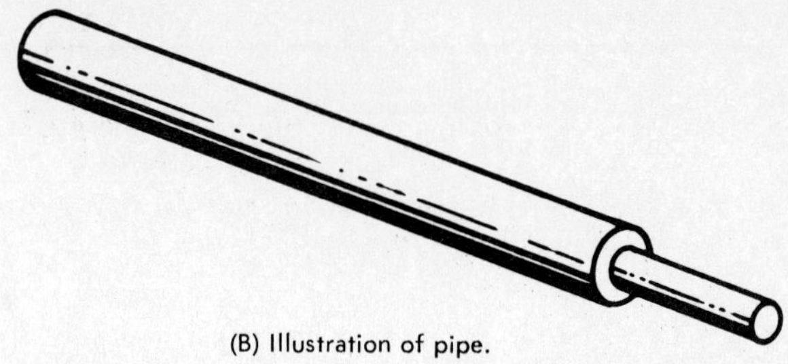

(B) Illustration of pipe.

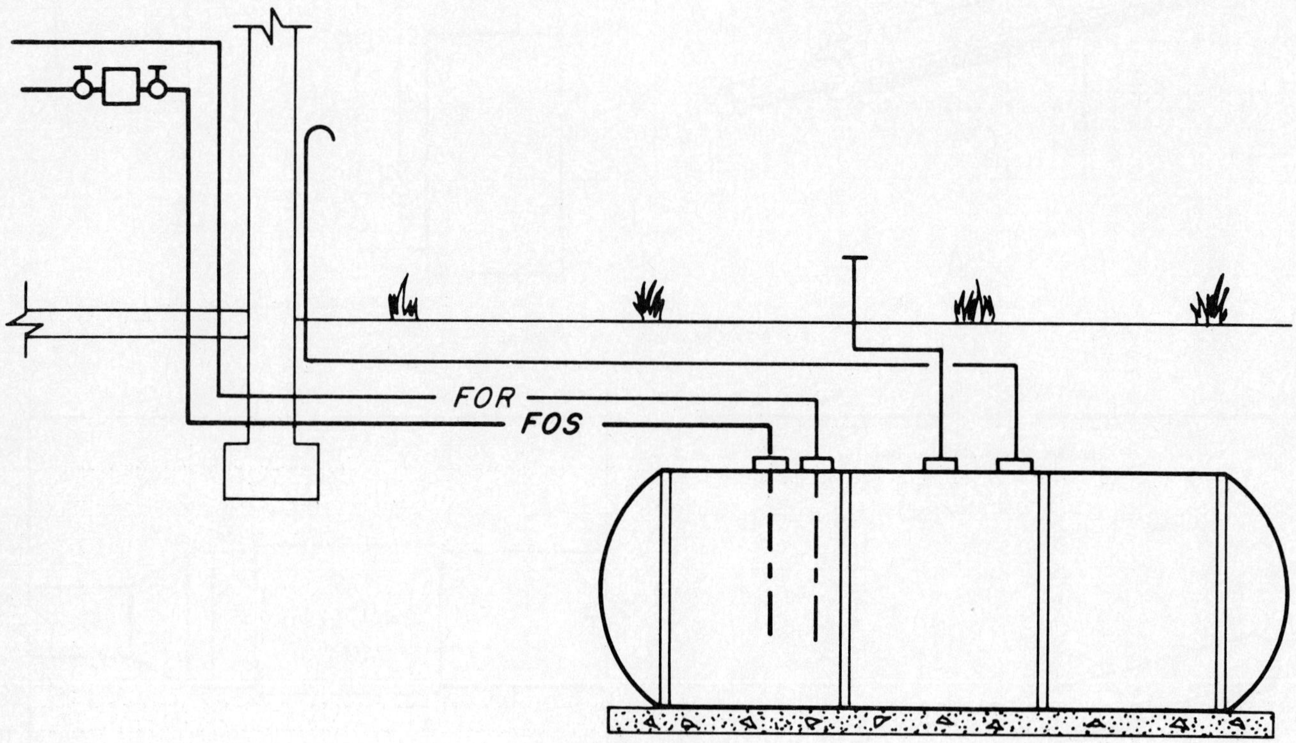

(C) Practical application.

Fig. 6-17. Fuel oil supply pipe.

(A) Recommended symbol.

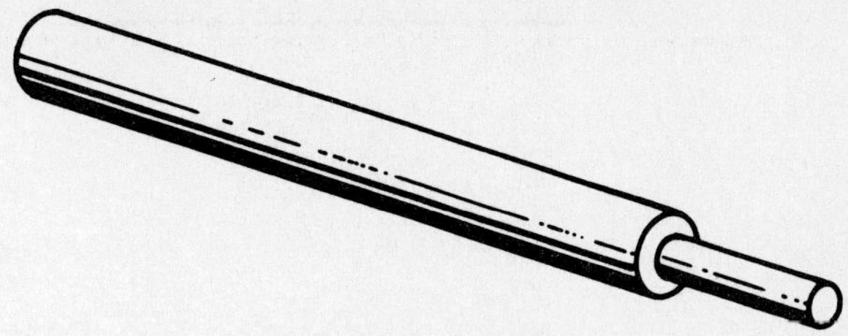

(B) Illustration of pipe.

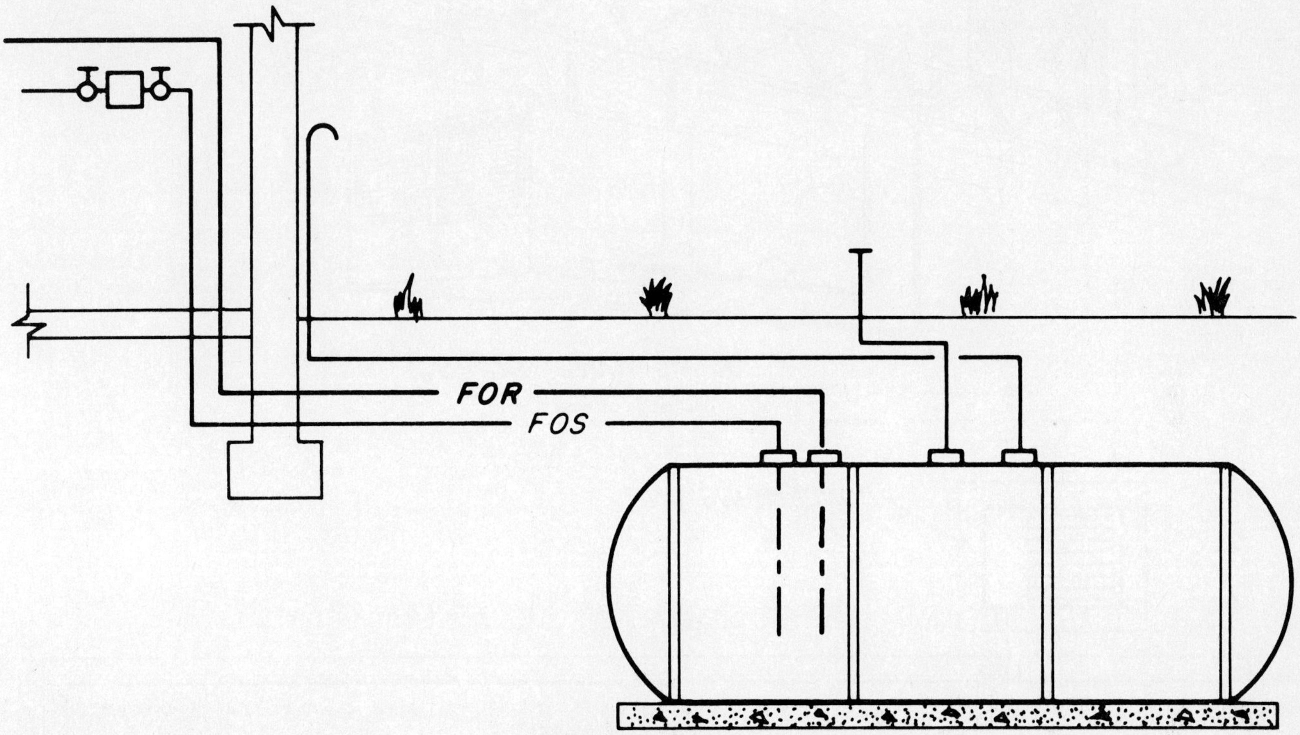

(C) Practical application.

Fig. 6-18. Fuel oil return pipe.

(A) Recommended symbol.

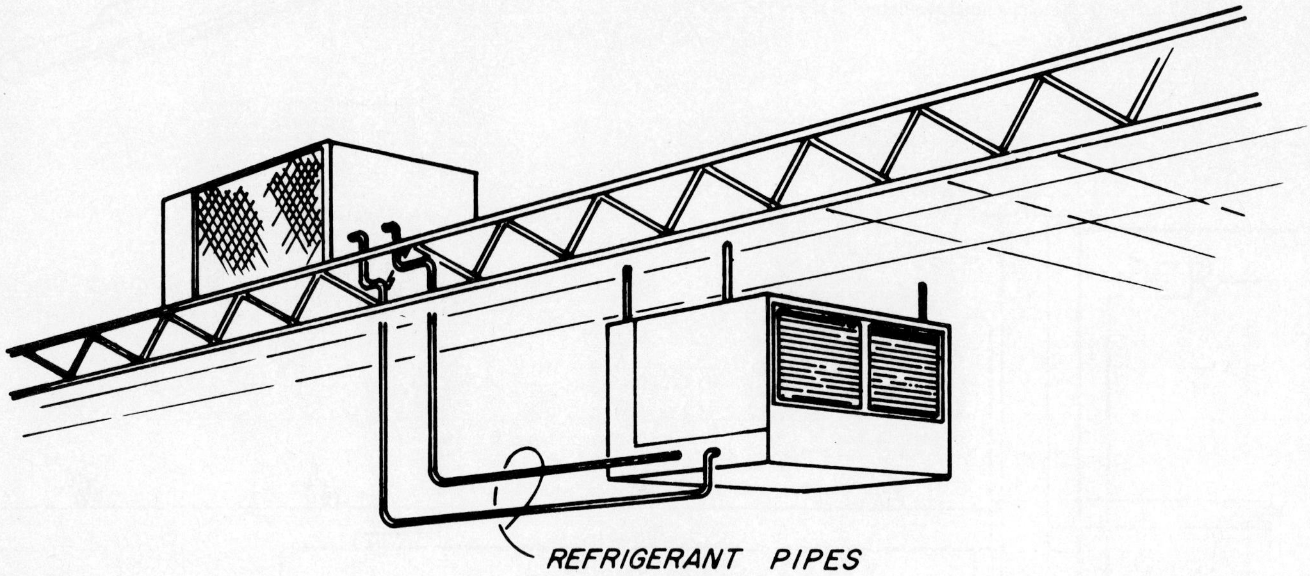

(B) Illustration of pipe.

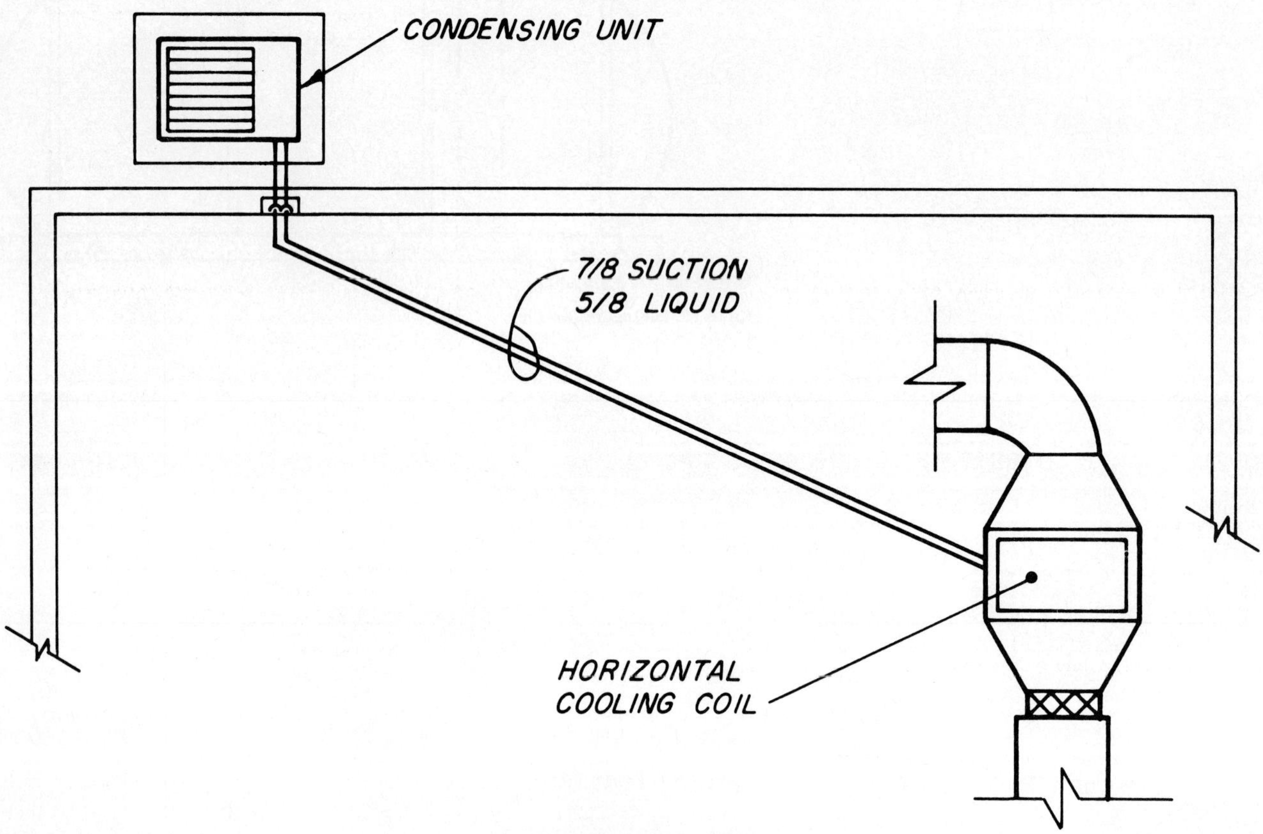

(C) Practical application.

Fig. 6-19. Refrigerant pipe.

(A) Recommended symbol.

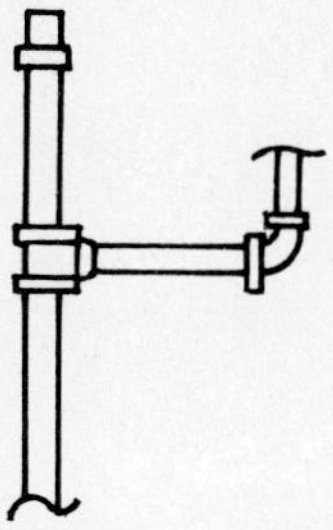

(B) Illustration of pipe rising.

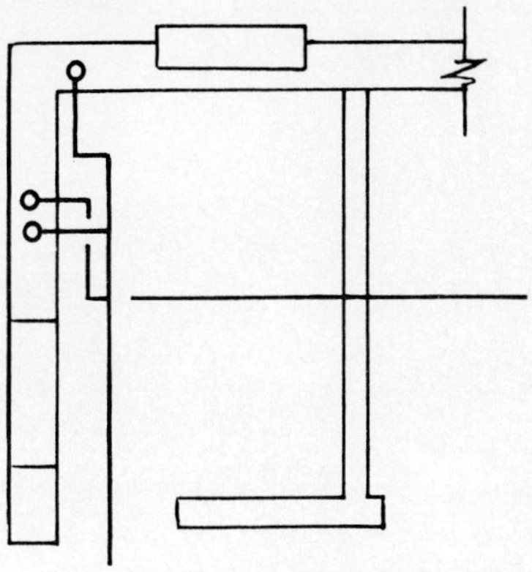

(C) Practical application.

Fig. 6-20. Pipe rising.

(A) Recommended symbol.

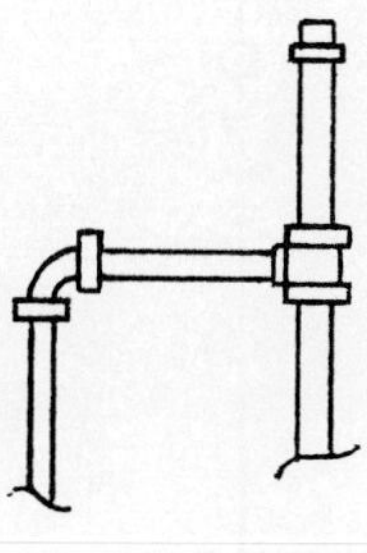

(B) Illustration of pipe turning down.

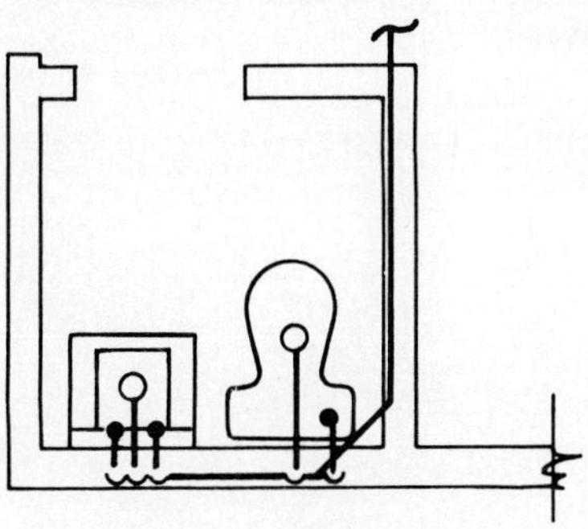

(C) Practical application.

Fig. 6-21. Pipe turning down.

(A) Recommended symbol.

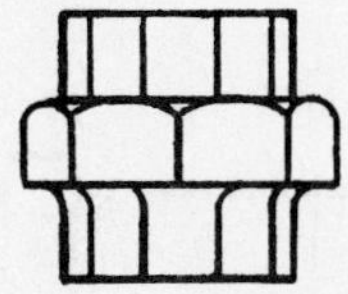

(B) Illustration of union.

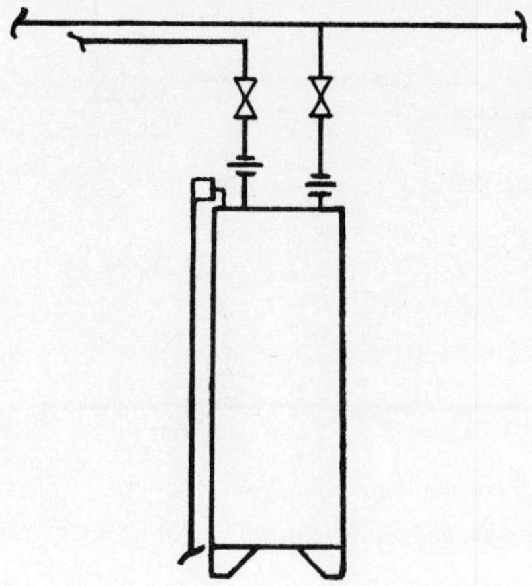

(C) Practical application.

Fig. 6-22. Union.

(A) Recommended symbol.

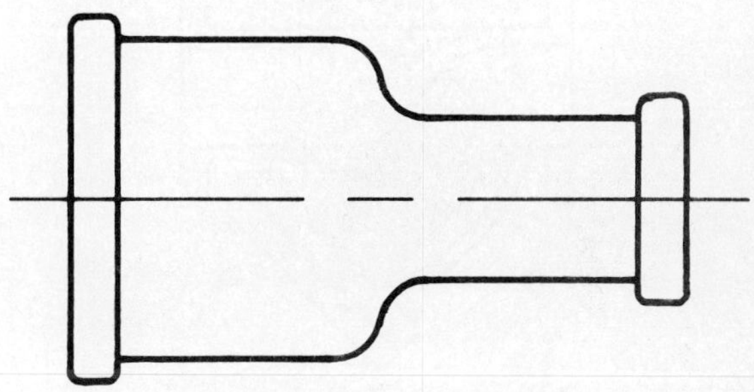

(B) Illustration of reducer-concentric.

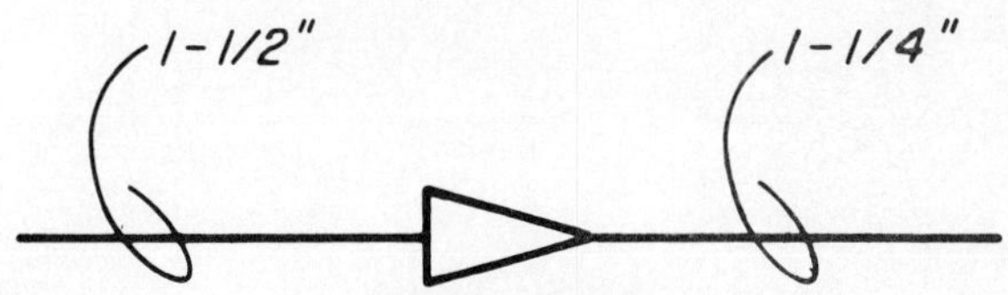

(C) Practical application.

Fig. 6-23. Reducer-concentric.

(A) Recommended symbol.

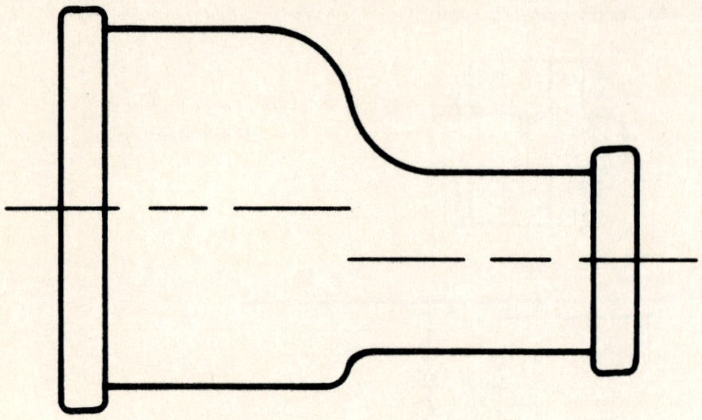

(B) Illustration of reducer eccentric.

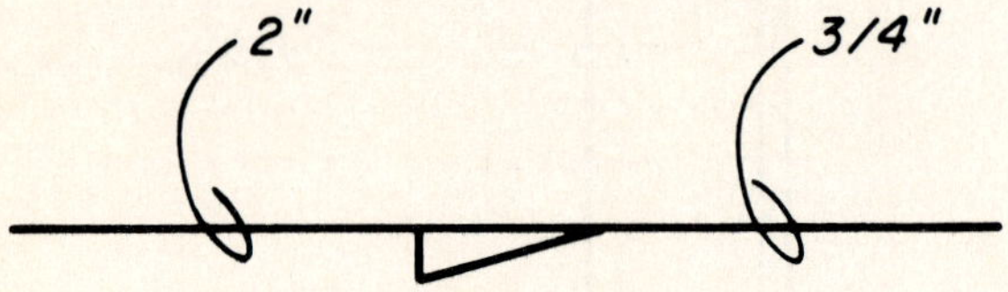

(C) Practical application.

Fig. 6-24. Reducer-eccentric.

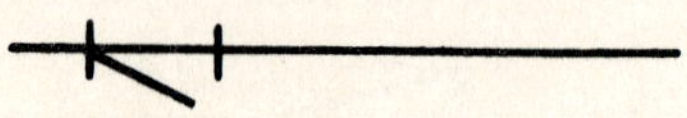

(A) Recommended symbol.

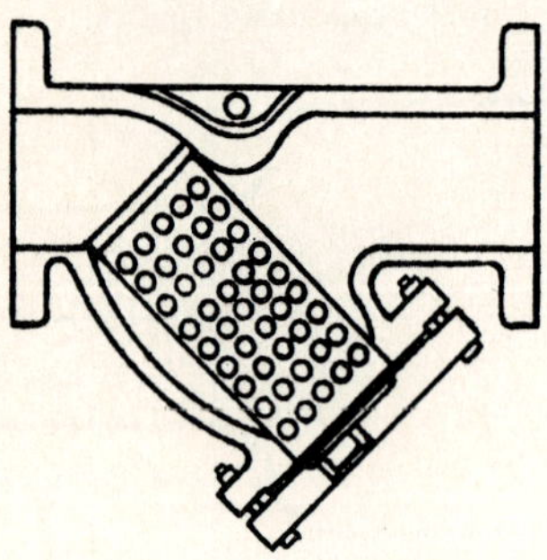

(B) Illustration of strainer.

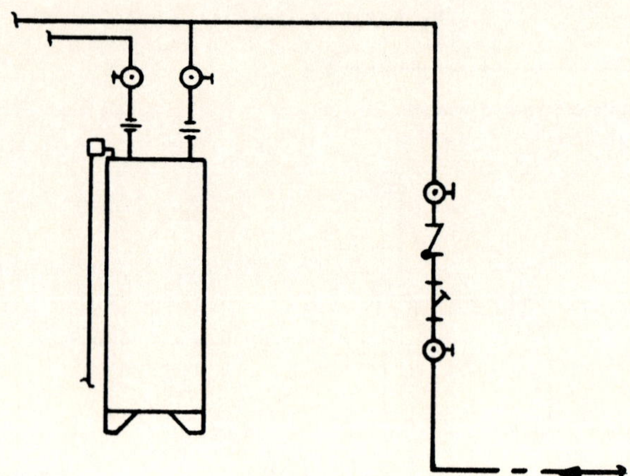

(C) Practical application.

Fig. 6-25. Strainer.

(A) Recommended symbol.

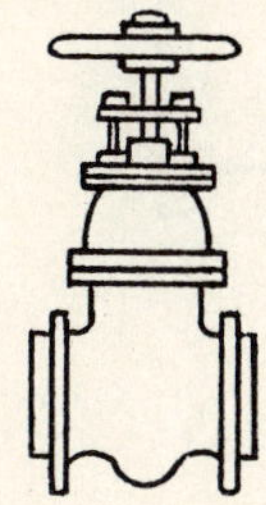

(B) Illustration of gate valve.

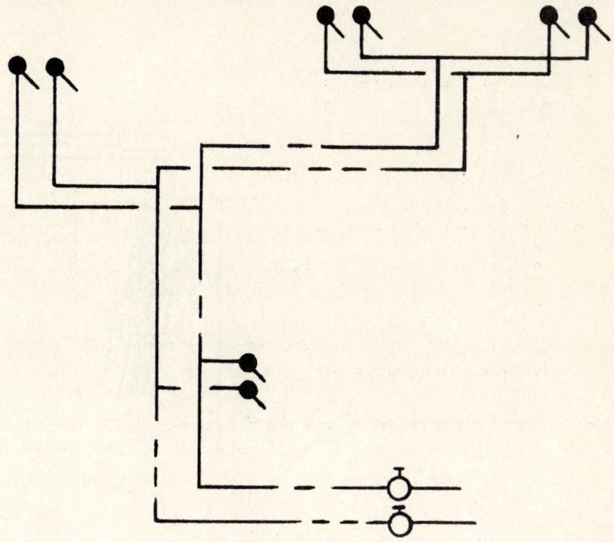

(C) Practical application.

Fig. 6-26. Gate valve.

(A) Recommended symbol.

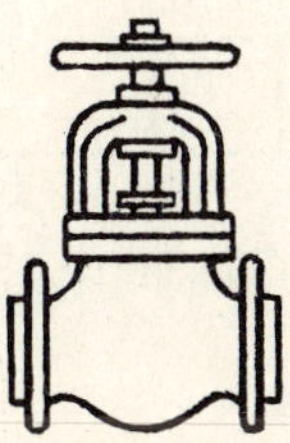

(B) Illustration of globe valve.

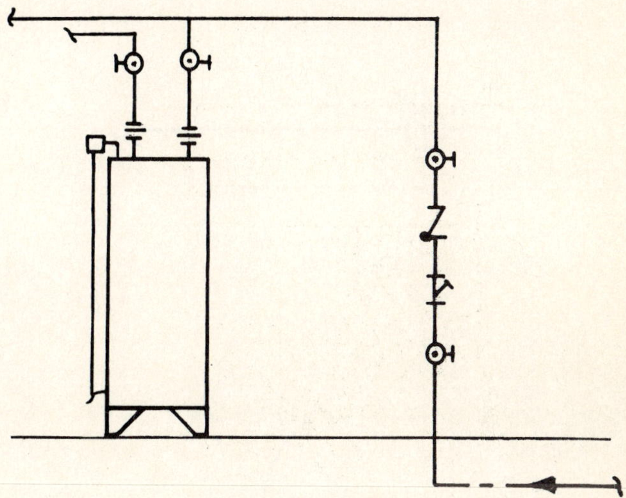

(C) Practical application.

Fig. 6-27. Globe valve.

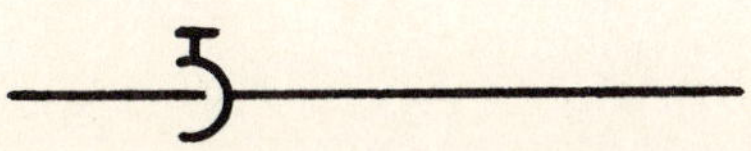

(A) Recommended symbol.

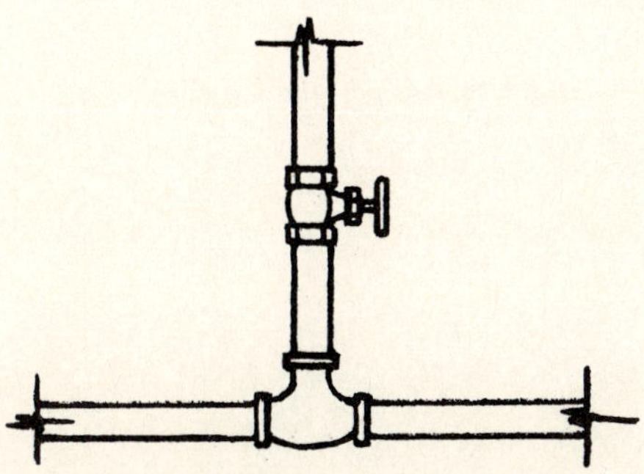

(B) Illustration of valve in riser.

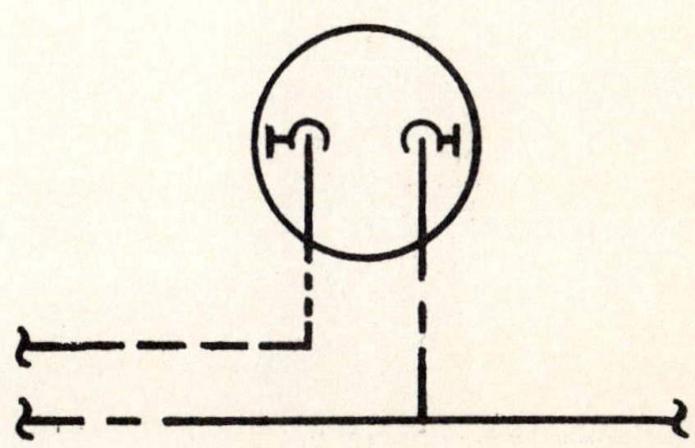

(C) Practical application.

Fig. 6-28. Valve in riser.

(A) Recommended symbol.

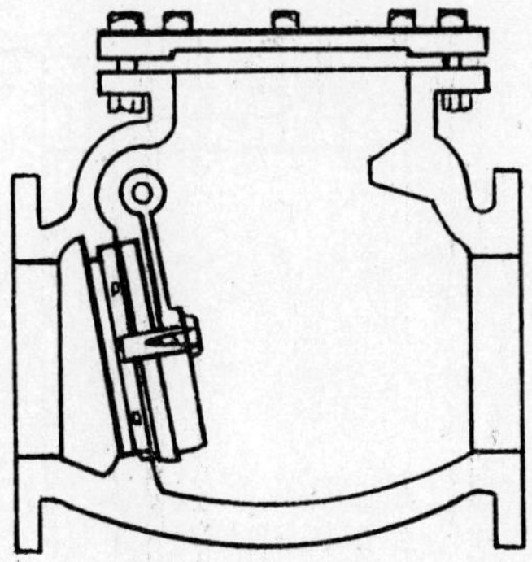

(B) Illustration of check valve.

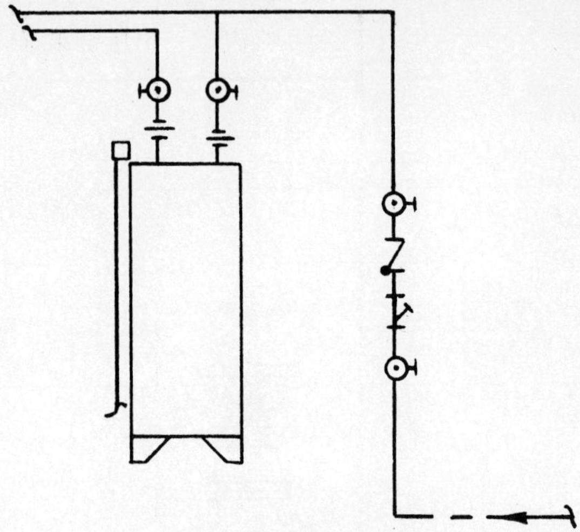

(C) Practical application.

Fig. 6-29. Check valve.

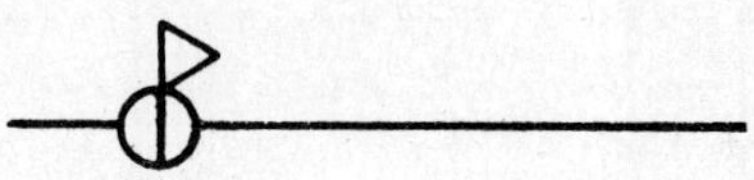

(A) Recommended symbol.

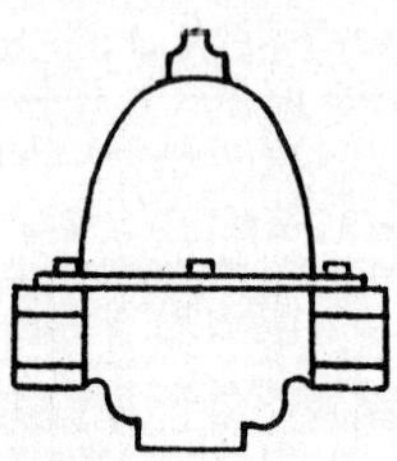

(B) Illustration of valve.

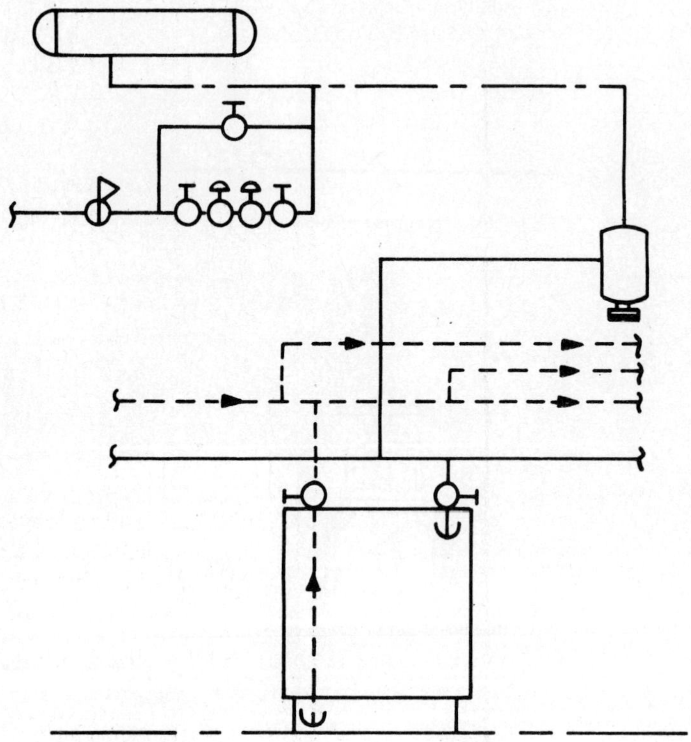

(C) Practical application.

Fig. 6-30. Pressure-reducing valve.

(A) Recommended symbol.

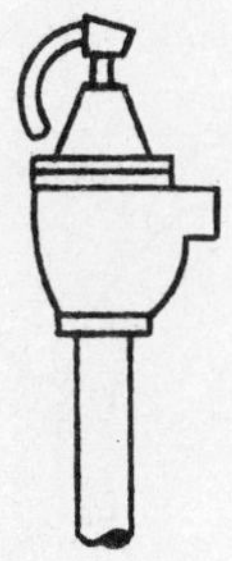

(B) Illustration of valve.

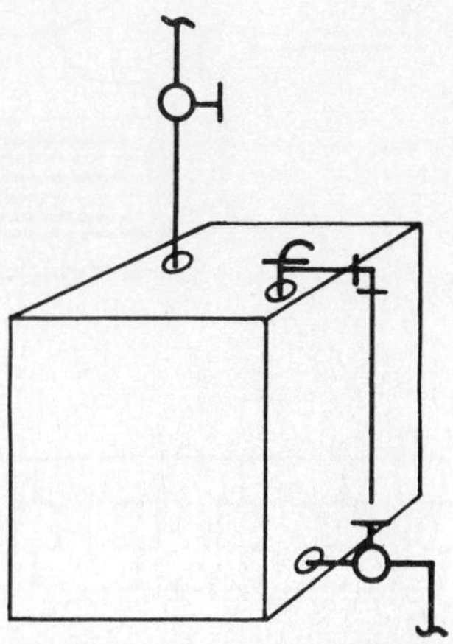

(C) Practical application.

Fig. 6-31. Pressure-relief valve.

(A) Recommended symbol.

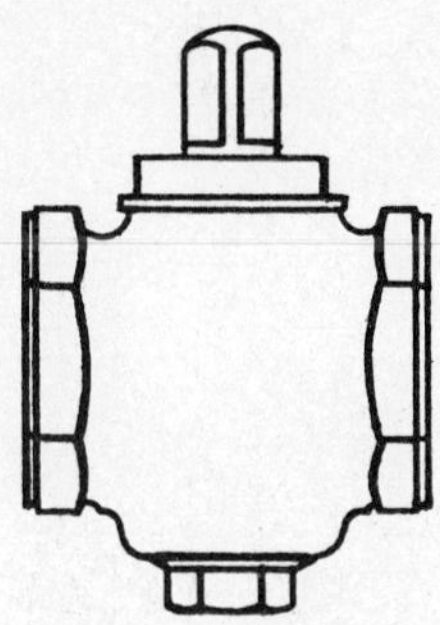

(B) Illustration of cock.

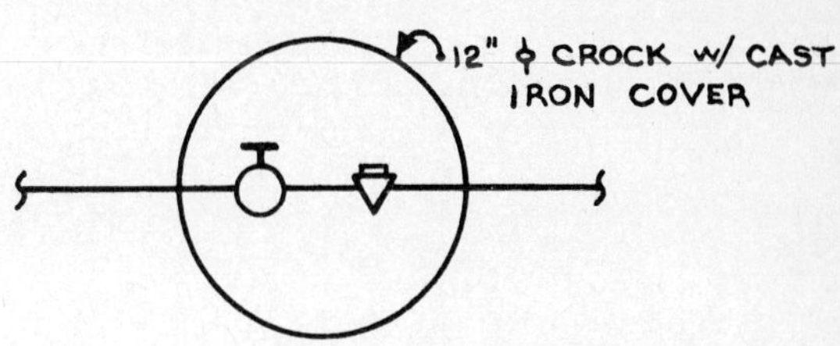

(C) Practical application.

Fig. 6-32. Square head cock.

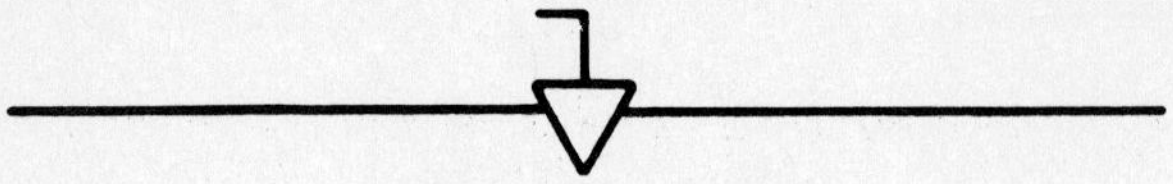

(A) Recommended symbol.

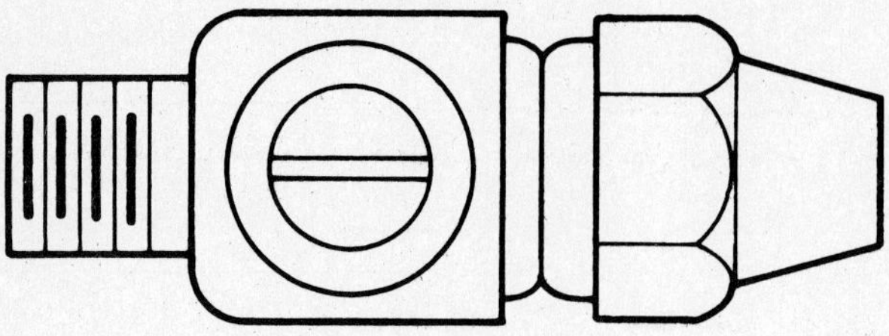

(B) Illustration of valve.

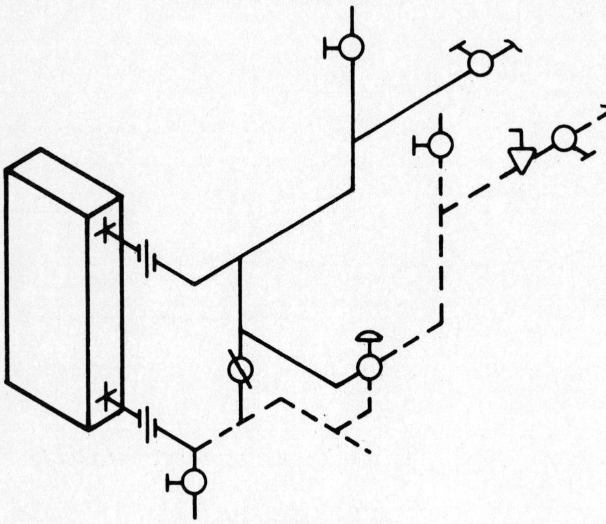

(C) Practical application.

Fig. 6-33. Balancing valve.

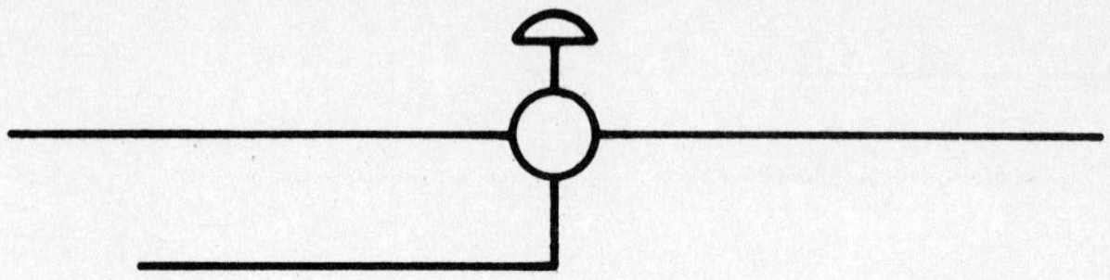

(A) Recommended symbol.

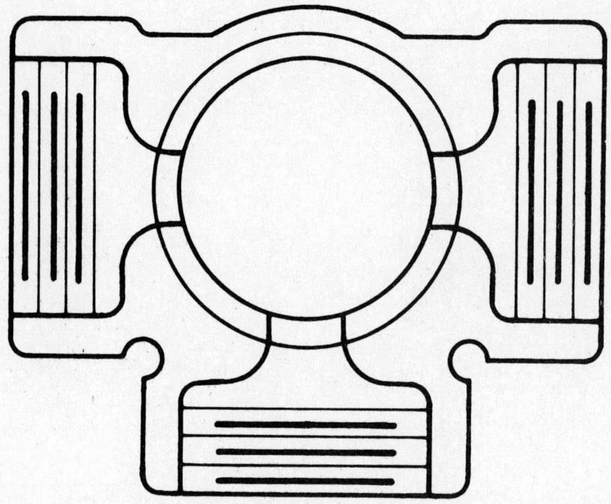

(B) Illustration of valve.

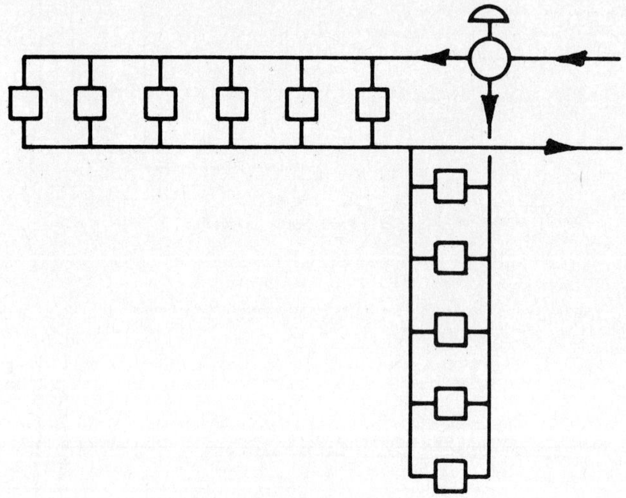

(C) Practical application.

Fig.6-34. Three-way control valve.

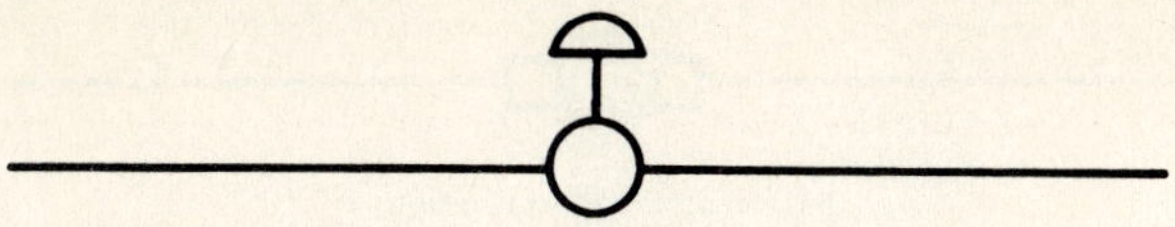

(A) Recommended symbol.

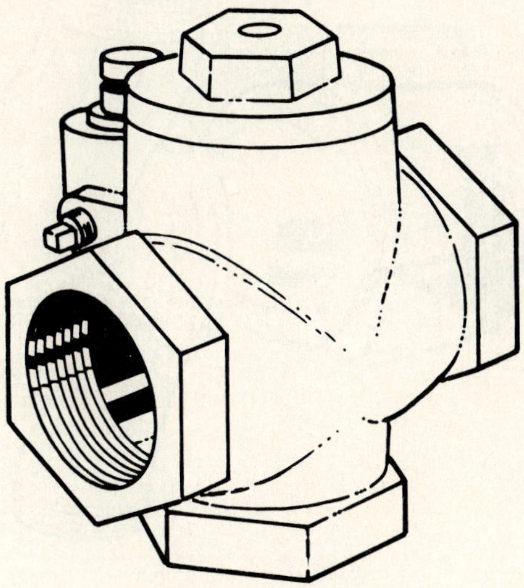

(B) Illustration of valve.

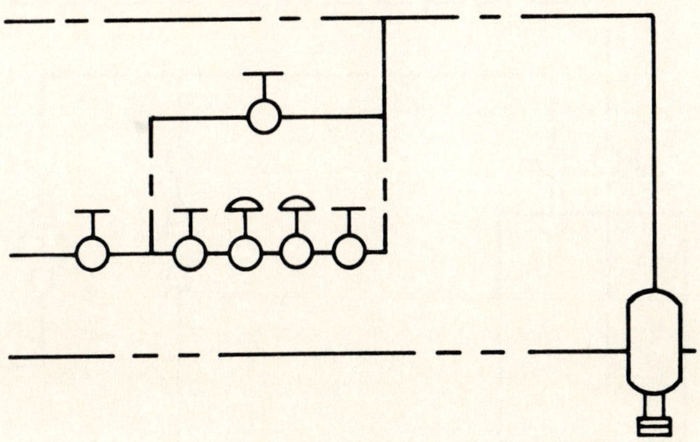

(C) Practical application.

Fig. 6-35. Two-way control valve.

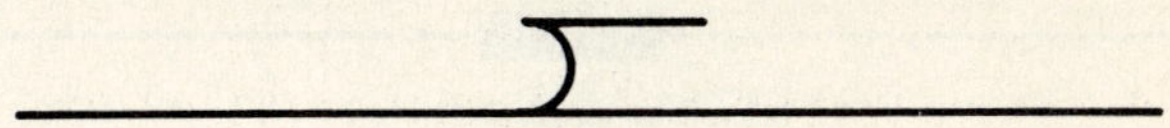

(A) Recommended symbol.

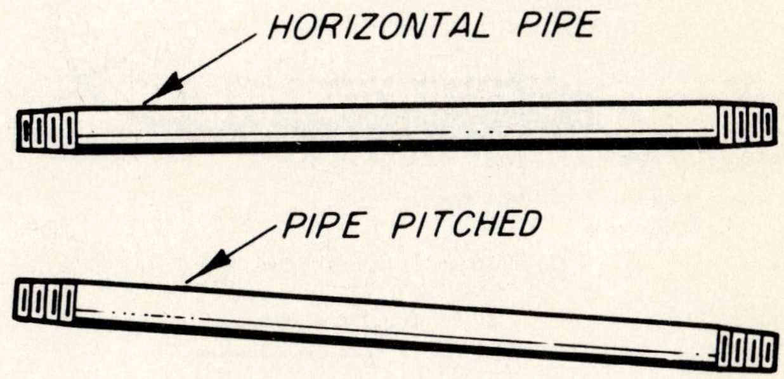

(B) Illustration of pipe pitched.

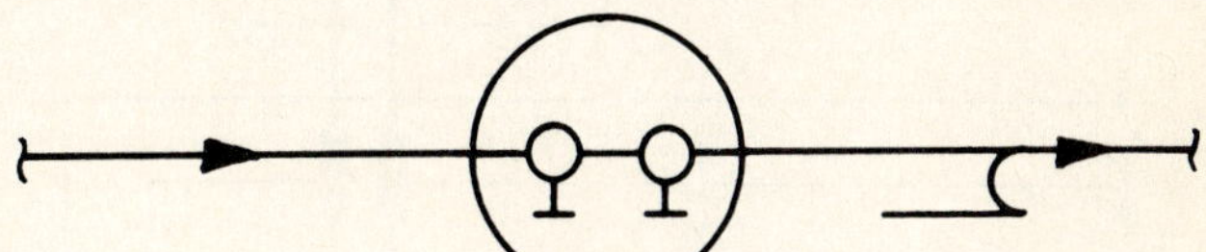

(C) Practical application.

Fig. 6-36. Pipe pitched minimum 1"/40'.

(A) Recommended symbol.

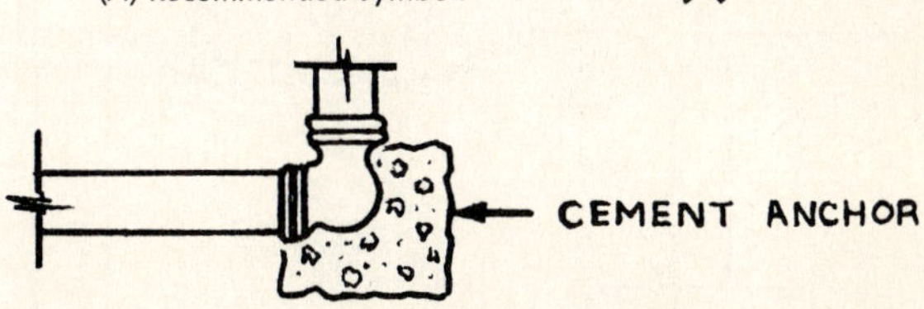

(B) Illustration of anchor.

Fig. 6-37. Anchor location.

(A) Recommended symbol.

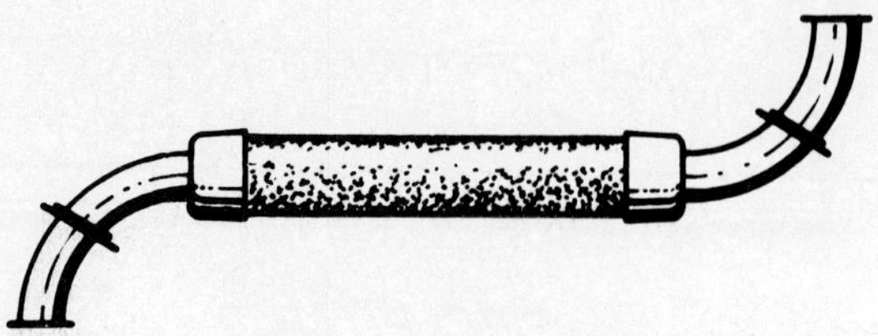

(B) Illustration of pipe connection.

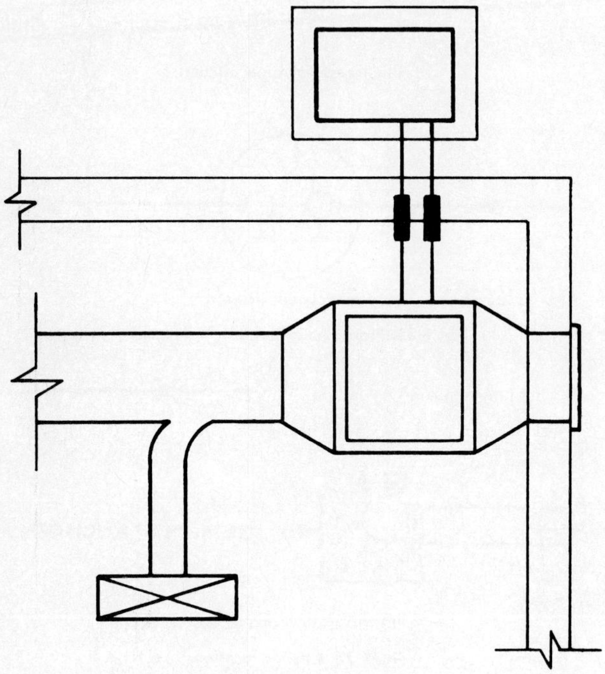

(C) Practical application.

Fig. 6-38. Flexible pipe connection.

(A) Recommended symbol.

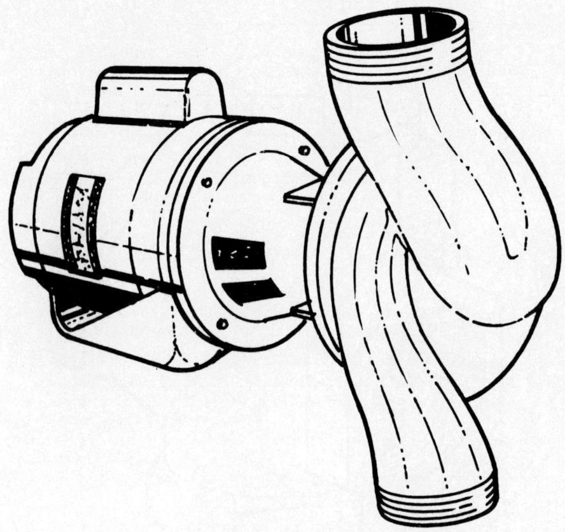

(B) Illustration of pump.

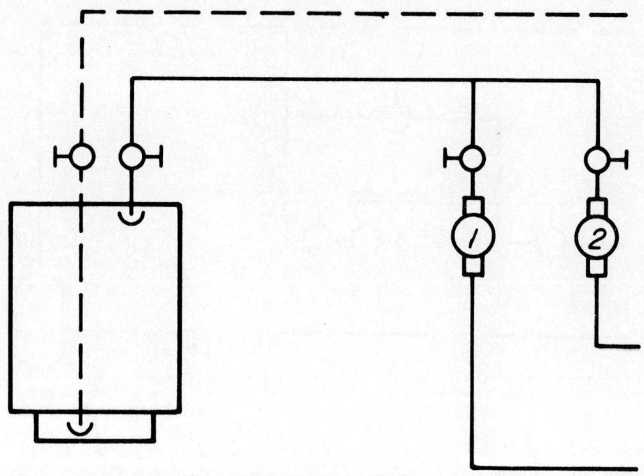

(C) Practical application.

Fig. 6-39. In-line pump.

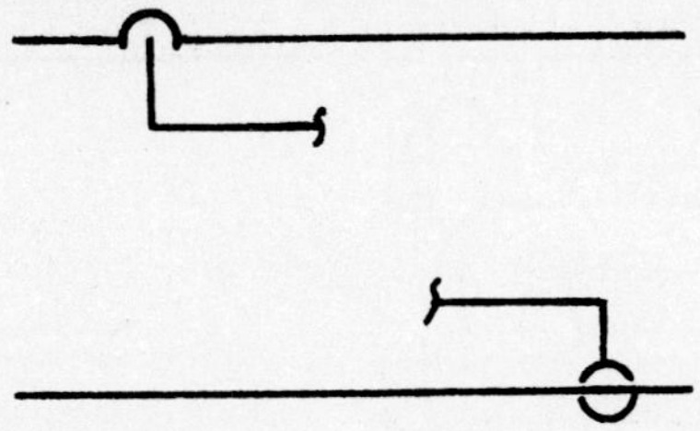

(A) Recommended symbol.

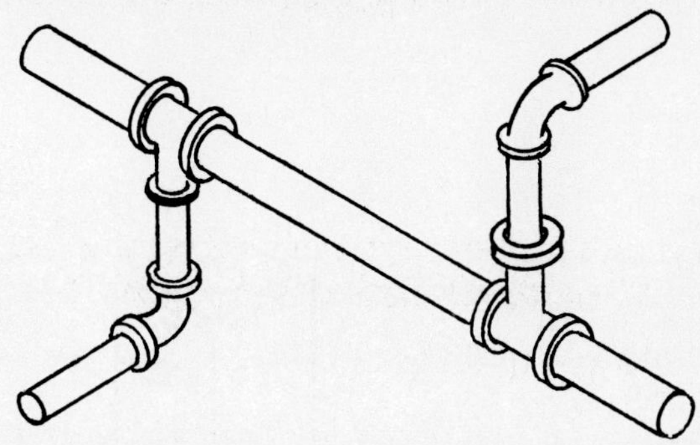

(B) Illustration of top and bottom takeoff.

(C) Practical application.

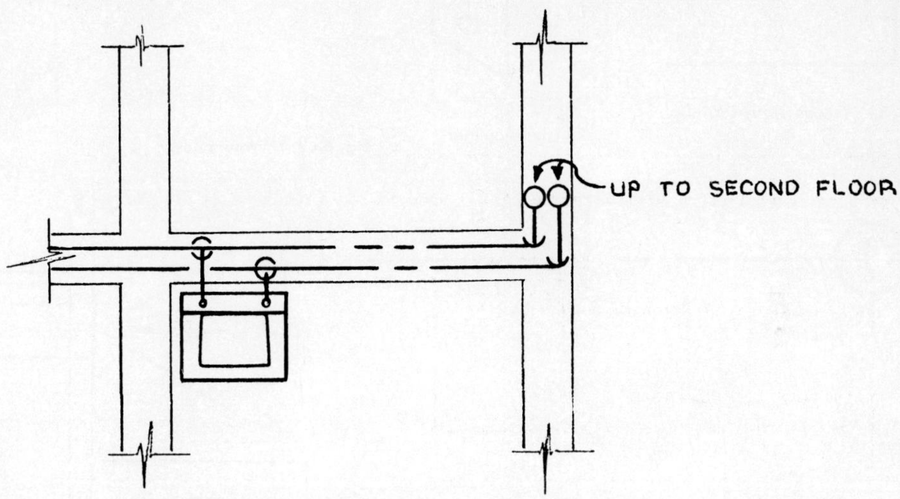

Fig. 6-40. Top and bottom takeoff.

(A) Recommended symbol.

(B) Illustration of gauge.

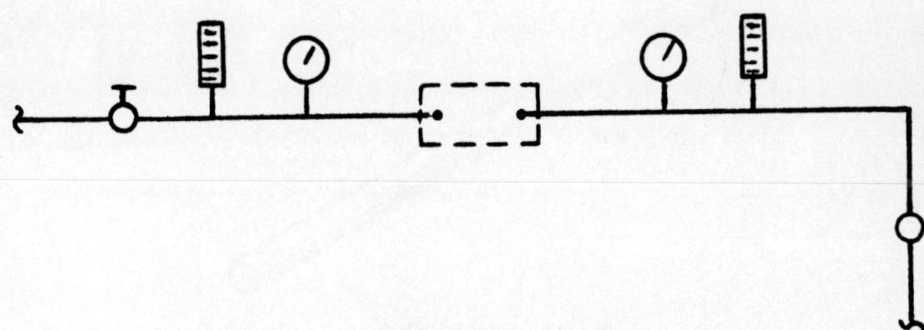

(C) Practical application.

Fig. 6-41. Pressure gauge.

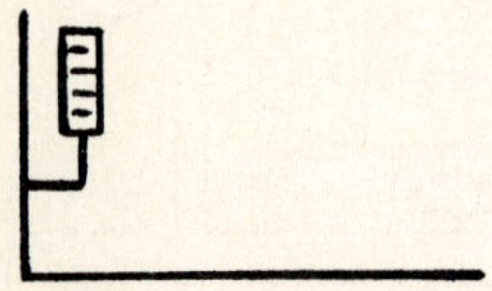

(A) Recommended symbol.

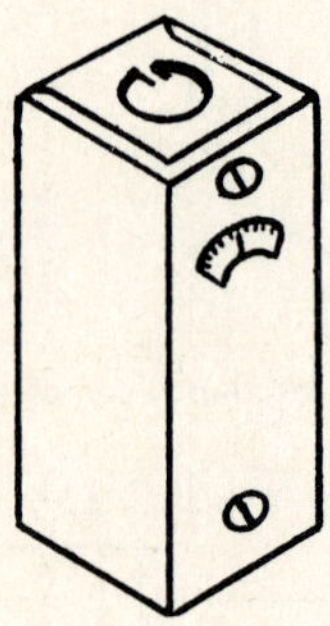

(B) Illustration of thermometer.

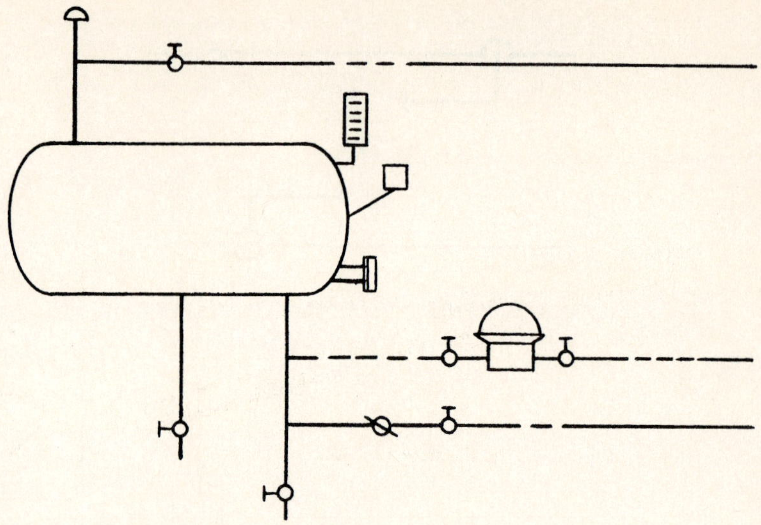

(C) Practical application.

Fig. 6-42. Thermometer.

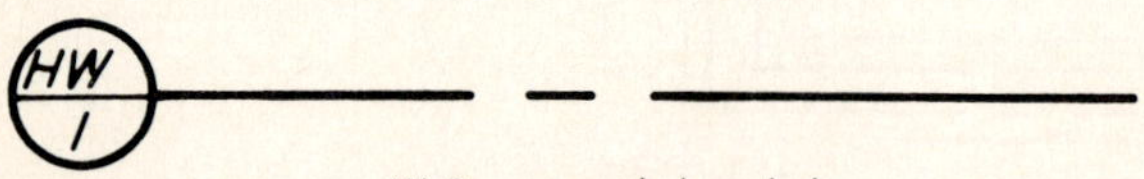

(A) Recommended symbol.

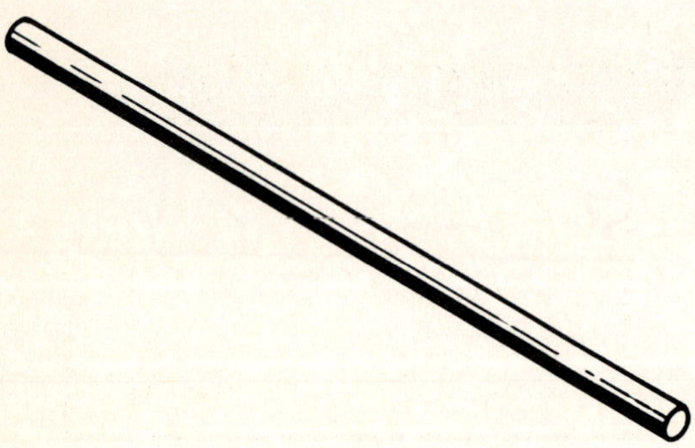

(B) Illustration of riser.

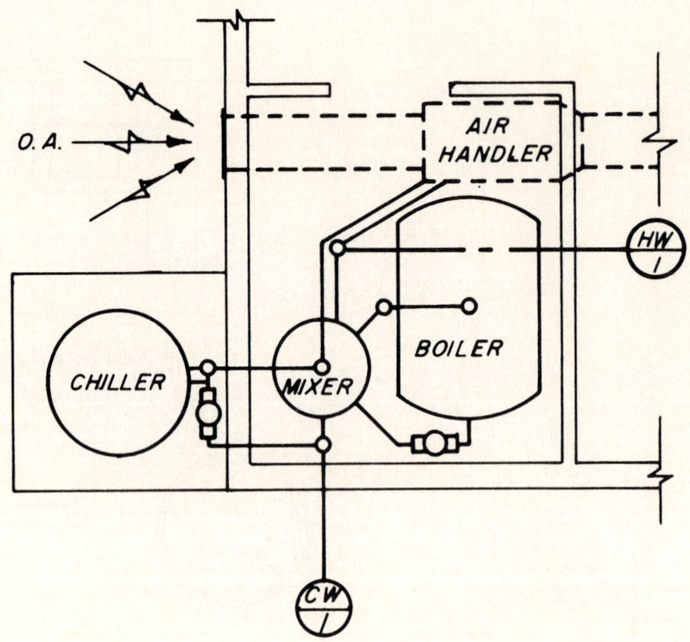

(C) Practical application.

Fig. 6-43. Hot-water riser.

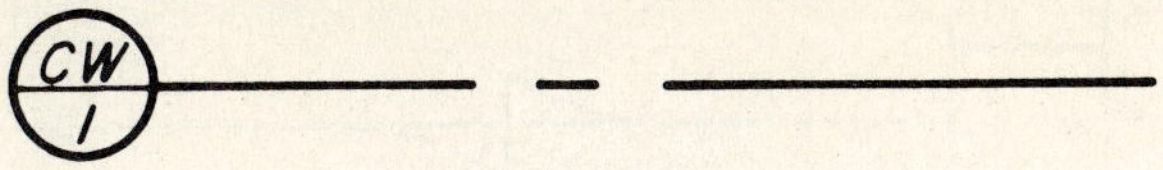

(A) Recommended symbol.

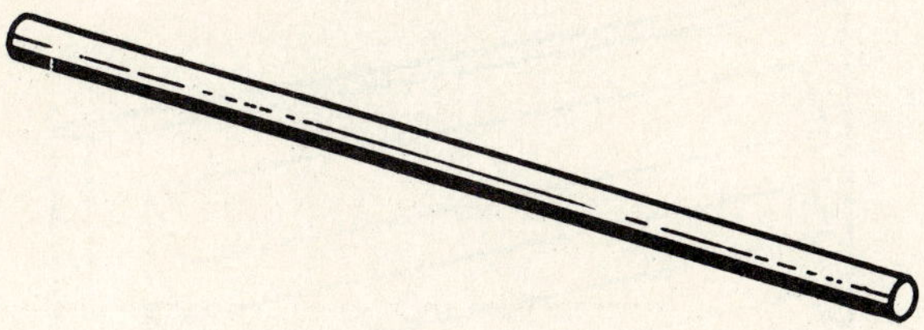

(B) Illustration of riser.

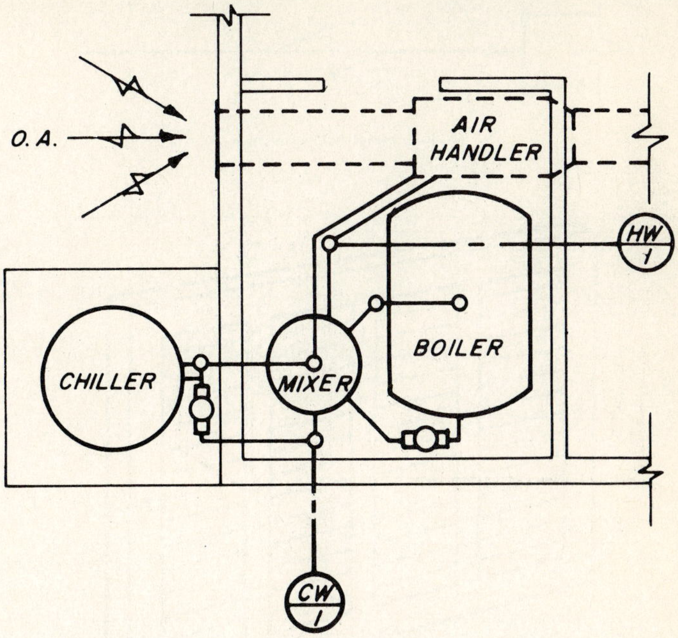

(C) Practical application.

Fig. 6-44. Chilled-water riser

(A) Recommended symbol.

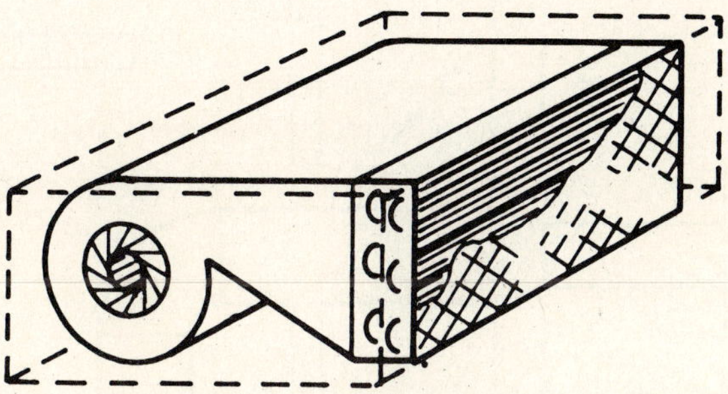

(B) Illustration of fan coil unit.

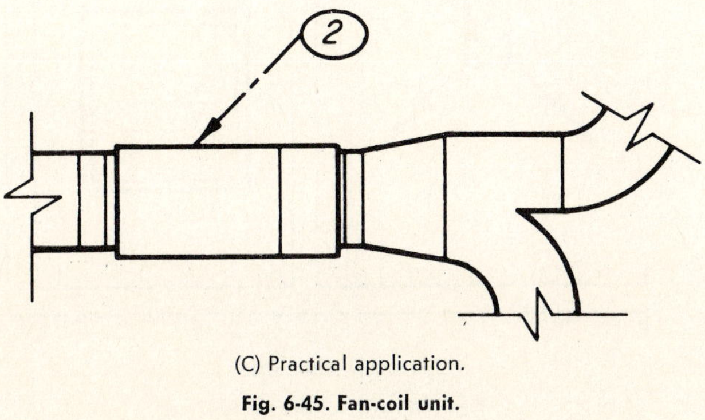

(C) Practical application.

Fig. 6-45. Fan-coil unit.

(A) Recommended symbol.

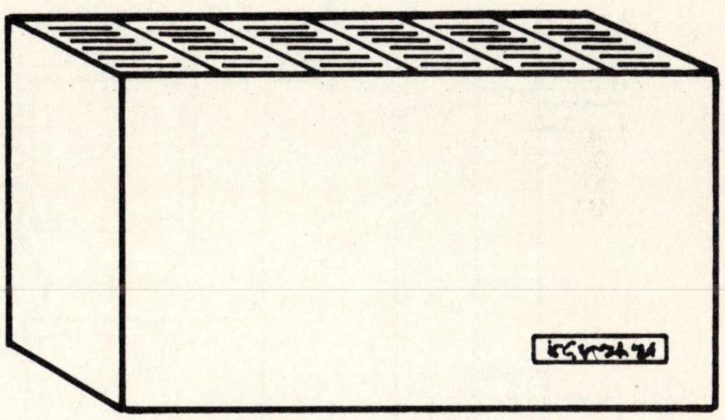

(B) Illustration of equipment.

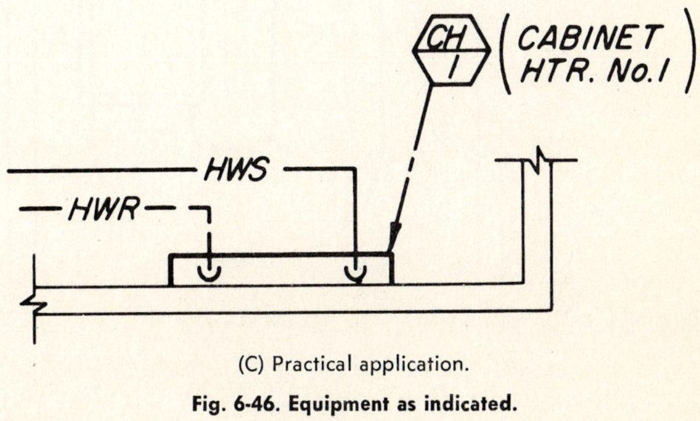

(C) Practical application.

Fig. 6-46. Equipment as indicated.

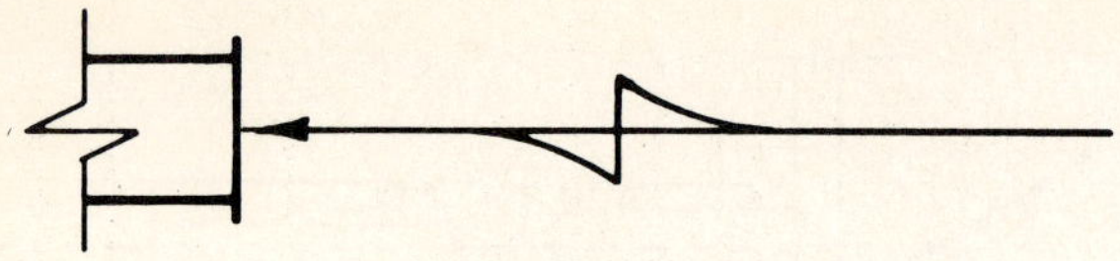

(A) Recommended symbol.

(B) Illustration of air into register.

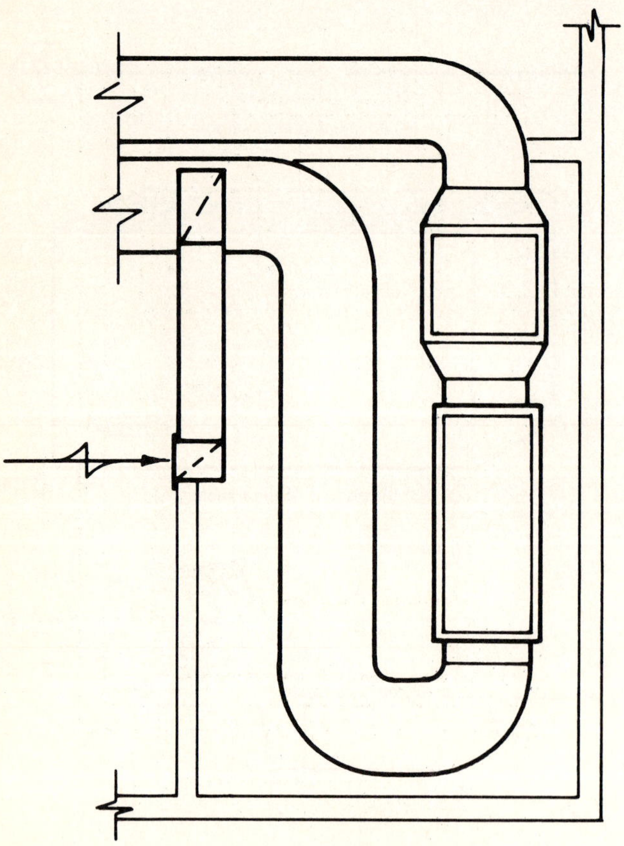

(C) Practical application.

Fig. 6-47. Air into register.

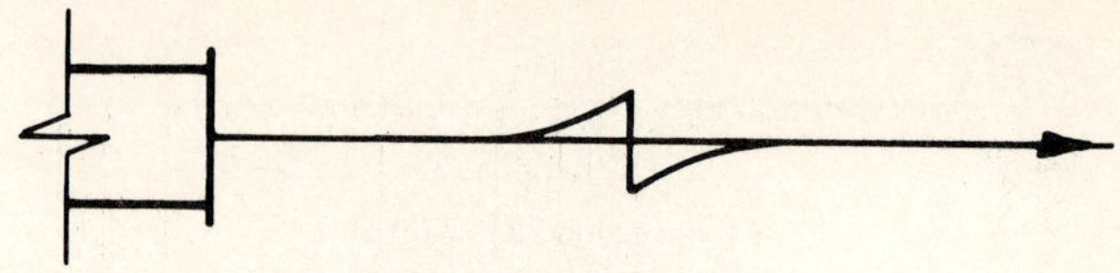

(A) Recommended symbol.

(B) Illustration of air out of register.

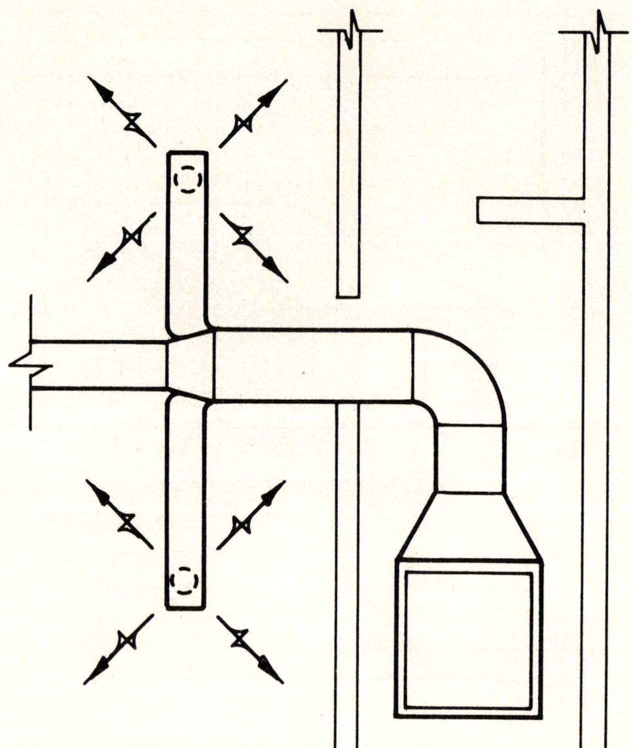

(C) Practical application.

Fig. 6-48. Air out of register.

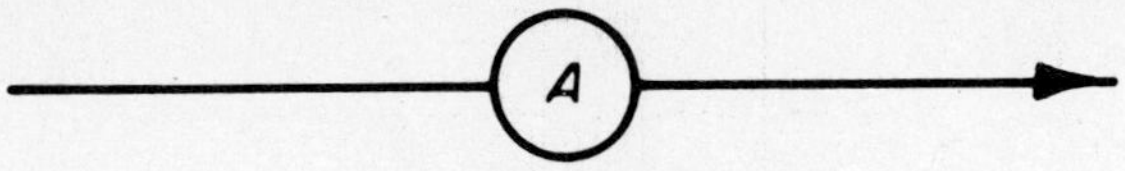

(A) Recommended symbol.

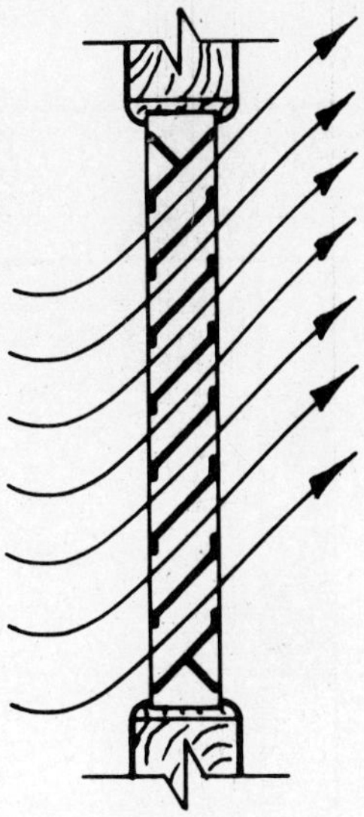

(B) Illustration of airflow.

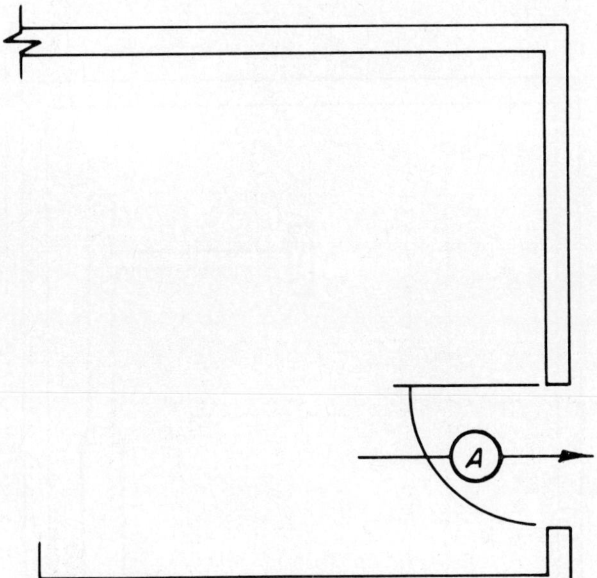

(C) Practical application.

Fig. 6-49. Airflow through undercut or louvered door.

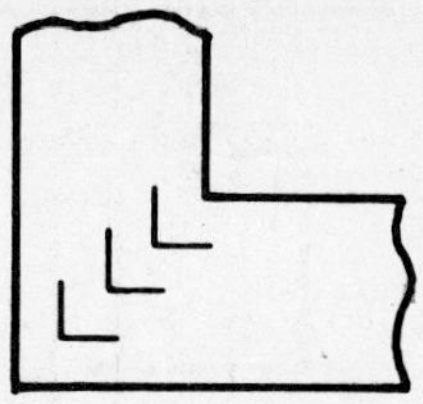

(A) Recommended symbol.

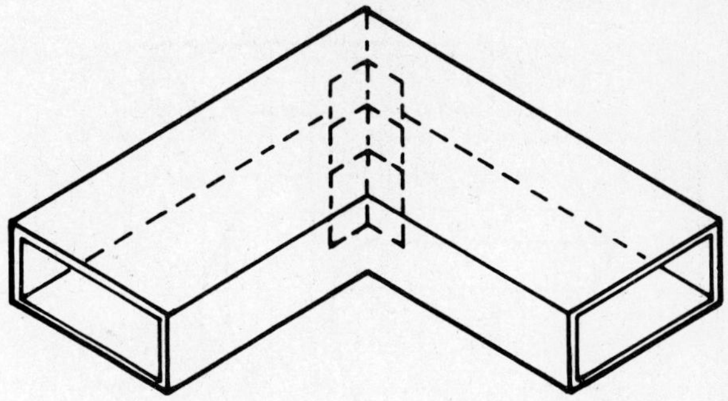

(B) Illustration of turning vanes.

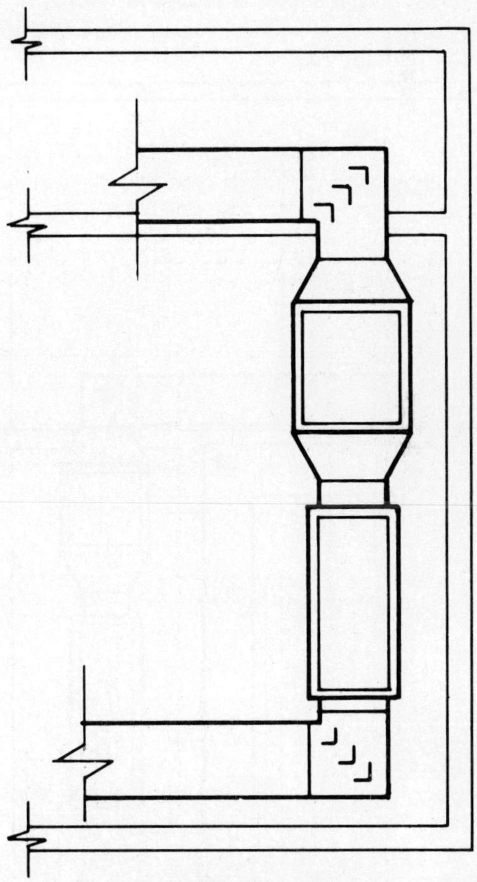

(C) Practical application.

Fig. 6-50. Turning vanes.

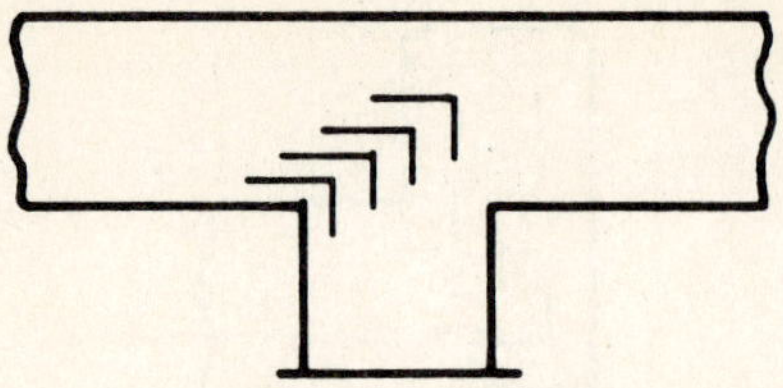

(A) Recommended symbol.

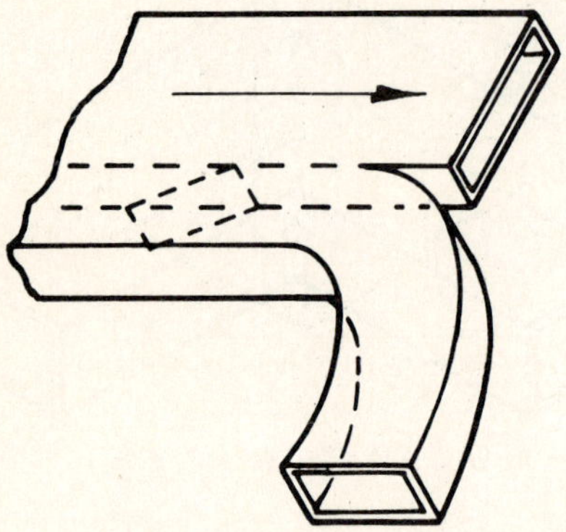

(B) Illustration of air extractor.

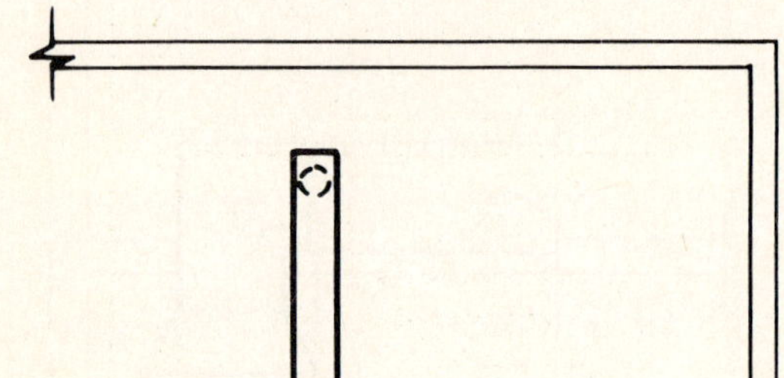

(C) Practical application.

Fig. 6-51. Air extractor.

(A) Recommended symbol.

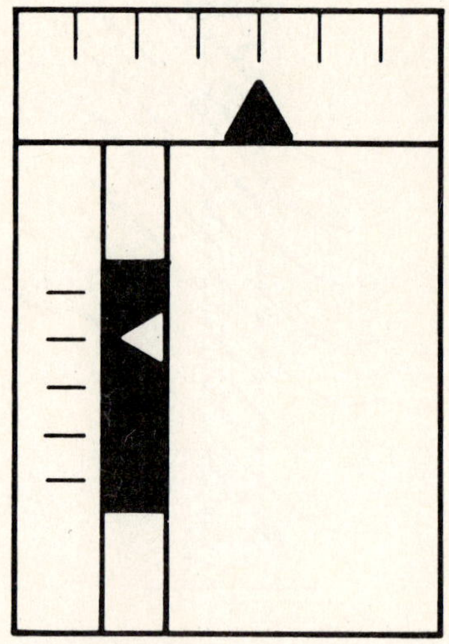

(B) Illustration of thermostat.

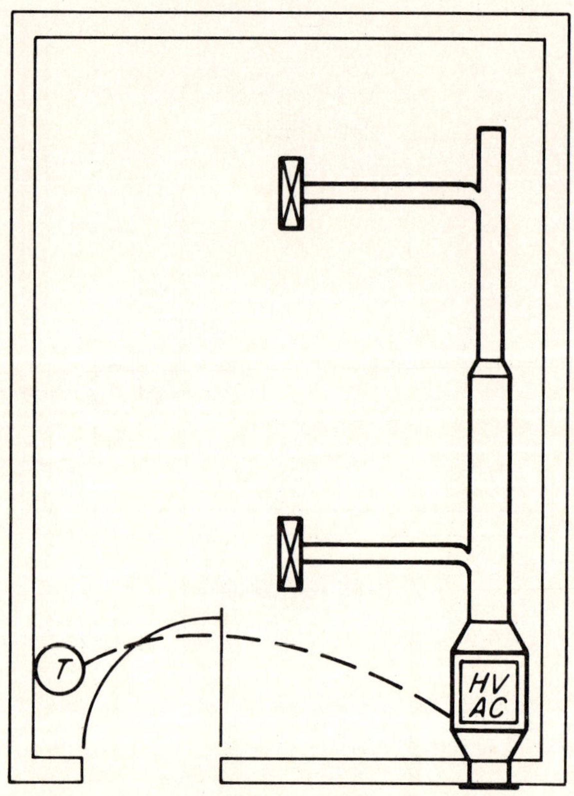

(C) Practical application.

Fig. 6-52. Various kinds of thermostats.

(A) Recommended symbol.

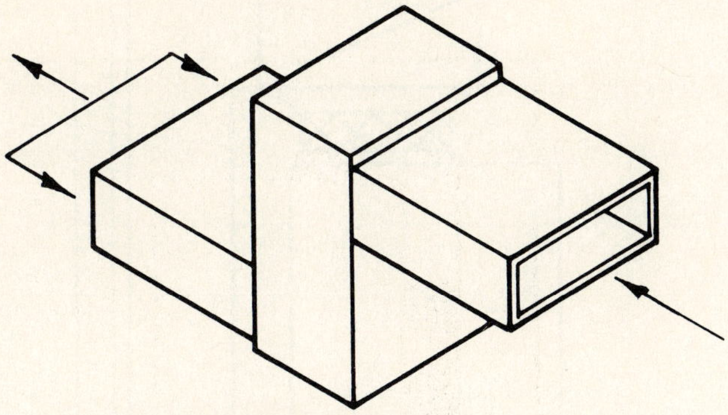

(B) Illustration of air duct.

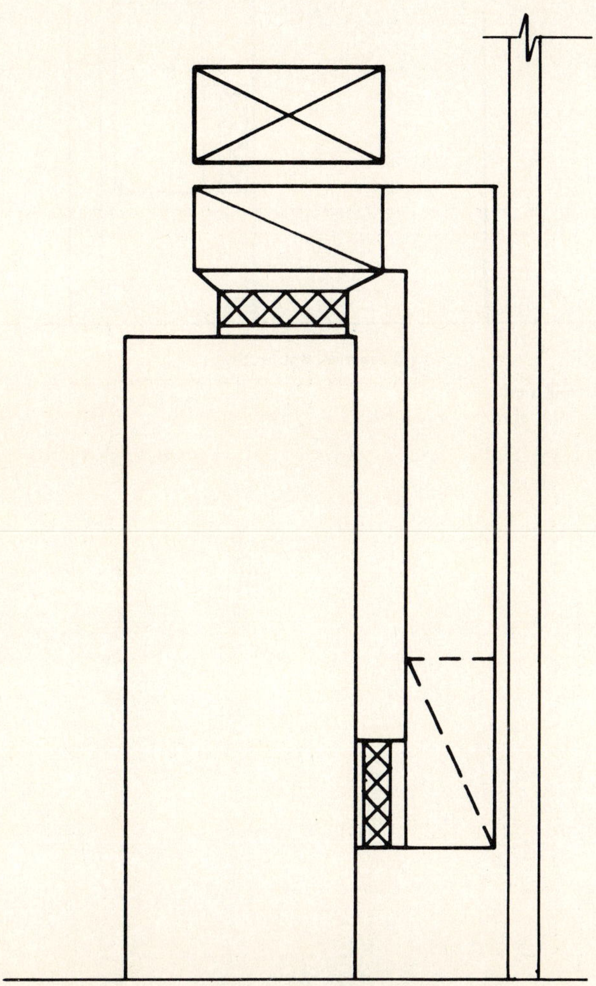

(C) Practical application.

Fig. 6-53. Supply-air duct system.

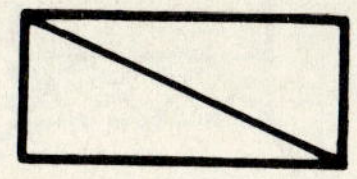

(A) Recommended symbol.

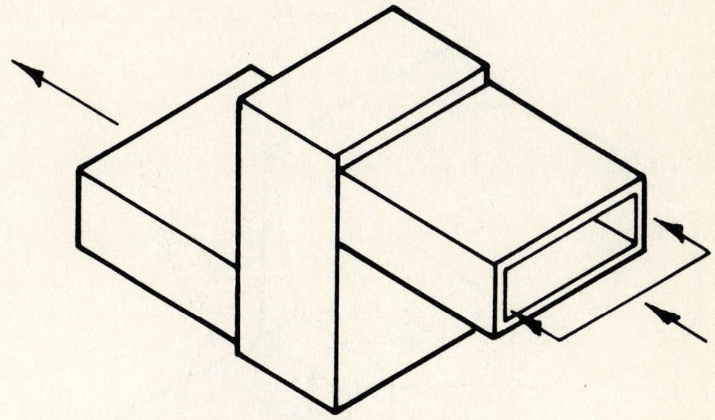

(B) Illustration of air or exhaust duct.

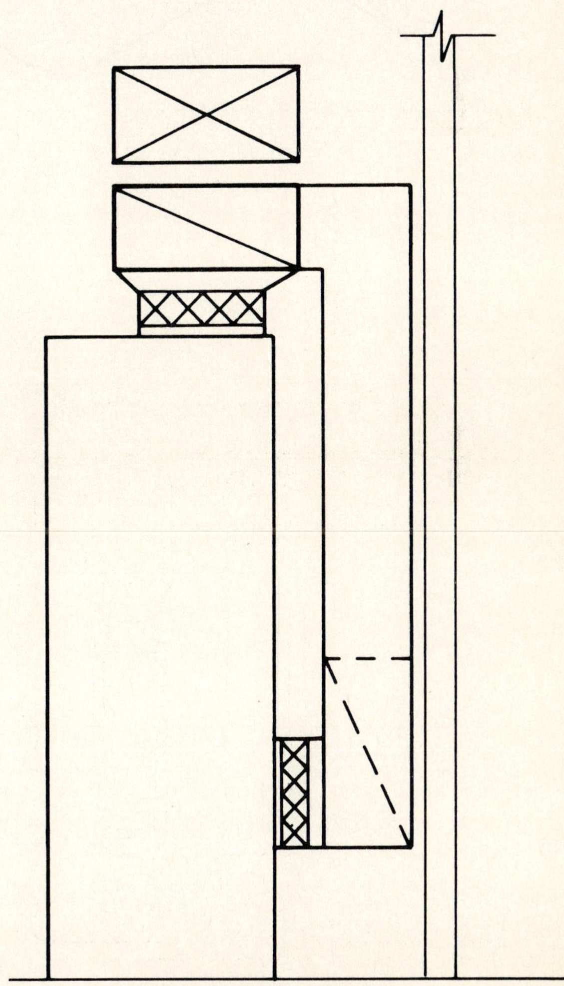

(C) Practical application.

Fig. 6-54. Return-air or exhaust duct system.

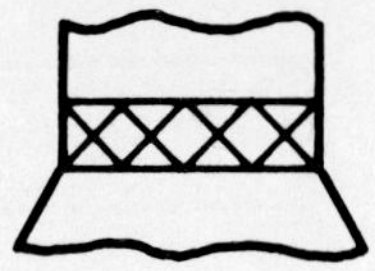

(A) Recommended symbol.

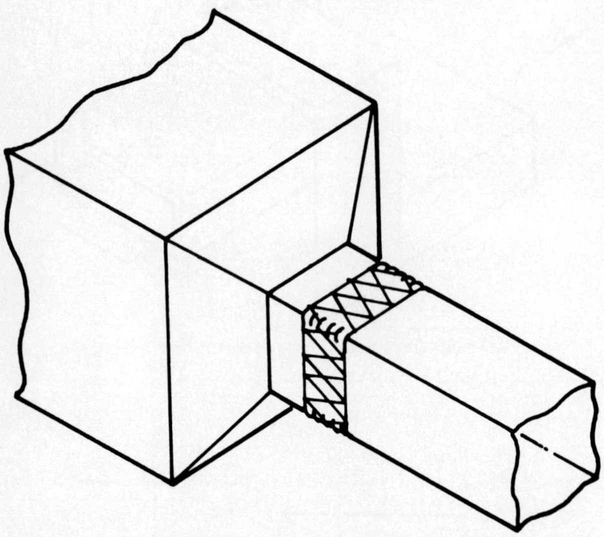

(B) Illustration of duct connection.

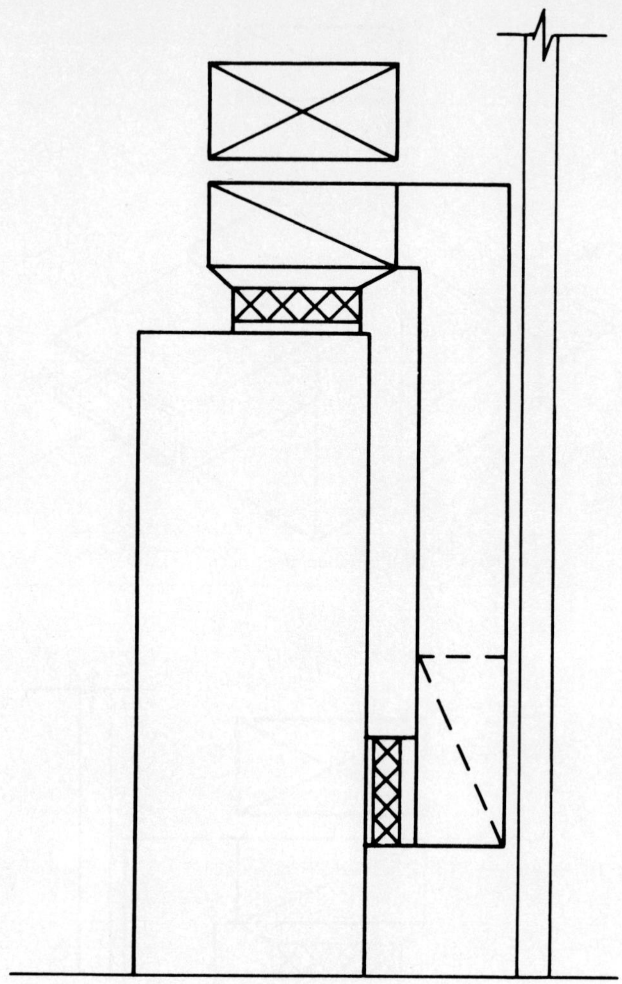

(C) Practical application.

Fig. 6-55. Flexible duct connection.

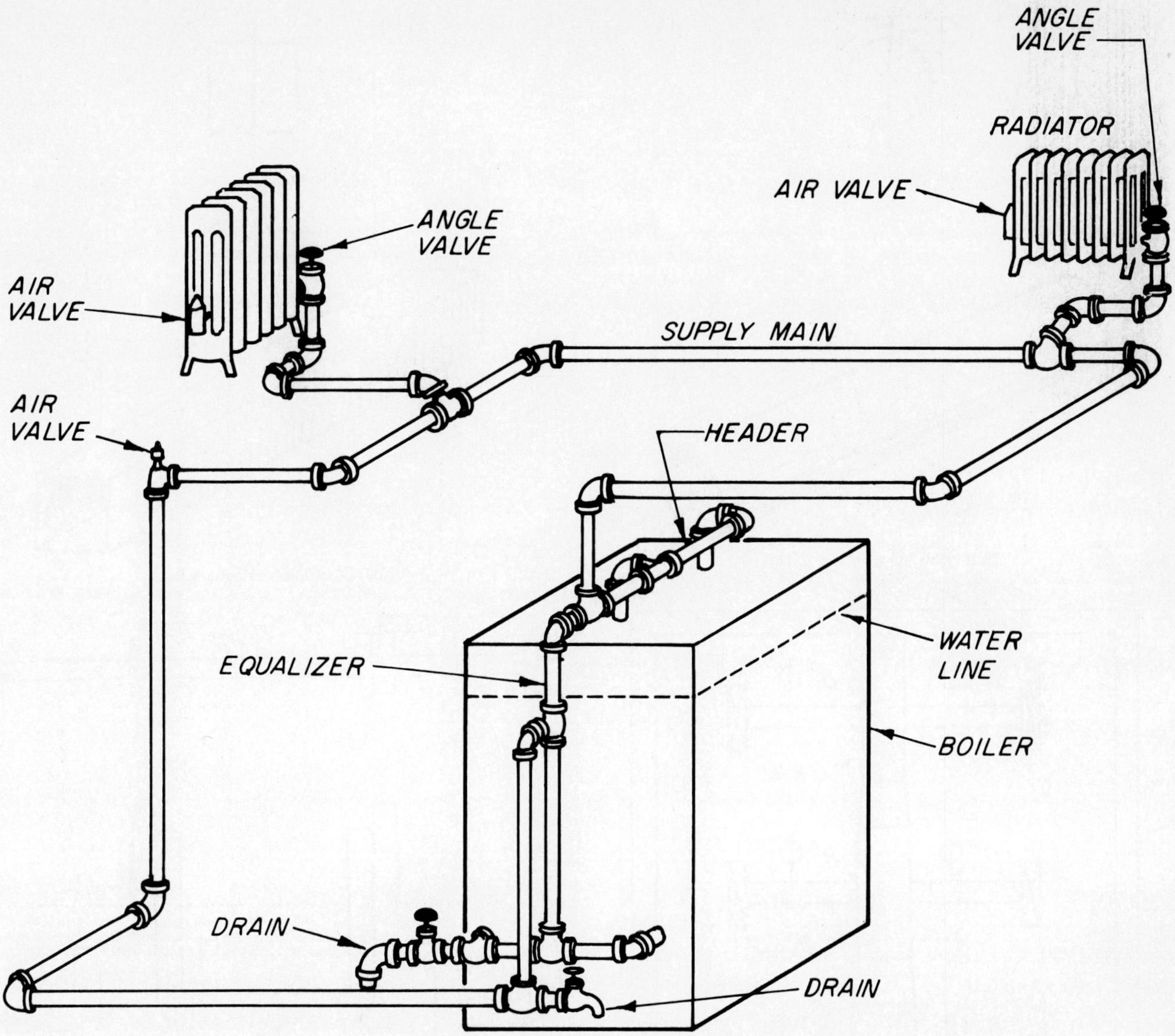

Fig. 6-56. Pictorial drawing of a heating system.

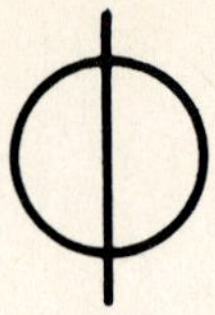

(A) Recommended symbol.

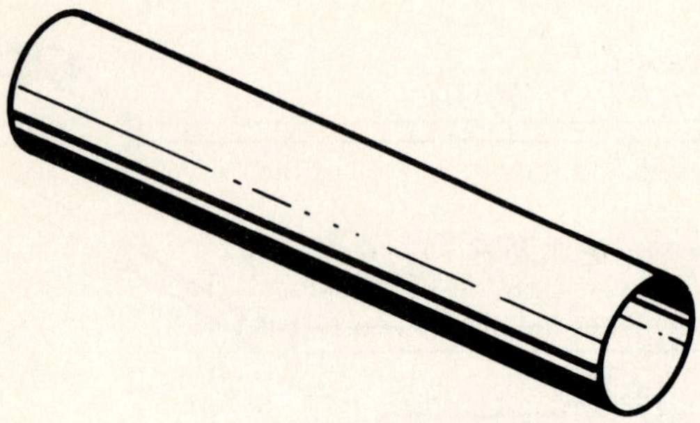

(B) Illustration of round duct.

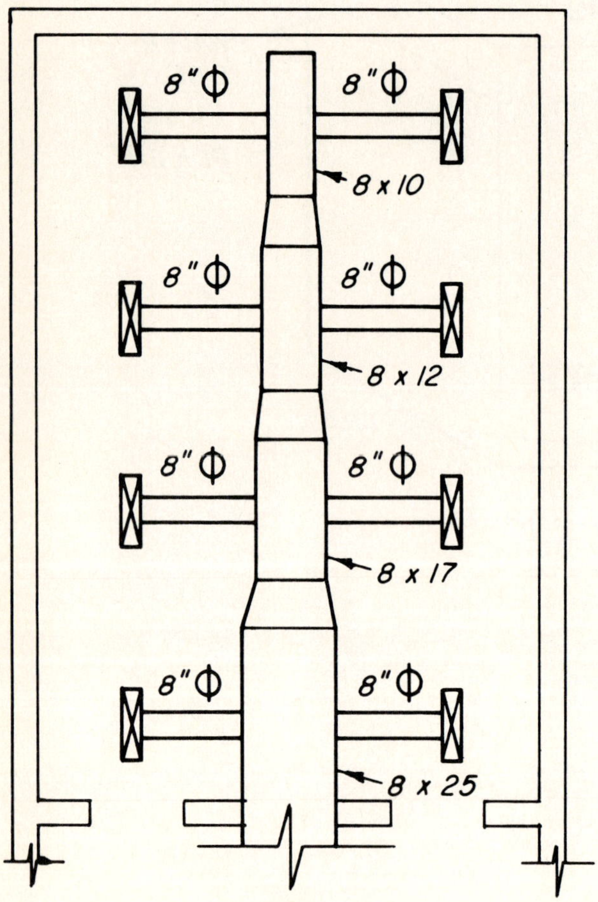

(C) Practical application.

Fig. 6-57. Round duct.

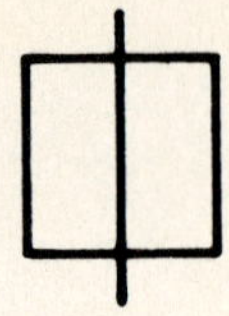

(A) Recommended symbol.

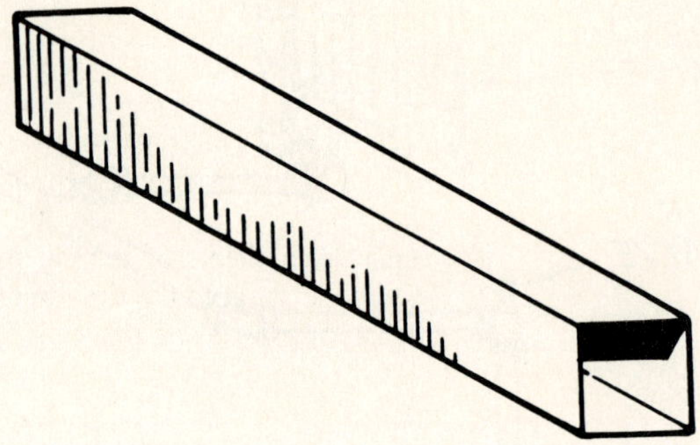

(B) Illustration of square duct.

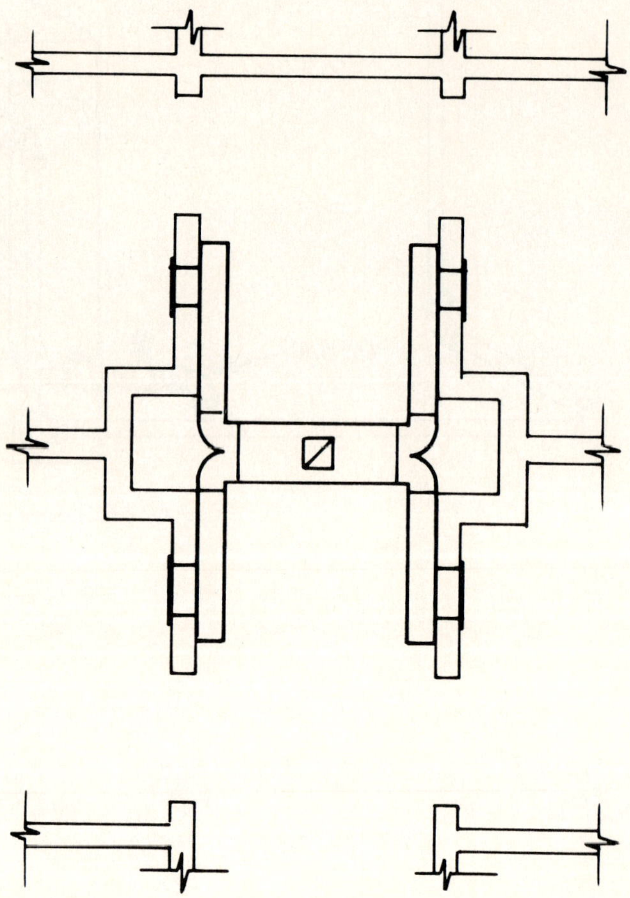

(C) Practical application.

Fig. 6-58. Square duct.

SUMMARY

The purpose of a duct drawing is to show the location and size of the ducts, diffusers, and grilles and to show other necessary information required in a heating, cooling, and ventilating air system.

Symbols are used to simplify the draftsman's work and to save time.

Most symbols used on HVAC drawings are those adopted by the USASI.

A good procedure to follow in learning HVAC symbols is to first learn the basic form and then apply the variations for obtaining different meanings.

UNIT 6 — ASSIGNMENT

1. Clean any dust off your drawing board with the dust brush. If necessary, also wash the board using soap and water on a damp cloth. This should remove any other debris or stains.
2. Prepare two sheets of tracing paper by drawing border lines as described in Unit 2.
3. Fasten one sheet to the drawing board and turn to Fig. 6–4. Draw all of the symbols that are listed, as well as lettering their meanings. Next, letter the abbreviations listed and their meanings.
4. Tape the second sheet of tracing paper to your drawing board and turn to Fig. 6–56. The usefulness of the symbols you drew in No. 3 above is clear if you consider the time it would take to draw Fig. 6–56 exactly as shown. Redraw this figure, using HVAC symbols to simplify the drawing. Be sure to retain the exact meaning.
5. Answer the following True or False questions.

	True	False
a. The HVAC draftsman is always required to have a detailed knowledge of duct design.	____	____
b. One purpose of an HVAC drawing is to show air outlets, including the volume and velocity of air from each.	____	____
c. HVAC symbols are used to simplify drawings, but they take more time to draw.	____	____
d. Although the ANSI has adopted a set of symbols, each firm may use its own modifications.	____	____
e. It isn't necessary to use a symbol list on a drawing because the meanings of the symbols are obvious.	____	____

UNIT 7

Duct Design and Layout

OBJECTIVES: To show how air-duct systems are shown on HVAC drawings

To teach the student how to prepare a working drawing from an engineer's design sketch

GENERAL

Air-distribution systems carry the air from the fan-coil unit to the space to be conditioned, and then back to the fan-coil unit. In designing such a system, the designer usually decides upon the combination of fans, duct routes, outlets, and controls that is the simplest and that will result in the best system.

DUCT SYSTEMS

Ducts are available in a variety of shapes and sizes, but the most-used shapes will be circular, rectangular, or square. The size of the ducts will depend on the amount of air to be carried through them.

Duct systems are divided into two general categories: one, those designed for areas in which the cold season is the major consideration, such as northern climates; two, those designed for areas in which the hot season is the major consideration, such as the Southern Hemisphere. Generally, duct systems that supply air at or near floor level are best for areas where the heating system is the major concern; this, of course, is due to the fact that warm air rises. Overhead systems are usually best for areas where the cooling system is the major concern since cool air falls. Of course, there are exceptions to the preceding statements. Many times it is not practical to install ductwork from below because of the construction of the building, economy reasons, etc.; the reverse is also true—that is, it may be impractical to install the ductwork overhead because of the ceiling structure or a similar reason. The following example will serve to illustrate how such an exception can come about.

The floor plan of the Branch Bank in Appendix D was designed for an area where the heating system is the major concern. However, because of limited space within the building, the construction budget, and other reasons, the designer determined that the heating system would be better if it were installed in the attic space above the conditioned space.

DESIGN PROCEDURE—BRANCH BANK

The designer calculates the heat loss for the heating system and the heat gain for the cooling system. Then HVAC equipment is selected and located in the attic space. Supply and return air ducts are sized to carry the proper volume of air and are then sketched as shown in Fig. 7-1. Then the draftsman takes the engineer's sketch and turns it into a neat, easily understood working drawing.

DRAFTING PROCEDURE

1. The border lines and title block are drawn on an appropriate-size sheet of tracing paper.
2. The outline of the building structure is traced from the architectural floor plans, and all walls, partitions, doors, windows, and the vault are shown. This background is drawn with a light solid line and with lead no softer than 4H. Exact architectural details are not necessary—just an outline or background of the building floor plan.
3. The physical dimensions of the Trane oil-fired furnace and its related cooling coil are found in the Trane Catalog, and then both are drawn on

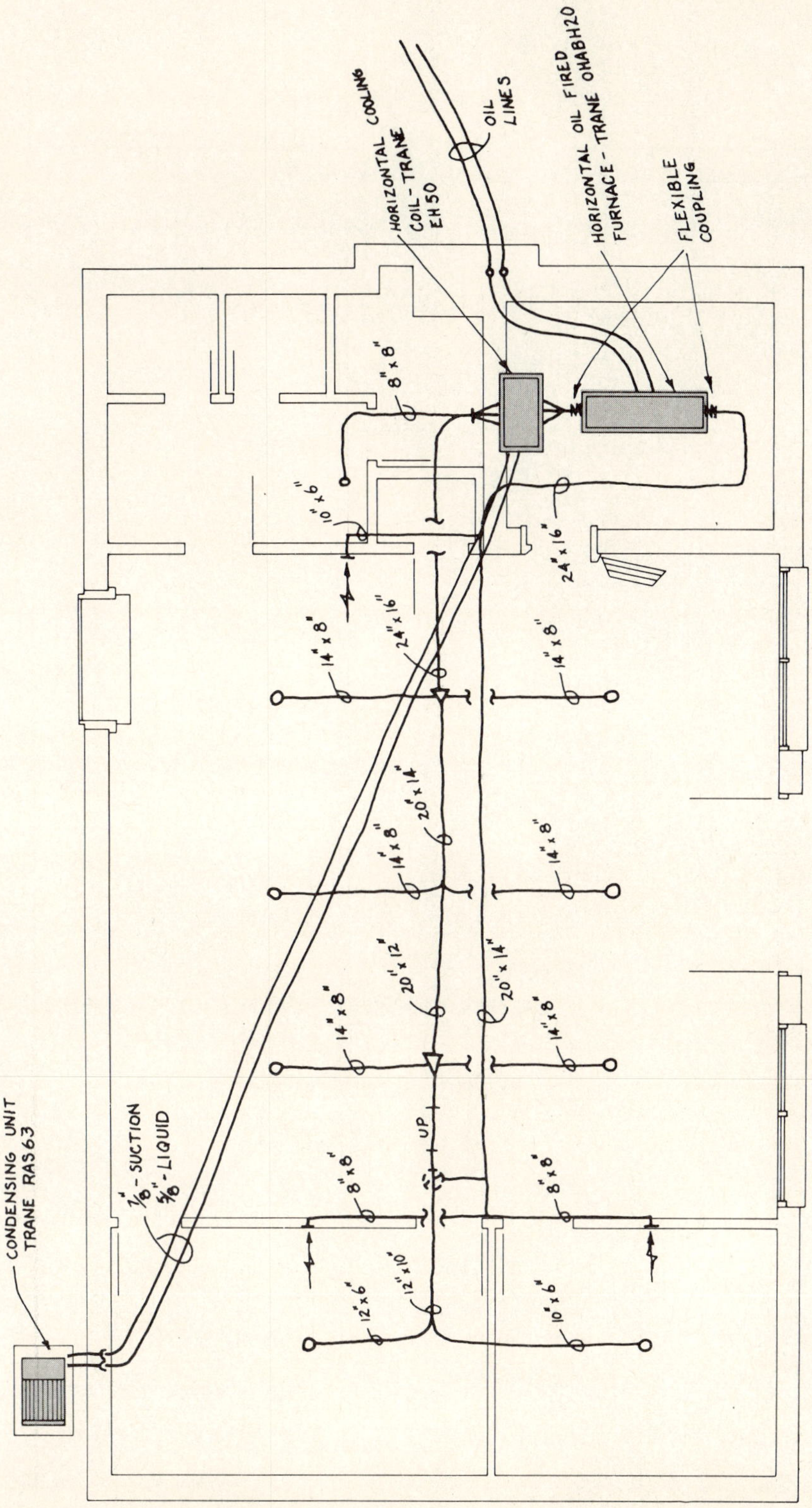

Fig. 7-1. Engineer's sketch of an HVAC system before drafting.

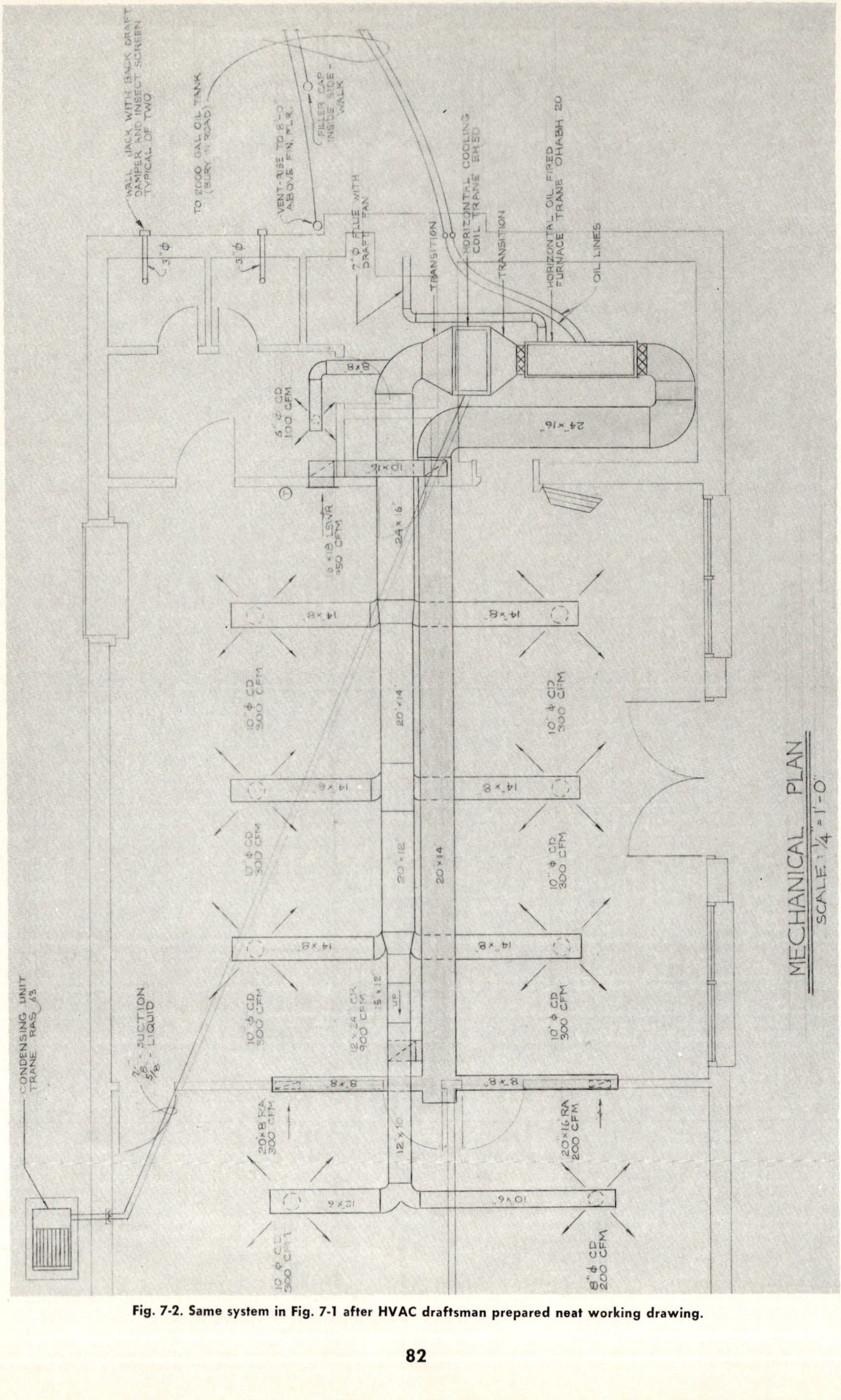

Fig. 7-2. Same system in Fig. 7-1 after HVAC draftsman prepared neat working drawing.

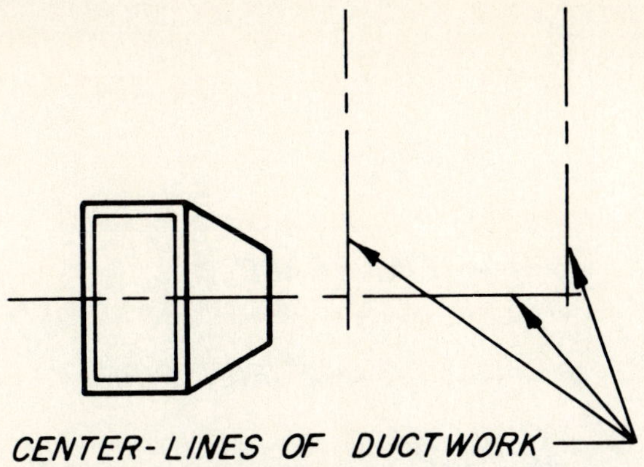

Fig. 7-3. Illustration of center line marked to aid the draftsman in preparing two-line duct drawings.

the floor plan in the space outlined by the designer. These two pieces of equipment, along with all other HVAC equipment on this drawing, are drawn with the solid heavy line and with 2H lead or softer. Looking at the manufacturer's catalog, the draftsman finds that a 7-inch-round flue is required for the oil-fired furnace; this is drawn next and routed to the nearby chimney flue.

Before reading further, look at the completed HVAC drawing in Fig. 7-2. Locate the building background, the Trane OHABH20 furnace, and the horizontal cooling coil. The remaining procedure is as follows:

4. The physical dimensions of the Trane RAS63 condensing unit are found in the manufacturer's catalog, and this piece is drawn on the tracing paper where indicated in the engineer's sketch.

   A note about the location of the condensing unit before proceeding: it is best to locate the condensing unit as close to the cooling coil as possible. However, in this case, the unit had to be located on the opposite side of the building because of limited outside space on the side next to the cooling coils.
5. The suction and liquid lines are drawn from the condensing unit to the cooling coil in the duct system.
6. A light center line is drawn from the cooling-coil housing, along the general route of the one-line duct diagram shown on the engineer's sketch. An illustration of this light center line appears in Fig. 7-3.
7. With the center line as a guide, the ductwork is drawn to scale. For example, the first supply-duct dimension is 24 by 16 inches; the first dimension (24 inches) always means the width of the duct, and the second dimension (16 inches) always means the depth of the duct. Therefore, the draftsman measures 12 inches (scaled at ¼ inch = 1 foot) on each side of the center line. This gives the required 24-inch width as indicated on the engineer's sketch, and this first section of duct is drawn in with the light solid line at first. The 90-degree short elbow from the cooling-coil transition to the section of duct previously drawn is drawn on the tracing paper with a ductwork template. If a ductwork template is not available, this 90-degree elbow may be drawn by using a circle template.
8. The remaining duct sizes are measured on the drawing and then drawn to the appropriate size according to the given scale. All elbows, takeoffs, etc., are constructed on the drawing with a duct template. After the entire system has been laid out with a light line and is found to be correct, the draftsman darkens these lines into a heavy solid line.
9. After all linework is completed, all notes, dimensions, etc., as indicated on the engineer's sketch are lettered in a neat manner. Refer again to Fig. 7-2 and study this draftsman drawing very carefully.

More details of the drawing shown in Fig. 7-2, including the supply-air diffuser/register and the return-air grilles, will be described later in this section.

Notice that oil lines are shown on the drawing as feeding the horizontal oil-fired furnace but are "cut" because space on the drawing does not permit showing further details. However, a detail such as that shown in Fig. 7-4 would clarify the drawing and leave little doubt as to what is required.

The floor plan of the office building shown in Fig. 7-5 shows another example of an HVAC duct system—this time on a larger scale. Regardless of the size of the system, basically the drafting is laid out and performed as described earlier in the unit. The reader should study this drawing very carefully, noticing all details.

SUMMARY

Air-distribution systems carry air from the air-handling unit to the space to be conditioned and then back to the air-handling unit.

The duct shapes most often used on HVAC drawings are round, rectangular, and square.

Duct systems are divided into two general categories:

1. Those designed for areas where the heating system is the major consideration
2. Those designed for areas where the cooling system is the major consideration

Overhead ducts are generally best suited for systems where cooling is the major consideration.

Underfloor ducts are generally best suited for systems where heating is the major consideration.

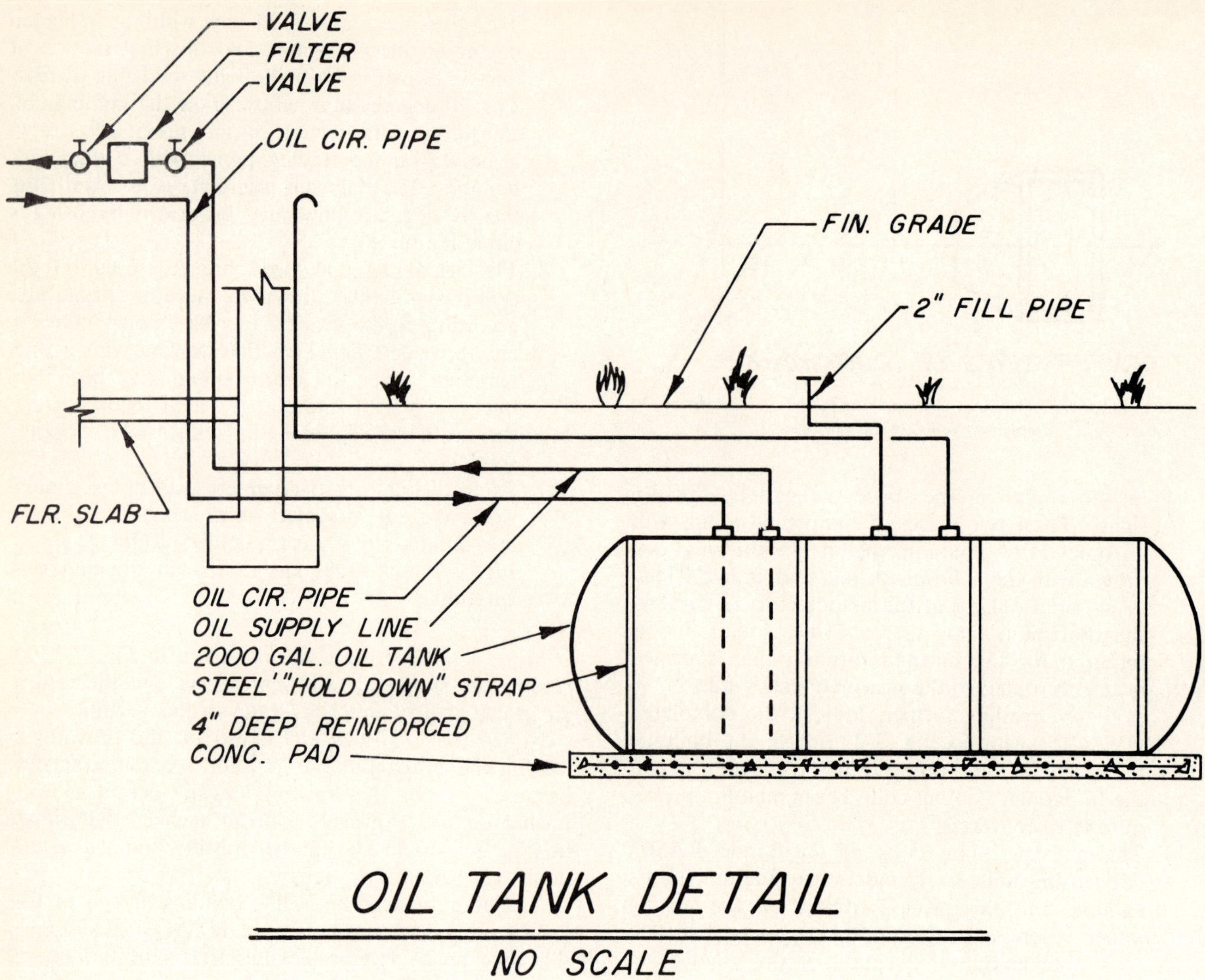

Fig. 7-4. Oil tank detail.

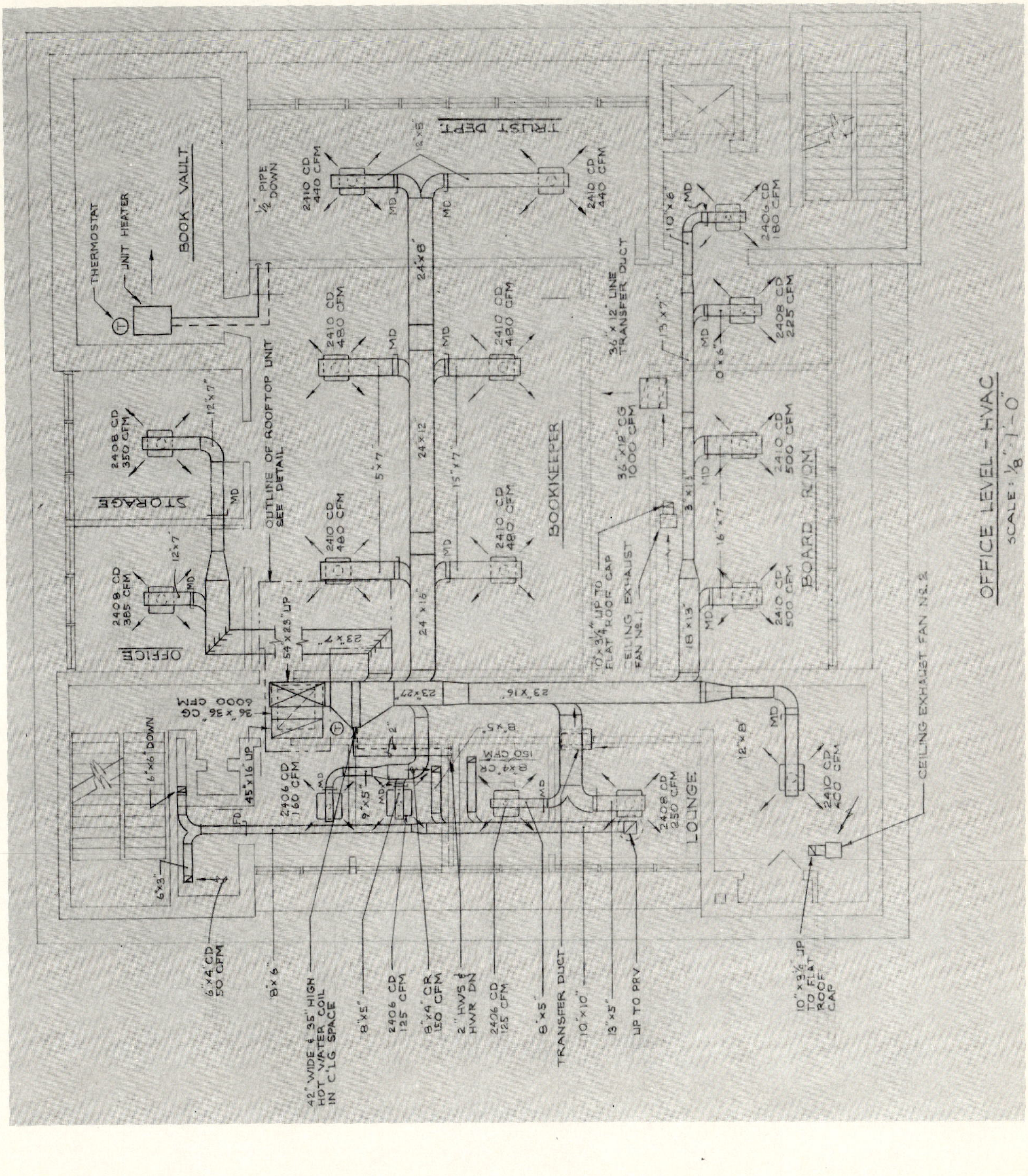

Fig. 7-5. Example of a HVAC duct system in a bank building.

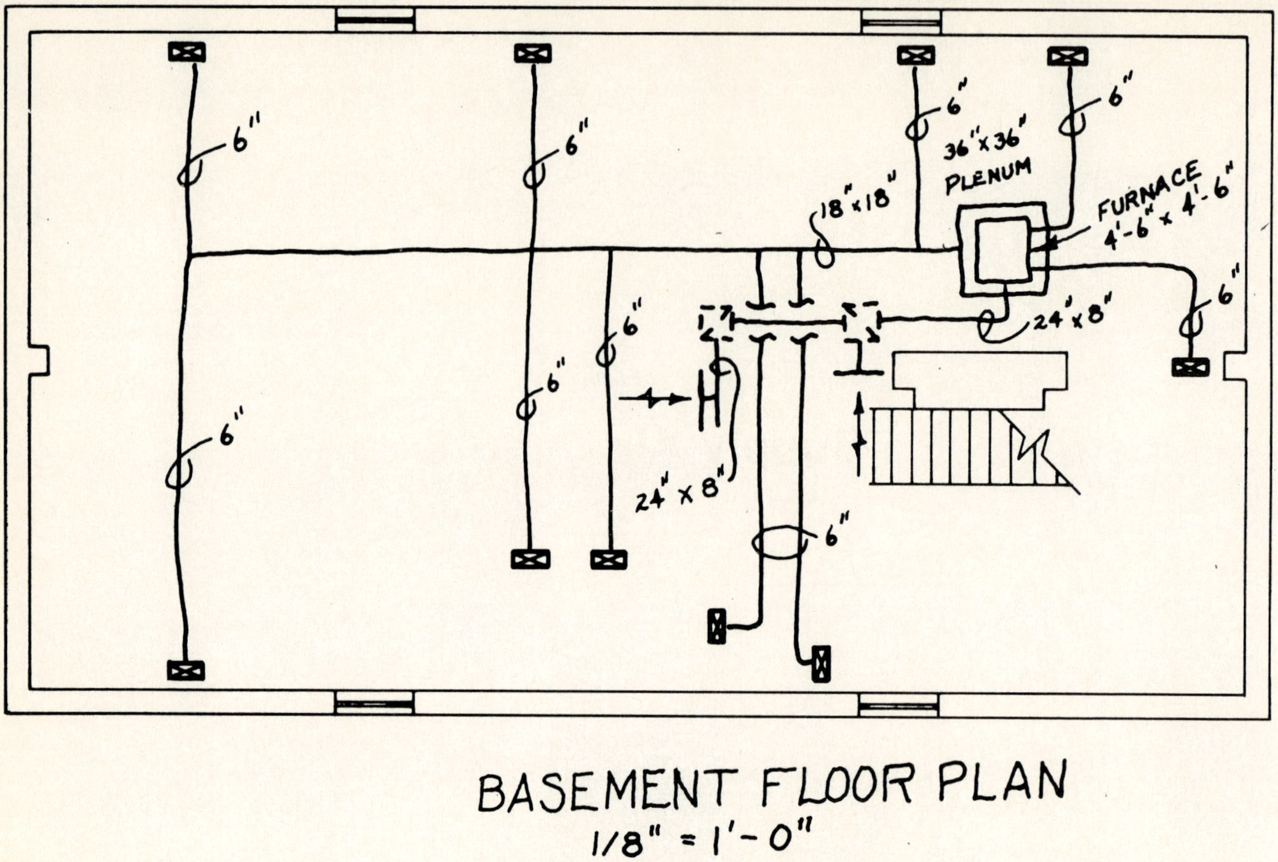

Fig. 7-6. Engineer's sketch for use in this assignment.

UNIT 7 — ASSIGNMENT

1. Prepare an 11-by-17-inch sheet of tracing paper with border lines as described in Unit 2.
2. Redraw the basement floor plan shown in Fig. 7–6, including stairway and furnace. If necessary, refer to the list in Unit 6 for the meaning of symbols.
3. Using the basement floor plan you just finished, draw a two-line duct system to scale, using the engineer's sketch in Fig. 7–6 as a guide.
4. Answer the following True or False questions.

	True	False
a. The most-used shapes of ducts are circular and triangular.	_____	_____
b. Dust systems are divided into three categories; there is a separate system for very cold, warm, and very hot climates.	_____	_____
c. Air-distribution systems carry air from the air-handling unit to the space to be conditioned, and then back to the air-handling unit.	_____	_____
d. If cooling is the major consideration, underfloor ducts are usually best.	_____	_____
e. The climate determines where the ductwork will be installed; the structure of the building makes no difference.	_____	_____

UNIT 8

Supply-Air Outlets

OBJECTIVES: To describe the various types of supply-air outlets

To show how these outlets are indicated on working drawings

GENERAL

Supply-air outlets are a major part of any air distribution system because they provide a means of distributing properly controlled air to a room or area. "Distributing properly controlled air" means that the supply-air outlet will deflect or diffuse the air, be adjustable for changing the airflow rate, avoid air noise, and be able to throw the conditioned air no less than ¾ the distance from the outlet to the opposite wall.

TYPES OF SUPPLY-AIR OUTLETS

Many types of supply-air outlets are available, some of which are described in the following paragraphs.

Supply-Air Outlet

A supply-air outlet is a floor, ceiling, or wall opening through which conditioned air is delivered to a room or area.

Ceiling Diffuser

The ceiling diffuser is a square, oval, circular, or semicircular facing device that covers the supply-air opening (usually in the ceiling) of a room or area. Most diffusers are adjustable for airflow direction and rate.

Grille

The grille is a covering for any opening through which air passes.

Register

The register is a covering for any opening through which air passes and has a built-in damper for controlling the air passing through it.

1. A fixed-louver register is a nonadjustable register in which the air pattern is factory set.
2. An adjustable-louver register is a register with adjustable bars for directing the air in several different patterns.

Fig. 8-1 gives examples of various kinds of supply-air outlets and the proper drawing symbol for each.

LOCATION OF SUPPLY-AIR OUTLETS

Supply-air outlets in the area to be conditioned must be located so that sufficient air is supplied to establish and maintain comfort conditions within the area. "Comfort conditions" are usually defined as a uniform air pattern that is free from hot or cold drafts.

The location of supply-air outlets are determined by the designer after he has taken the following factors into consideration.

1. The size and shape of the room or area to be conditioned
2. The decor of the area to be conditioned
3. The furniture placement within the area to be conditioned
4. The ceiling, wall, and floor finish of the area to be conditioned

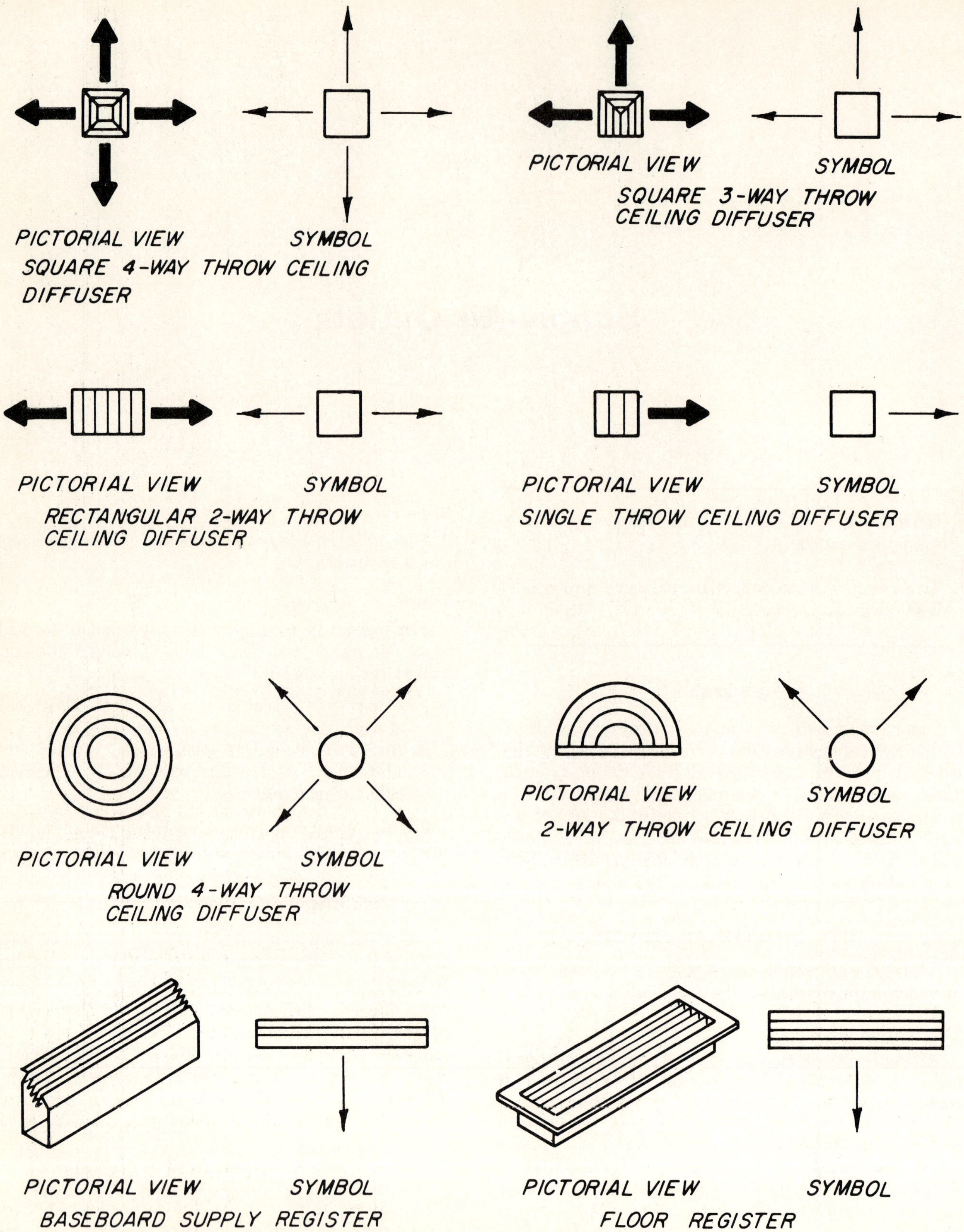

Fig. 8-1. Example of the most commonly used air outlets with pictorial and graphic views of each.

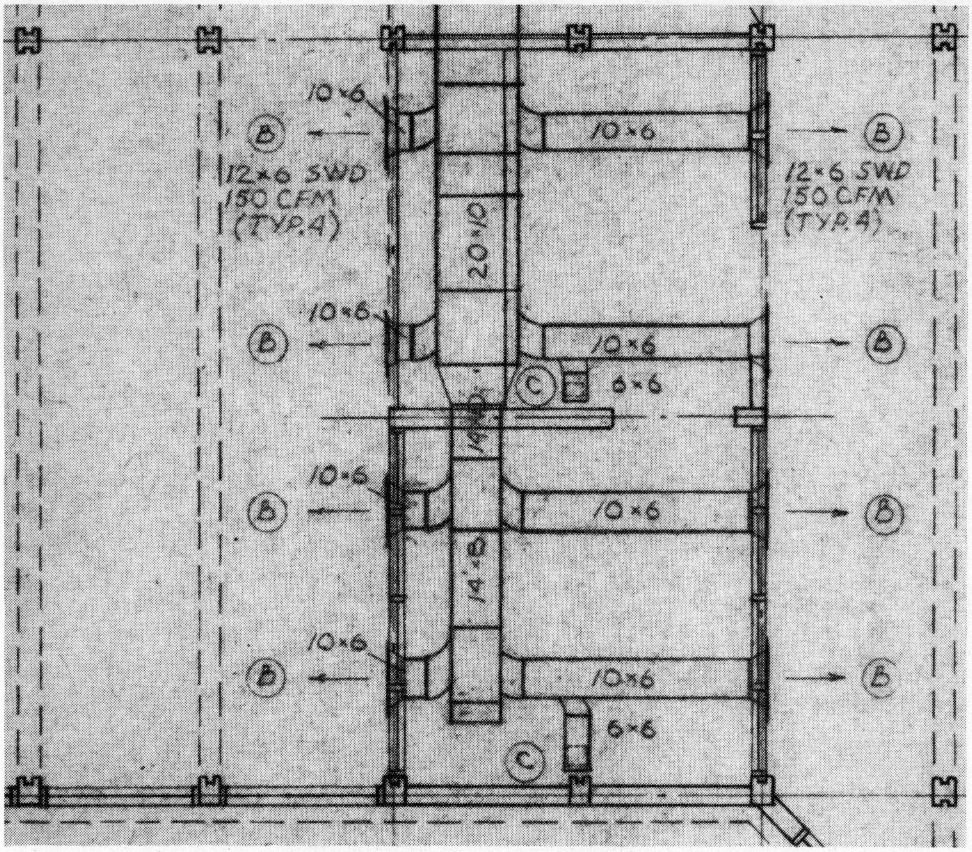

Fig. 8-2. Plan view of a supply-air duct system using high sidewall air outlets into the various areas.

5. The required budget—cost of outlets
6. Whether the system is designed for heating, cooling, or both. If both, the one that is used the most during the year should be given the greatest consideration
7. The total air quantity required for the area under consideration
8. The total load and draft conditions of the area
9. The diffusion or spread pattern required within the space to be conditioned
10. The level of noise allowed in the space to be conditioned

From the preceding list, we can say that supply-air outlets are located according to the shape, size, usage, and load concentration of the area, and according to whether the system is used in a cooling or heating application or both.

Although the engineer or designer usually selects and locates the supply-air outlets, the HVAC draftsman should have a general knowledge of how it is accomplished. The basic steps are as follows:

1. Study the architectural drawings to determine the size, shape, and use of each room or area.
2. Determine the air quantity for each area.
3. Select the outlet type and locate the outlets according to the requirements and conditions in the area under consideration.

PRACTICAL APPLICATION

The examples of HVAC systems in Figs. 8-2 through 8-10 were taken from actual working drawings as designed by professional engineers and drawn by HVAC draftsmen. The location of the supply-air outlets are given as well as the direction of flow for each. A careful analysis of each will enable the HVAC draftsman to become familiar with air-outlet locations.

SUMMARY

Supply-air outlets may be located at low wall, high wall, baseboard, floor or ceiling positions, depending upon area conditions.

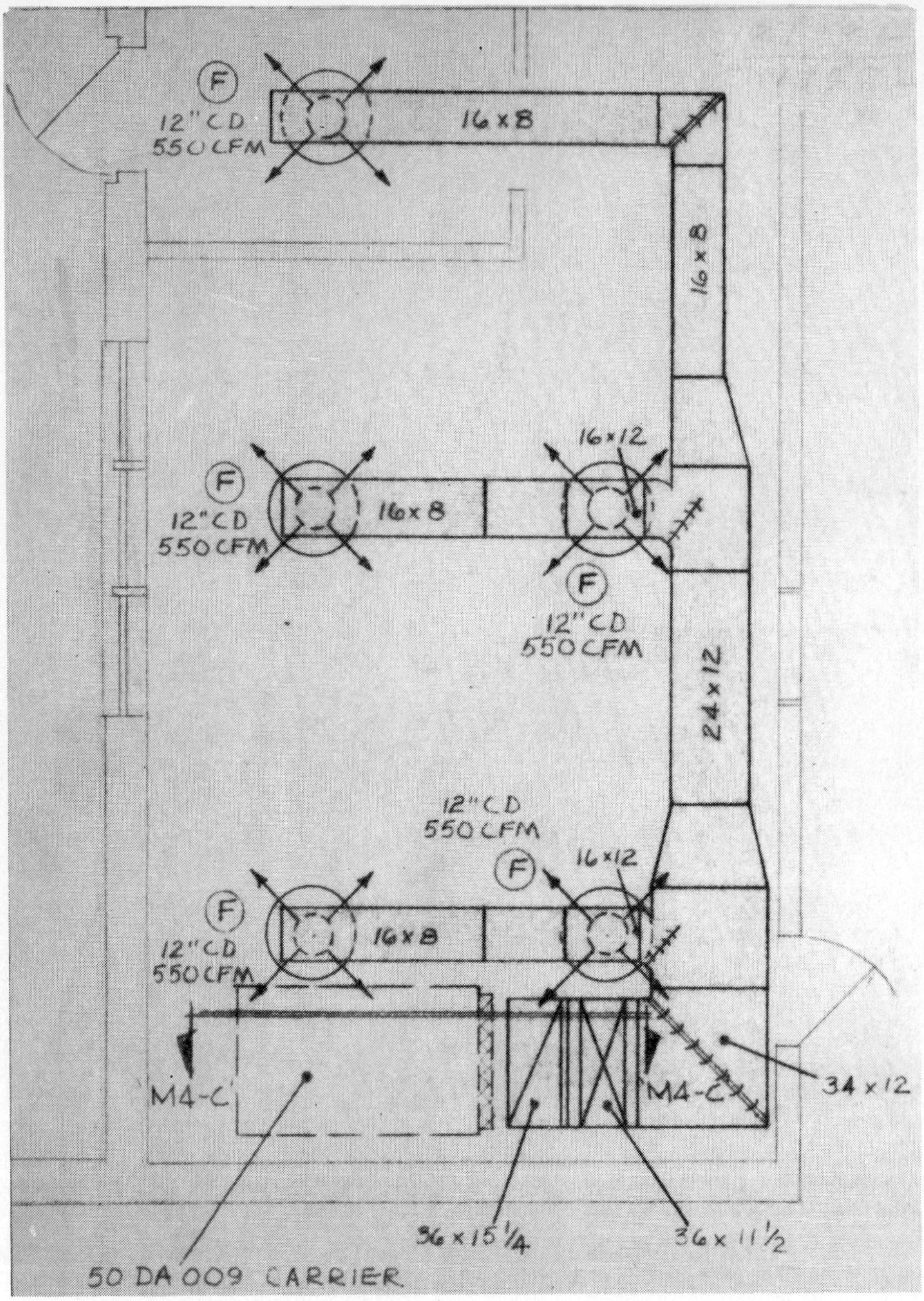

Fig. 8-3. Plan view of another air-distribution system showing the heating and cooling equipment on the roof of the building; the duct system uses ceiling diffusers with four-way throw.

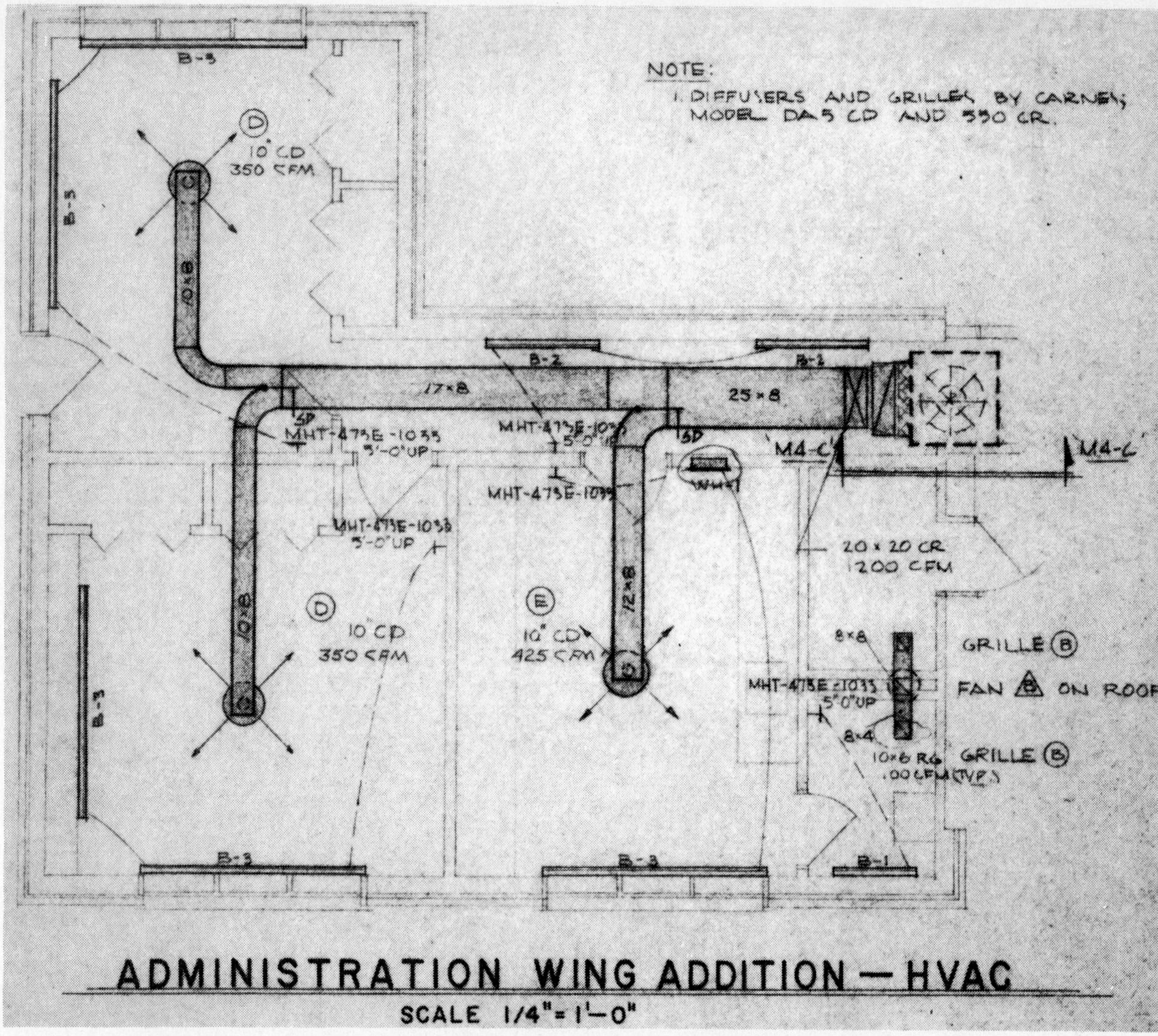

Fig. 8-4. This shows an example of a small air-conditioning system for a group of offices—again with four-way throw diffusers recessed in the ceiling; the equipment designated as B-2, B-3, etc., are electric baseboard heaters, and the Ts designate thermostats to control the heaters.

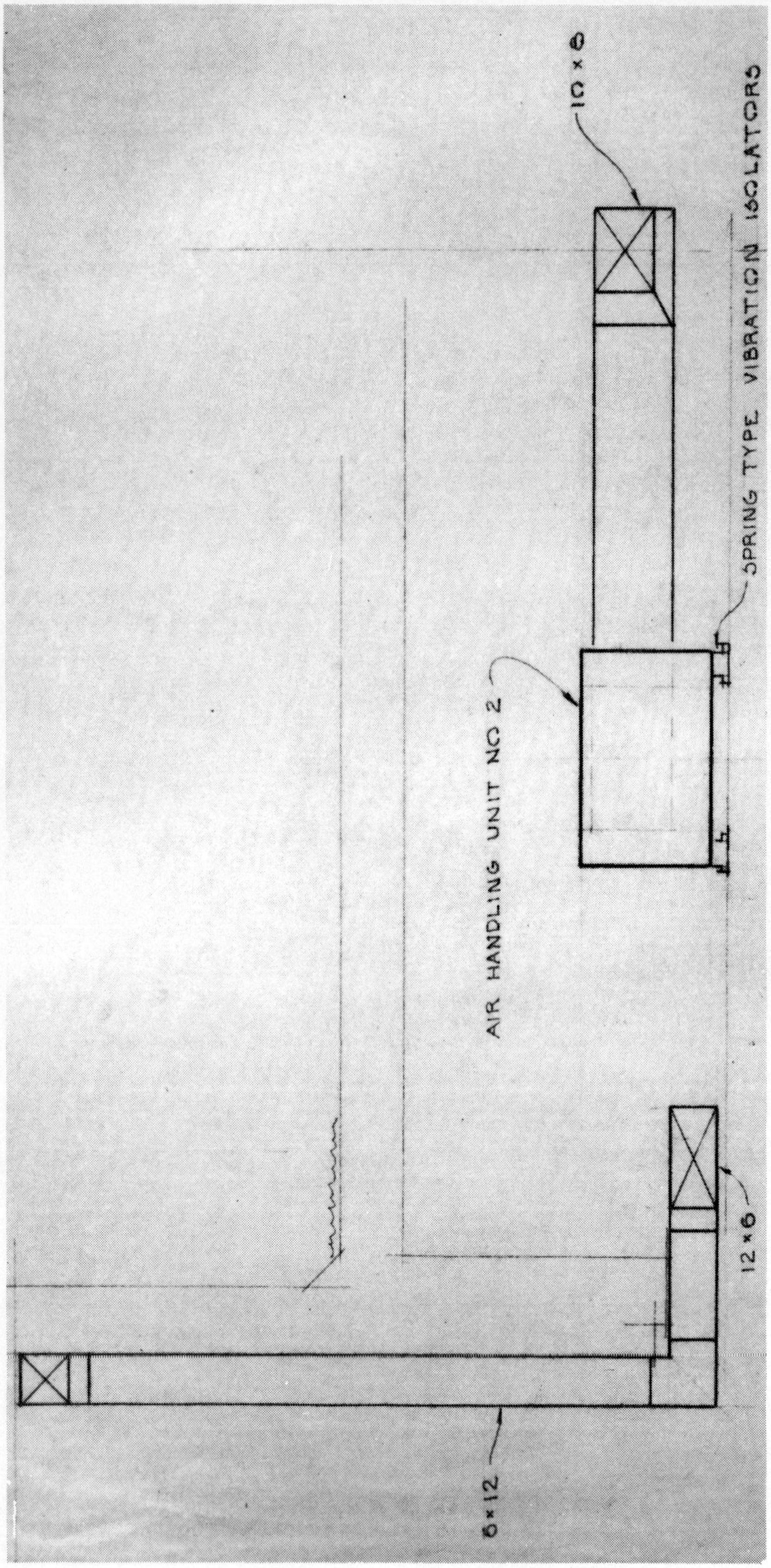

Fig. 8-5. A section view of an air-handling unit with a supply air duct.

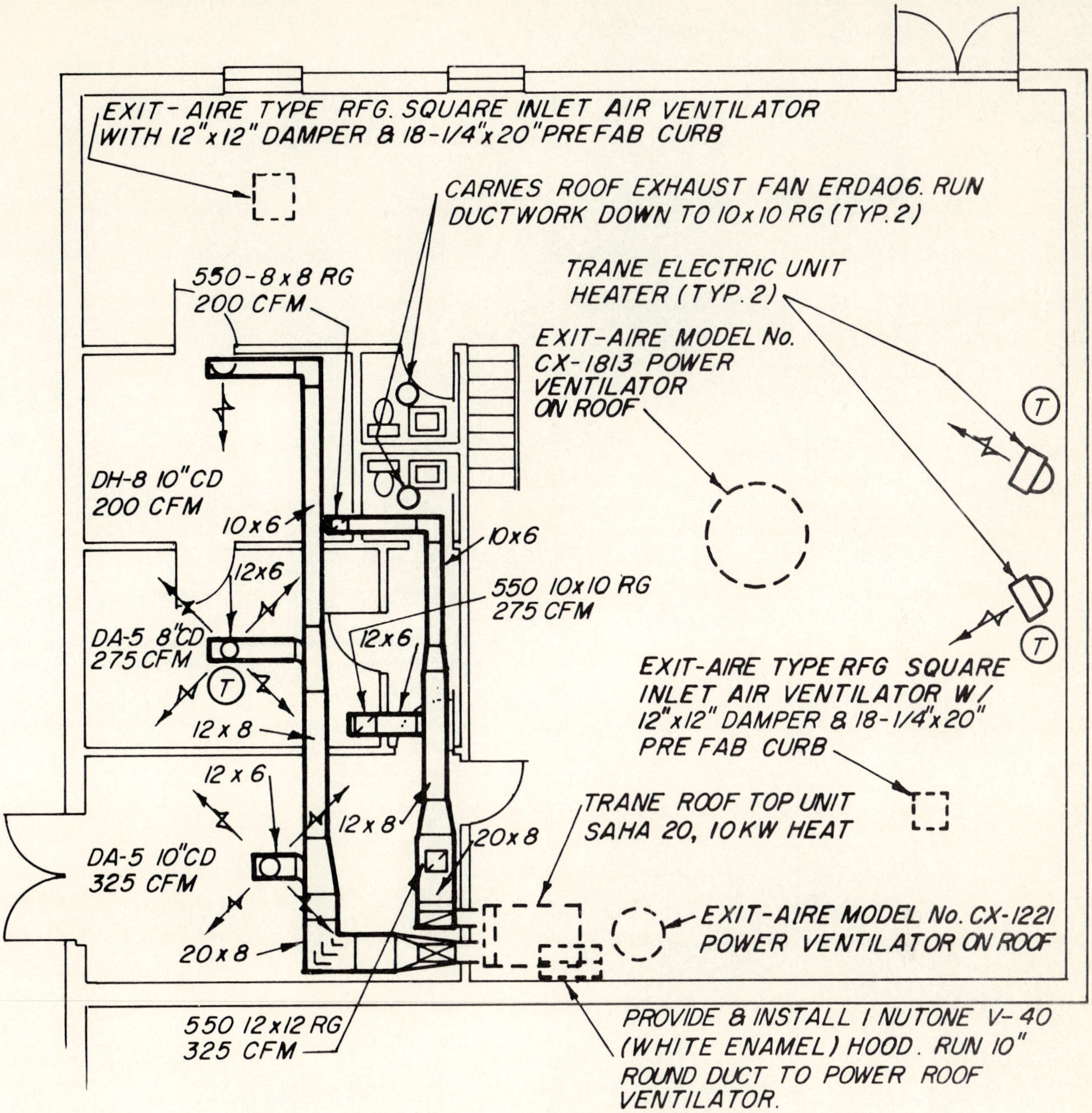

Fig. 8-6. An entire plan view of a small building showing all details of construction.

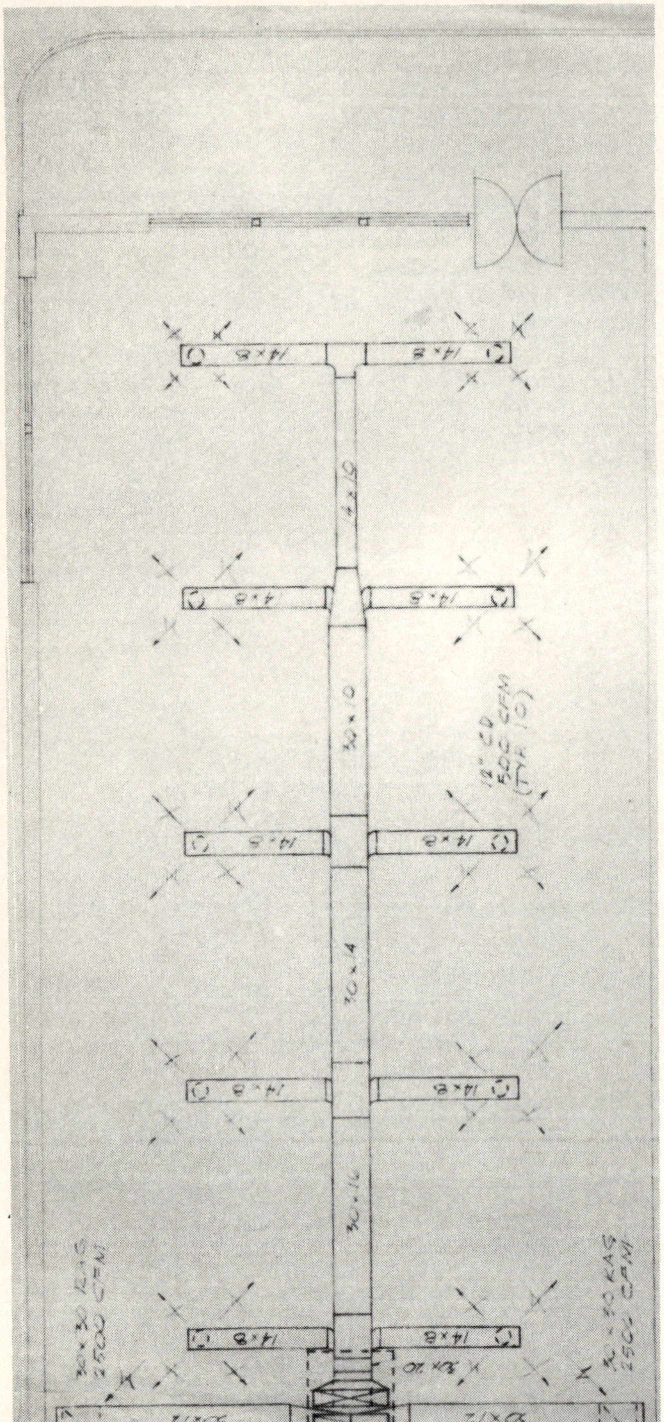

Fig. 8-7. A plan view of another store building showing HVAC equipment, supply air ducts, return air ducts, and air outlets.

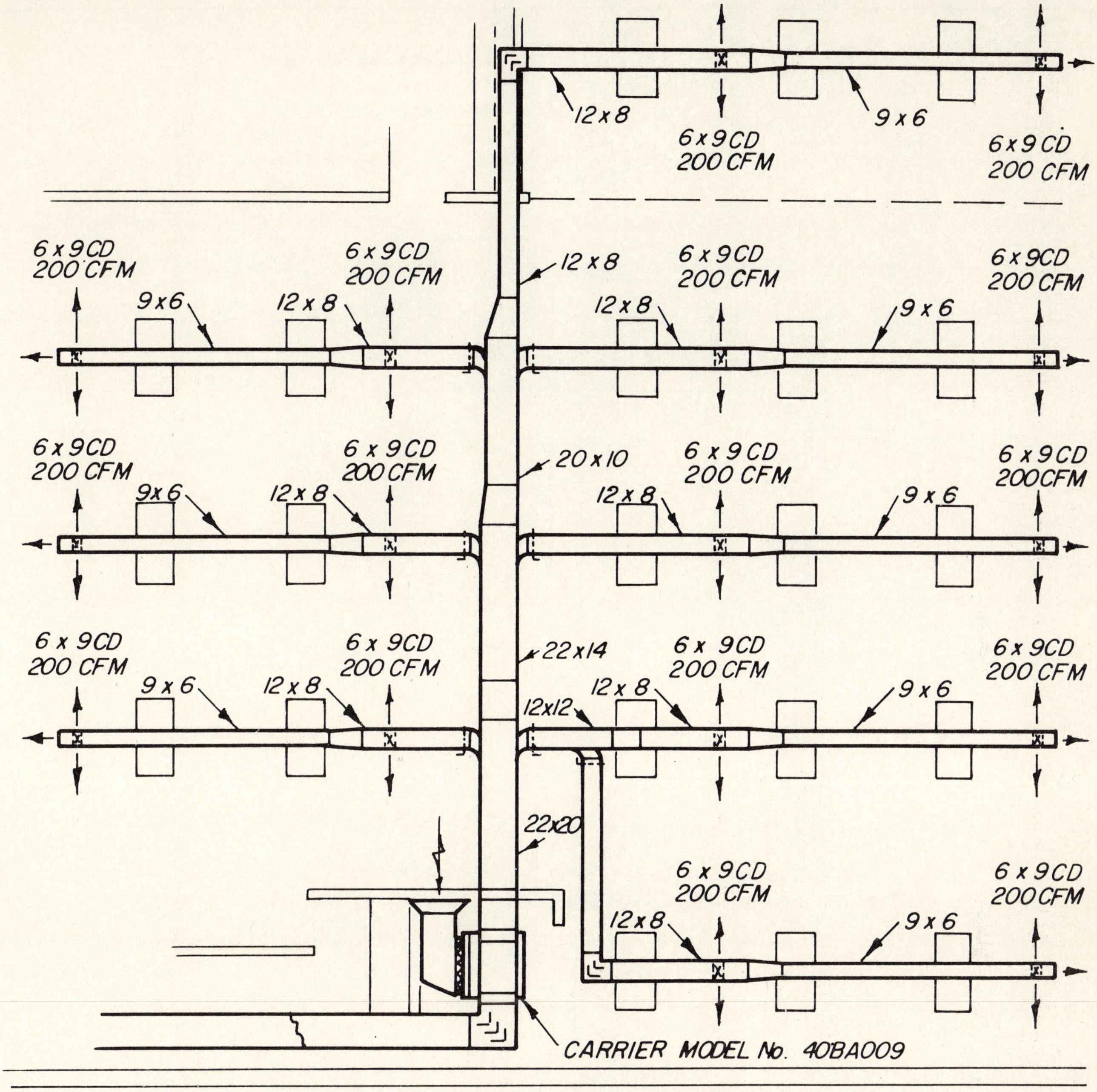

Fig. 8-8. A "check" set of drawings showing the air-distribution system for a large department store. Note the engineer found an error and corrected it; the draftsman will add this to the original drawings.

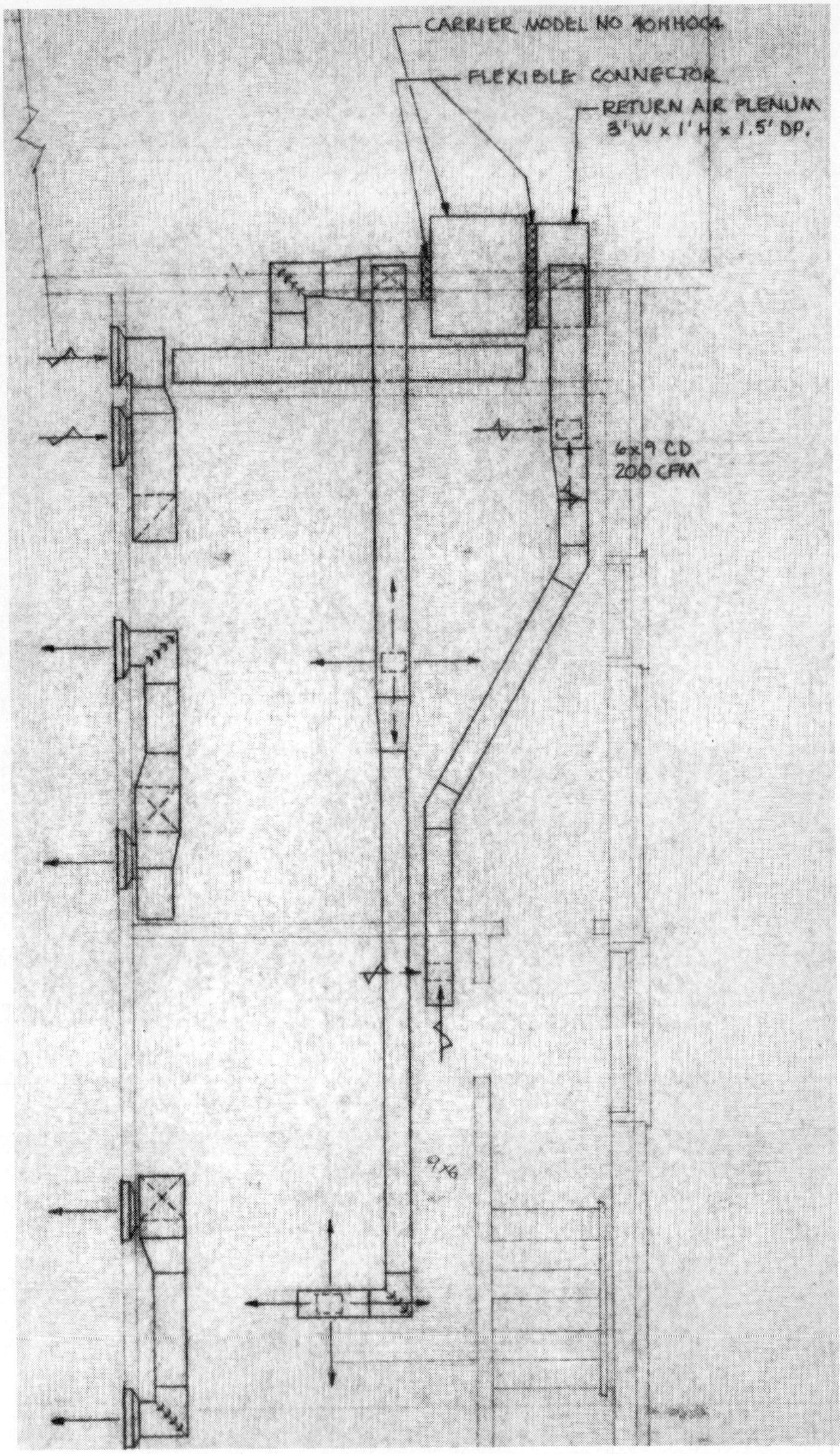

Fig. 8-9. A plan view of one complete air-conditioning zone along with ductwork for another system which has its equipment on another level of the building.

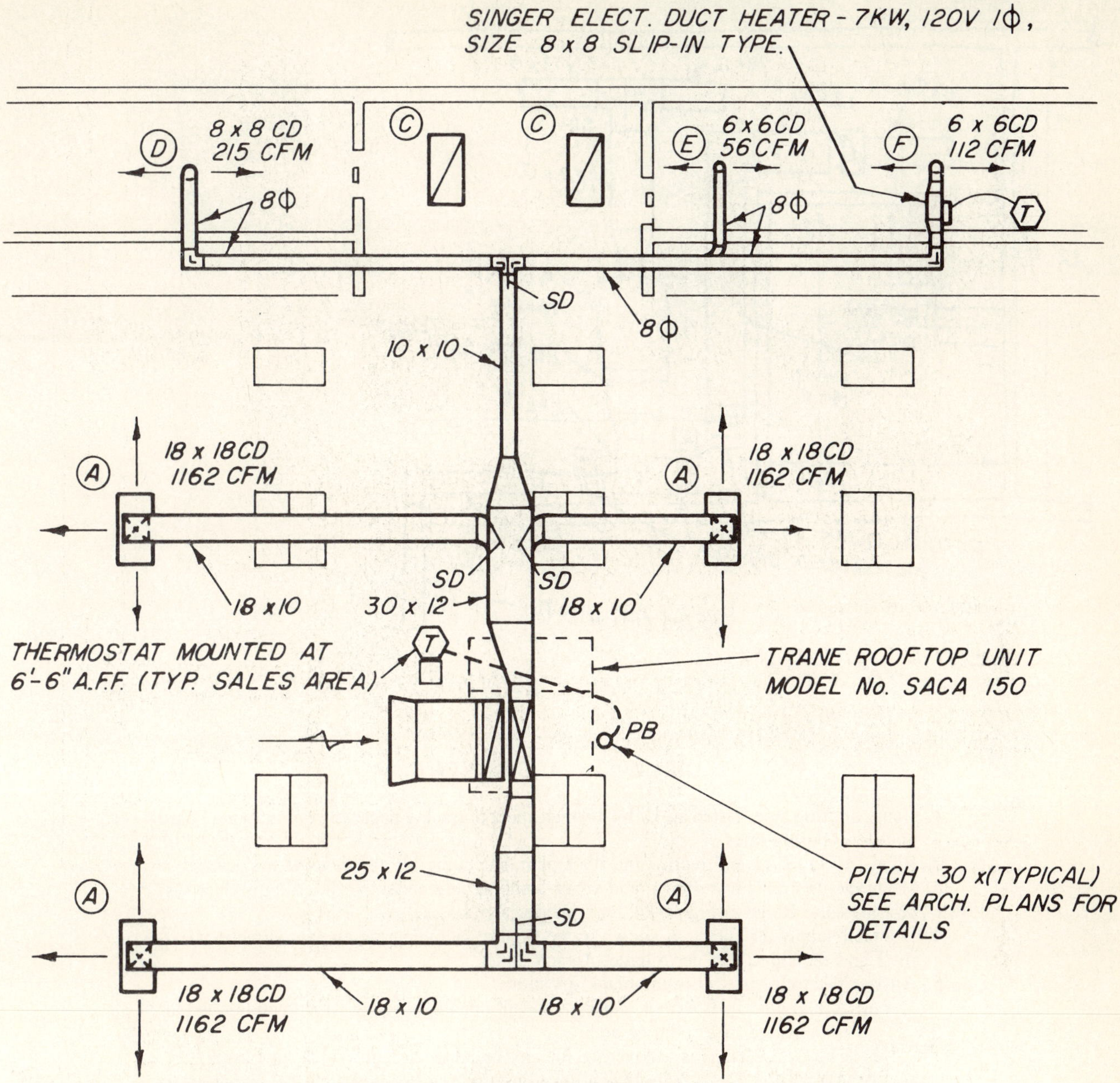

Fig. 8-10. A plan view showing rooftop HVAC equipment and related ductwork for a clothing store.

If the drawing scale permits, the supply outlets should be drawn to scale on the floor plans, and the shape of the outlet used—that is, square, circular, etc. If the drawing scale is too small for drawing the outlets to scale, all outlets should be exaggerated, but proportional in size. Thus, a 12-by-12-inch actual-size diffuser should appear twice the size of a 6-by-6-inch actual-size diffuser.

All air outlets should be drawn with the heavy solid line when the diffusers are visible in the drawing view. If they are not visible, as on the underside of a duct, they should be drawn with a heavy broken line.

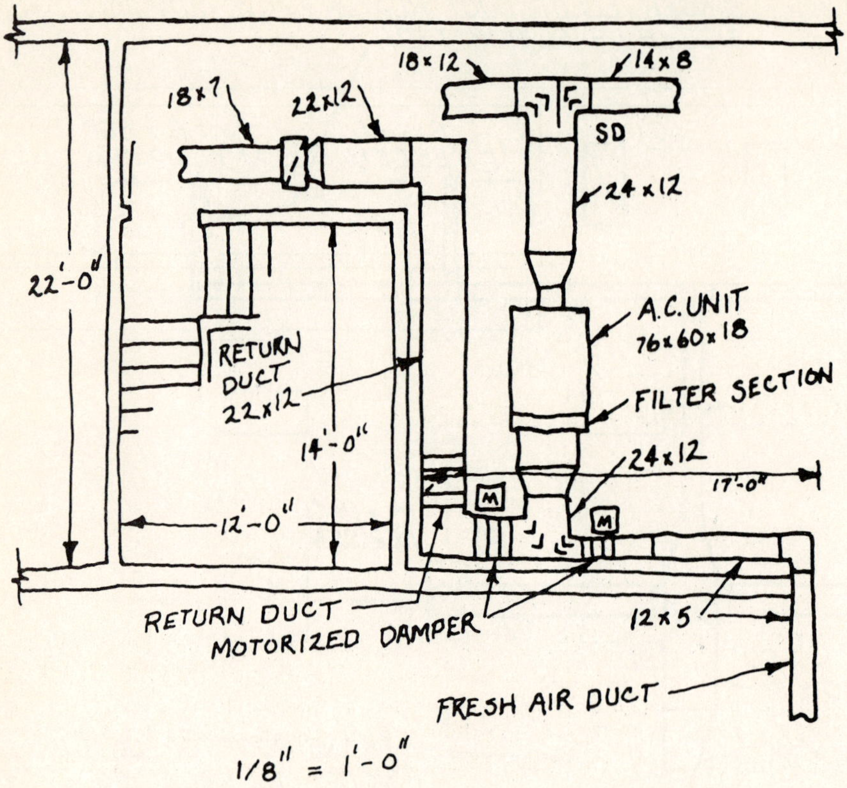

Fig. 8-11. Engineer's sketch for the assignment in this unit.

UNIT 8 — ASSIGNMENT

1. Using drafting tape, fasten an 11-by-17-inch sheet of tracing paper to your drawing board and draw border lines.
2. Refer to Fig. 8–11 and redraw the floor plan, including all doors, windows, etc.
3. Using the floor plan just drawn, add an air-handling unit and a two-line supply-duct system to scale. Refer to Fig. 8–11 for a guide.
4. Draw all supply-air outlets as shown in the sketch, including the air-flow directional arrows and the cubic feet per minute of each outlet.
5. Answer the following fill-in-the-blank questions.
   a. A __________ is a floor, ceiling, or wall opening through which conditioned air is delivered to a room or area.
   b. The __________ is a covering for any opening through which air passes.
   c. A(n) __________ register is a nonadjustable register in which the air pattern is factory set.
   d. The __________ is a square, oval, circular, or semicircular facing device that covers the supply-air opening of a room or area.
   e. A(n) __________ register is a register with adjustable bars for directing the air in several different patterns.

UNIT 9

Return-Air Outlets

OBJECTIVES: To describe the various types of return-air outlets

To show how these outlets are indicated on working drawings

GENERAL

The location of the return-air outlets is not as critical as the location of the supply-air outlets, but as a rule of thumb, the return-air outlets should be in the opposite location from the supply-air outlets. That is, if the supply-air outlets are located in the ceiling or high side-wall, the return-air outlets should be located in the floor or low side-wall. If the supply outlets are located on the outside wall of the room, the return-outlets should be located on an inside wall of the room—preferably on the opposite wall. For example, if the supply outlets were located in the ceiling and the return outlets were also located in the ceiling, there is danger of short-circuiting the air from the supply outlets directly to the return outlets (Fig. 9-1).

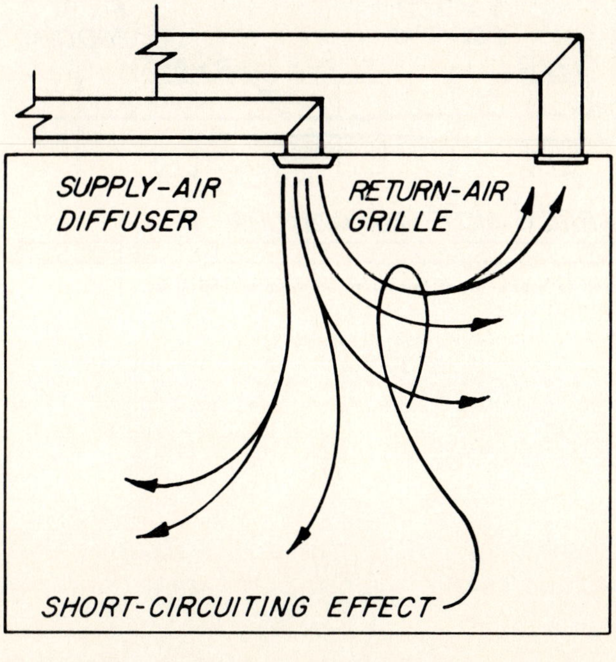

Fig. 9-1. Example of how poorly located air outlets can short-circuit air.

TYPES OF RETURN OUTLETS

If all supply-air outlets are located around the perimeter of a building and the air within the building is not "sealed" in each space, one centrally located return outlet is usually satisfactory. Transfer grilles, undercut doors, or a ceiling plenum can be used to ensure that all supply air returns to the one return outlet. A floor plan of this type of return-air system is shown in Fig. 9-2.

The return-air system for many multistory residences is located in the stairway, due to the fact that the cooler air flowing down the stairs will flow into the return outlet in the stairway; this minimizes cold-floor drafts. See Fig. 9-3.

Floor outlets are not recommended in commercial buildings because they are likely to become "dirt catchers." If low-return outlets are required, either baseboard or low side-wall return outlets would be a better choice.

Where conditioned space must be "sealed off," preventing one central-return outlet, the return-air outlets must be located in several areas and then transferred

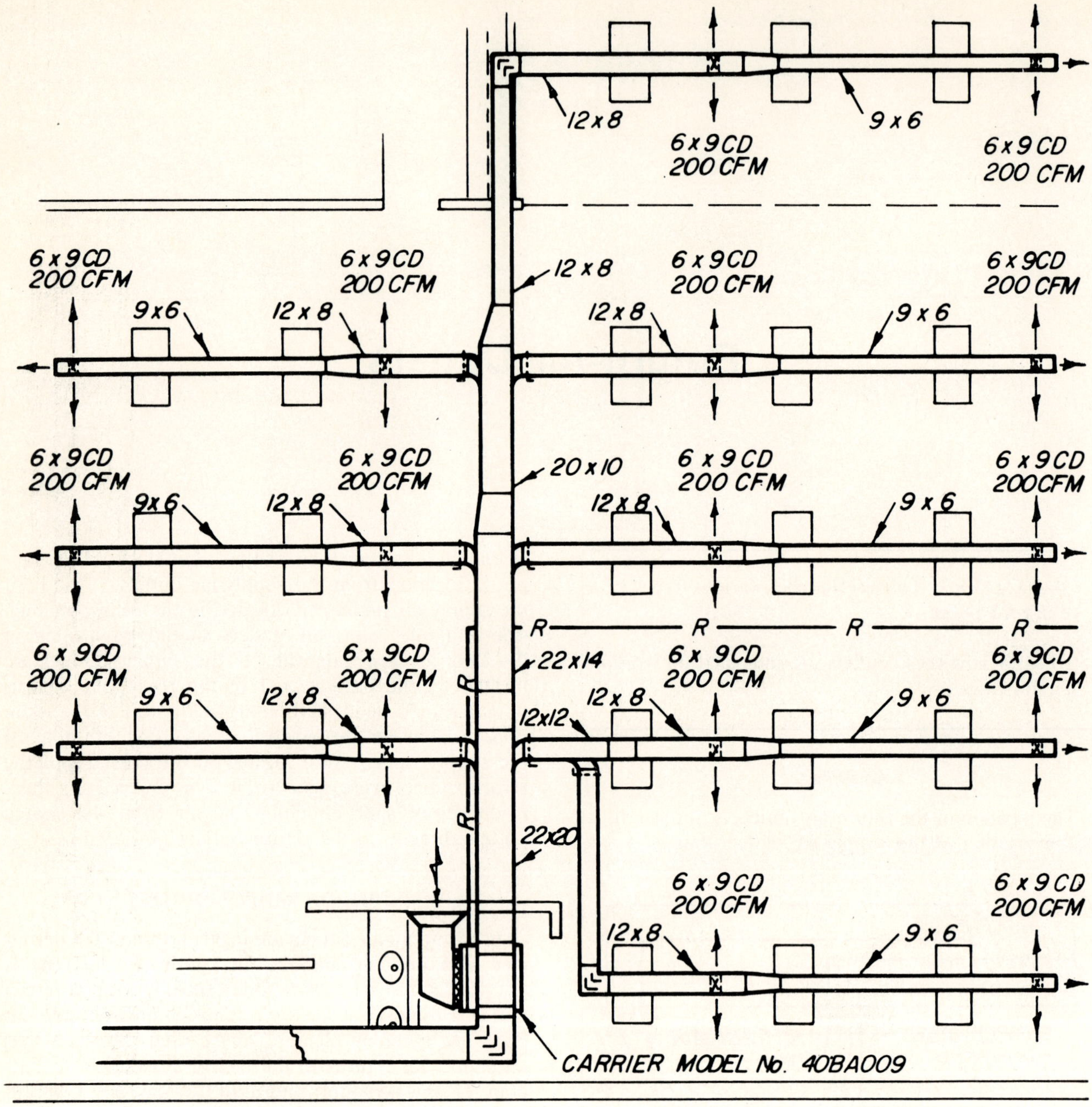

Fig. 9-2. A floor plan showing how one return-air outlet is used to handle all of the return air from several areas.

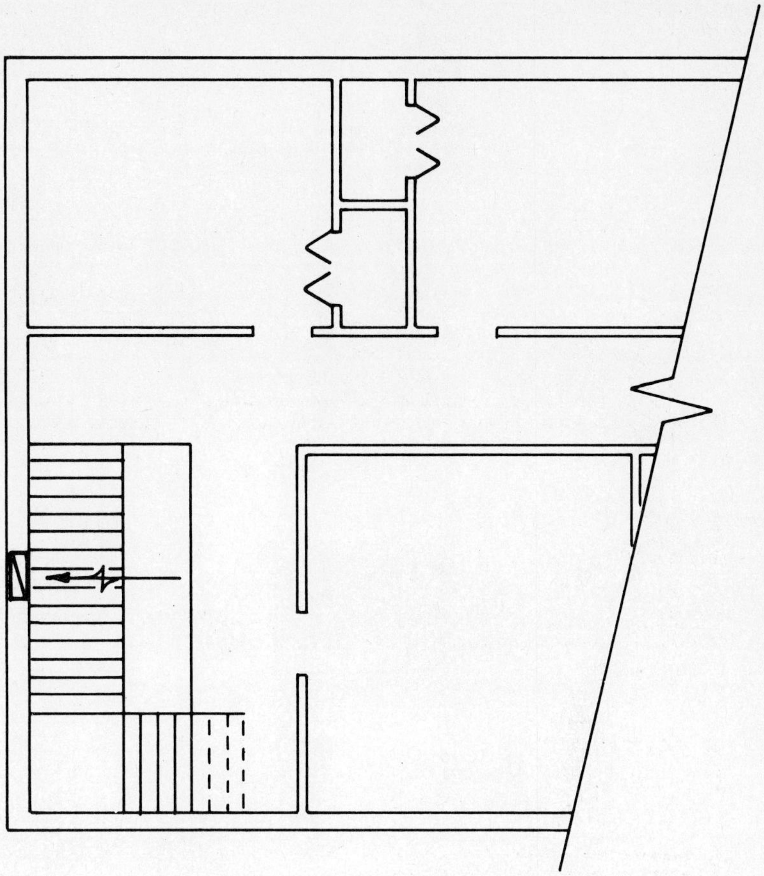

Fig. 9-3 A floor plan showing the return-air outlet located in the stairway.

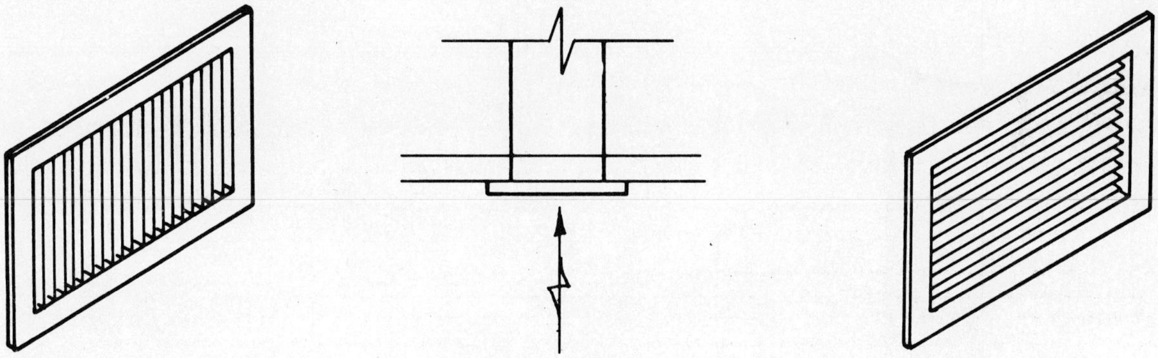

(A) Side-wall return-air outlet with vertical blades.

(B) Section of a ceiling-mounted return-air outlet.

(C) Side-wall return-air outlet with horizontal blades.

Fig. 9-4. Example of three types of return-air outlets.

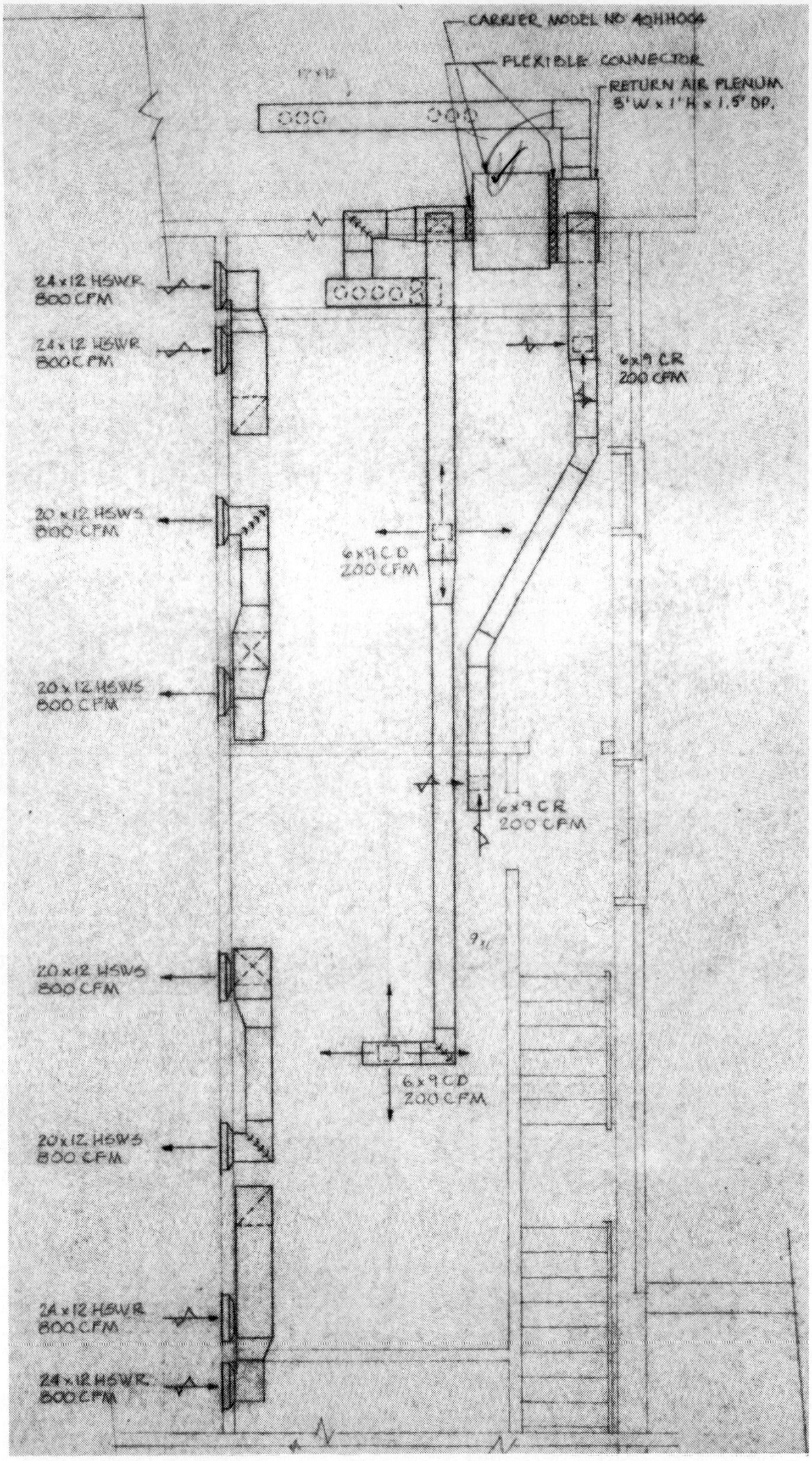

Fig. 9-5. Example of air-conditioning system with return air being carried back to the return-air plenum by means of duct work with return-air outlets.

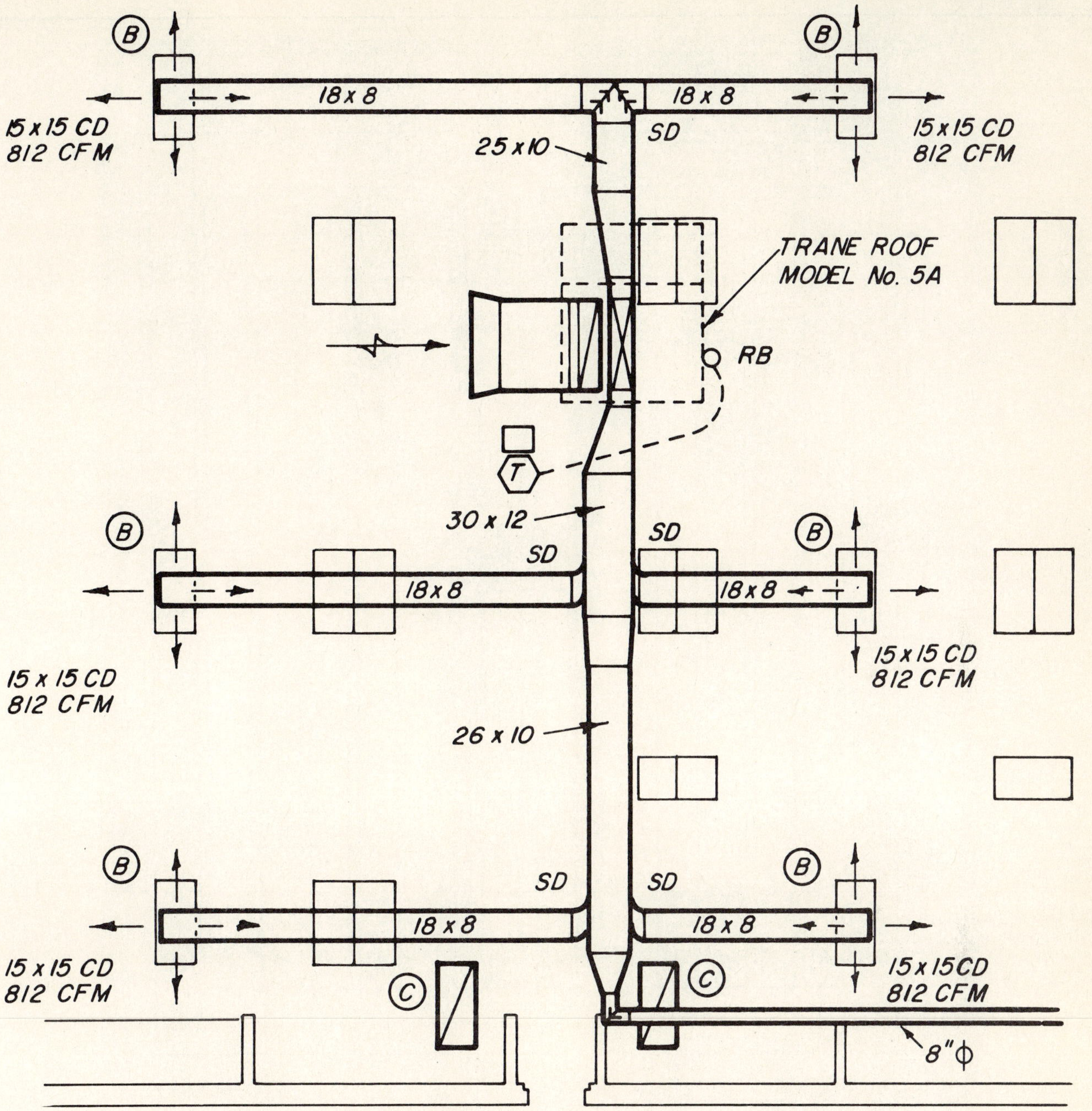

Fig. 9-6. A floor plan of an air-conditioning system using only one large outlet for the air return.

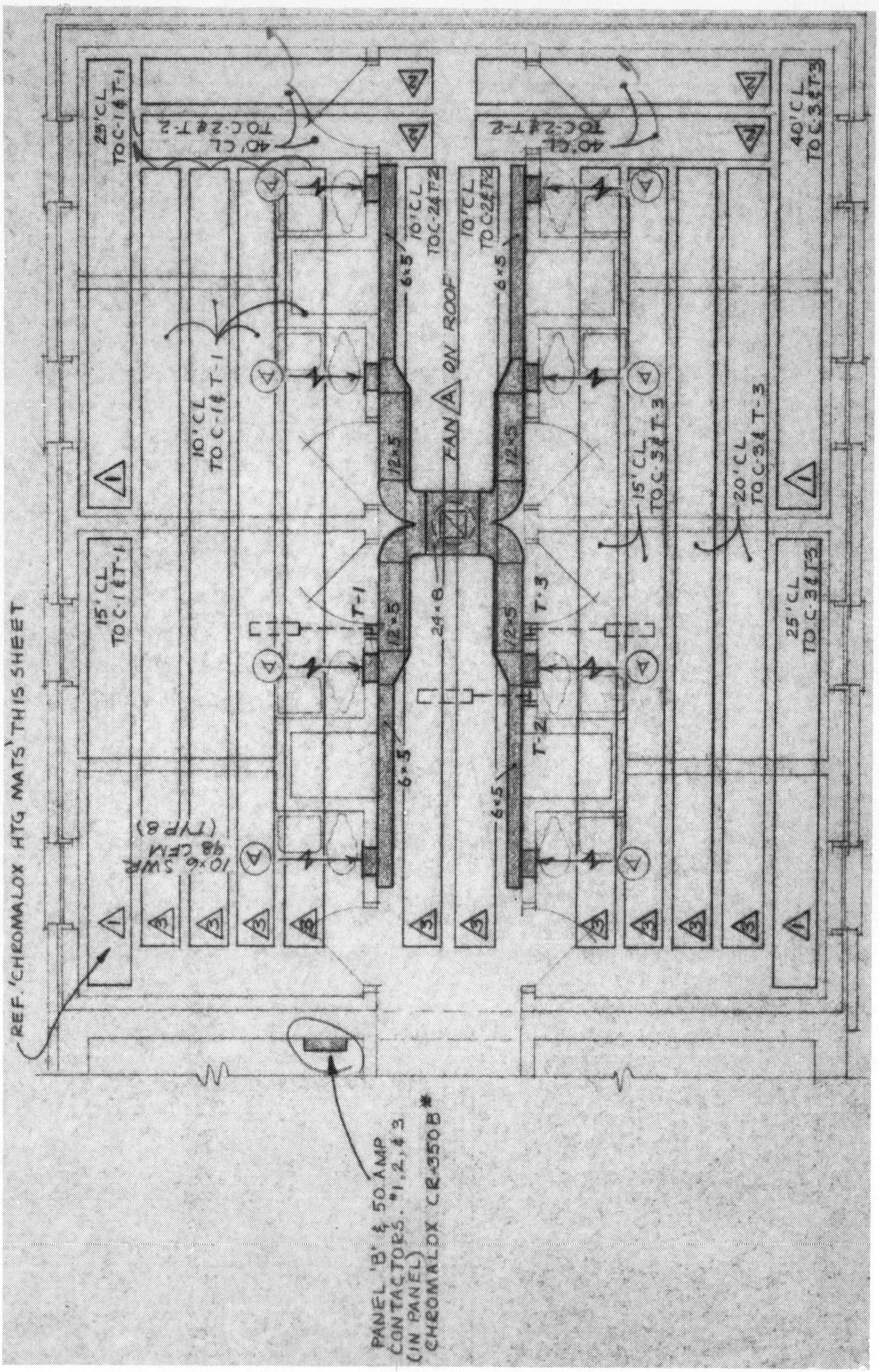

Fig. 9-7. A floor plan of an exhaust system using eight high side-wall return-air outlets.

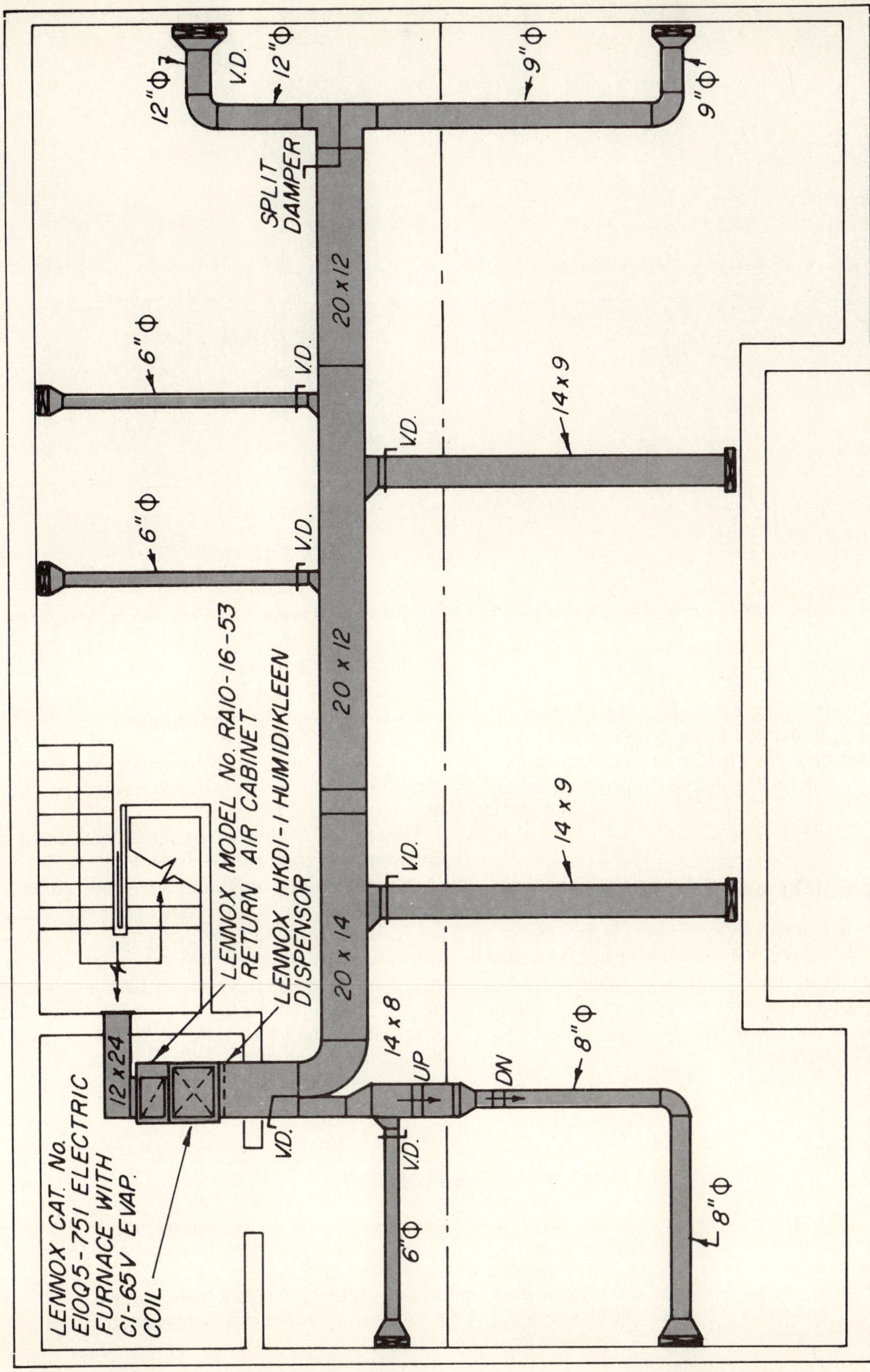

Fig. 9-8. Example of an air-conditioning system using a 12-by-24-inch duct to carry the return air from a grille, located in the stairwell, to the return-air cabinet.

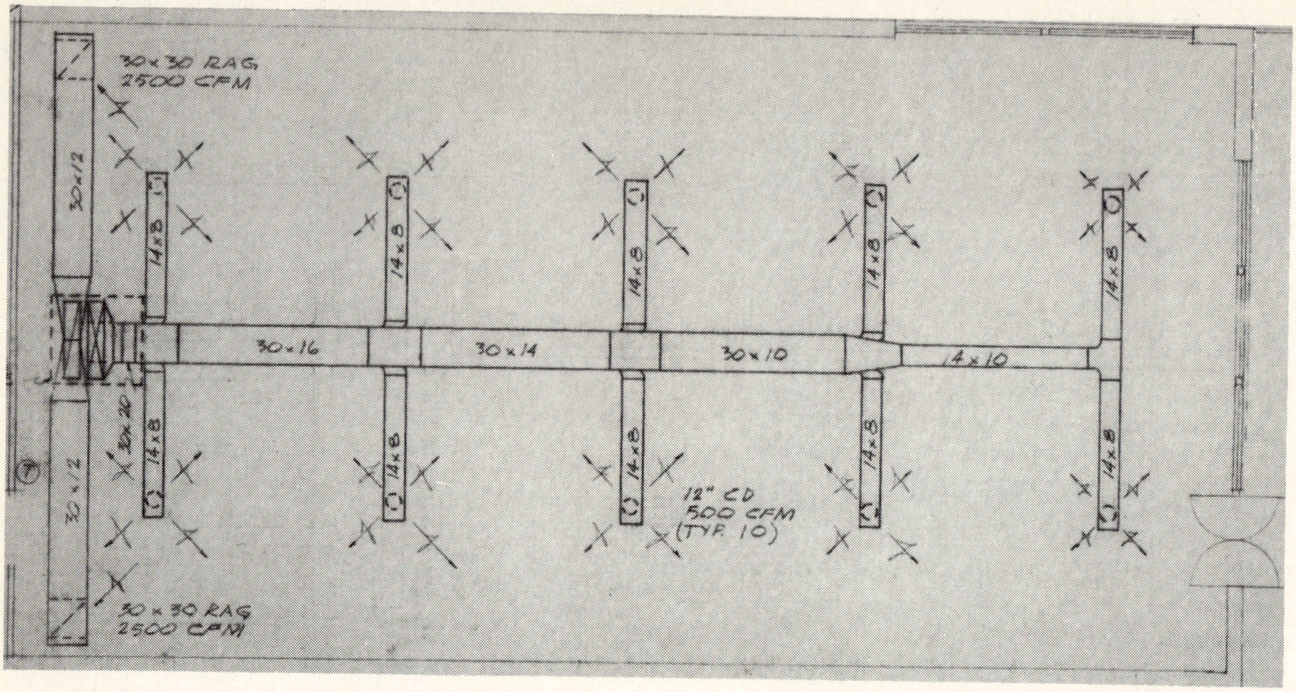

Fig. 9-9. A floor plan of a store building air-conditioning system using two 30-by-30-inch ceiling registers combined with ductwork for the return air.

by duct to the air-handling apparatus. An illustration of such a system is illustrated in Unit 7 (Fig. 7-2).

Fig. 9-4 shows different types of return-air outlets for use in both residential buildings and commercial buildings.

PRACTICAL APPLICATIONS

The examples of actual HVAC working drawings shown in Figs. 9-5 through 9-9 show various means of returning air in an air-conditioning system. They should be analyzed carefully in order to gain a better understanding of how return-air systems should be shown on a working drawing.

SUMMARY

Return-air outlets provide a means of returning the conditioned air to its original source so that it may be conditioned again.

Usually, return-air outlets should be located in a position opposite that of supply-air outlets.

Recessed, floor-mounted return-air outlets are not recommended in commercial buildings because they are likely to become "dirt catchers."

In some buildings, return-air outlets can be located in a central area. However, in buildings where all the supply air cannot return to one central outlet, the return-air outlets and their related ductwork must be located in several areas.

UNIT 9 — ASSIGNMENT

1. Locate the drawing of a supply-duct system that you completed as an assignment in Unit 8 and fasten it to your drawing board.
2. Referring to Fig. 9–10, draw a two-line return-air duct system.
3. Again, using Fig. 9–10 as a guide, draw all return-air grilles, airflow directional arrows, and the cubic feet per minute of each outlet. You now have both a supply-duct system and a return-air duct system on one page for comparison.

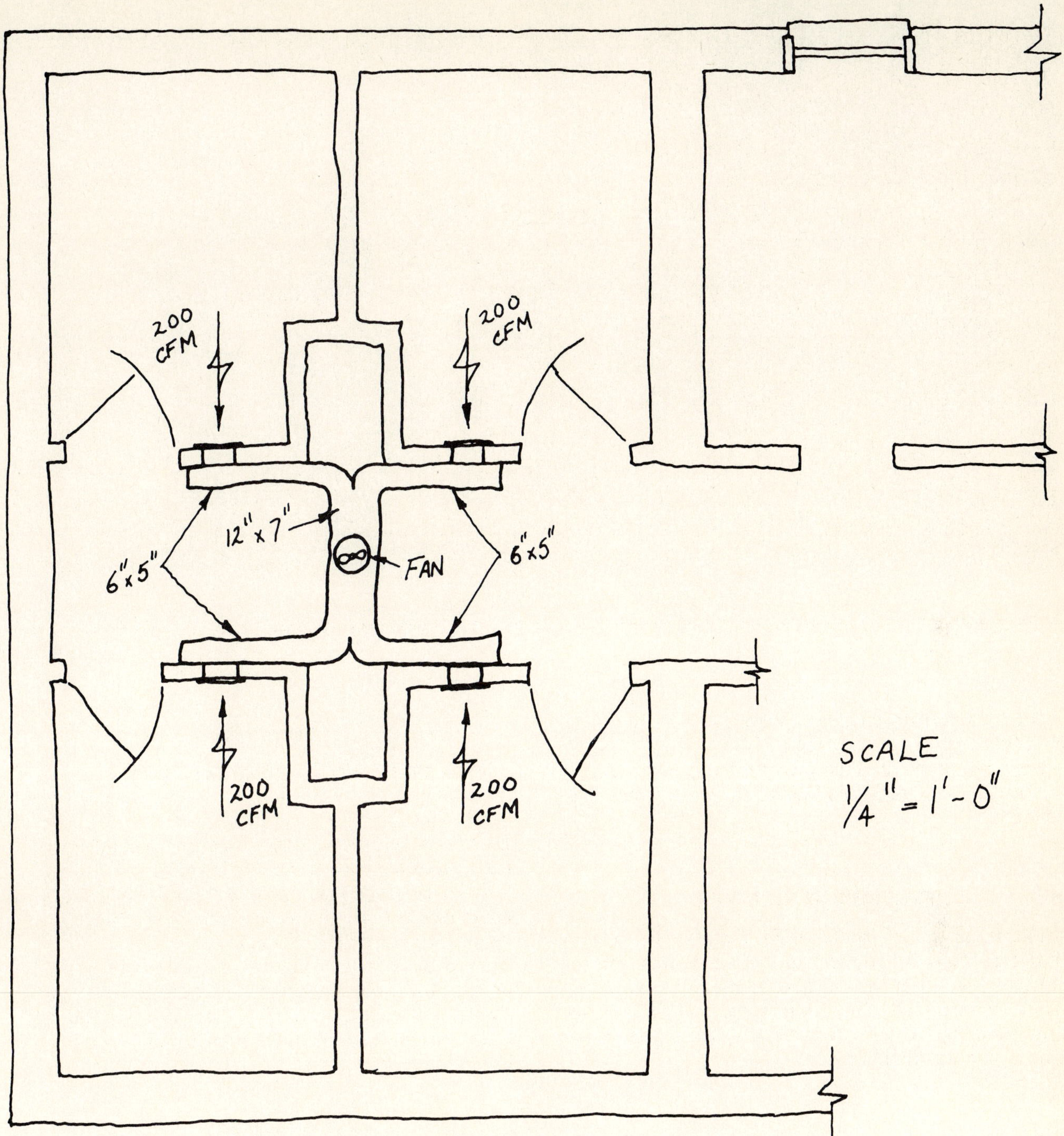

Fig. 9-10. Engineer's sketch of a ventilating system for the assignment in this unit.

4. Answer the following True or False questions.

	True	False
a. The location of the return-air outlets is more important than the location of the supply-air outlets.	____	____
b. As a rule-of-thumb, if the supply-air outlets are located in the ceiling, the return-air outlets should be located in the ceiling also.	____	____
c. Multi-story buildings usually have the return-air system in the stairway.	____	____
d. Baseboard or low side-wall return outlets are better than floor outlets in commercial buildings.	____	____
e. One central-return outlet is all that is ever needed.	____	____

SECTION III

Isometric Duct Drawing

Fundamentals of Isometric Drawing

OBJECTIVES: To teach the basic fundamentals of isometric drawings

To apply these basic fundamentals to HVAC working drawings

INTRODUCTION

Isometric drawing comes under the heading of pictorial drawing in that the objects are pictured in one view only. The use of isometric drawings is limited because they are more difficult to draw than multiview drawings, intricate parts cannot be pictured clearly, and they are usually more difficult to dimension. However, isometric drawings of air ducts and other HVAC equipment are very useful on HVAC working drawings for describing objects in highly complex systems and conveying information about them.

The orthographic drawing in Fig. 10-1 shows a plan view of a two-story building. The return-air duct is shown in the ceiling of the second floor where it receives air from four return-air grilles. The main duct

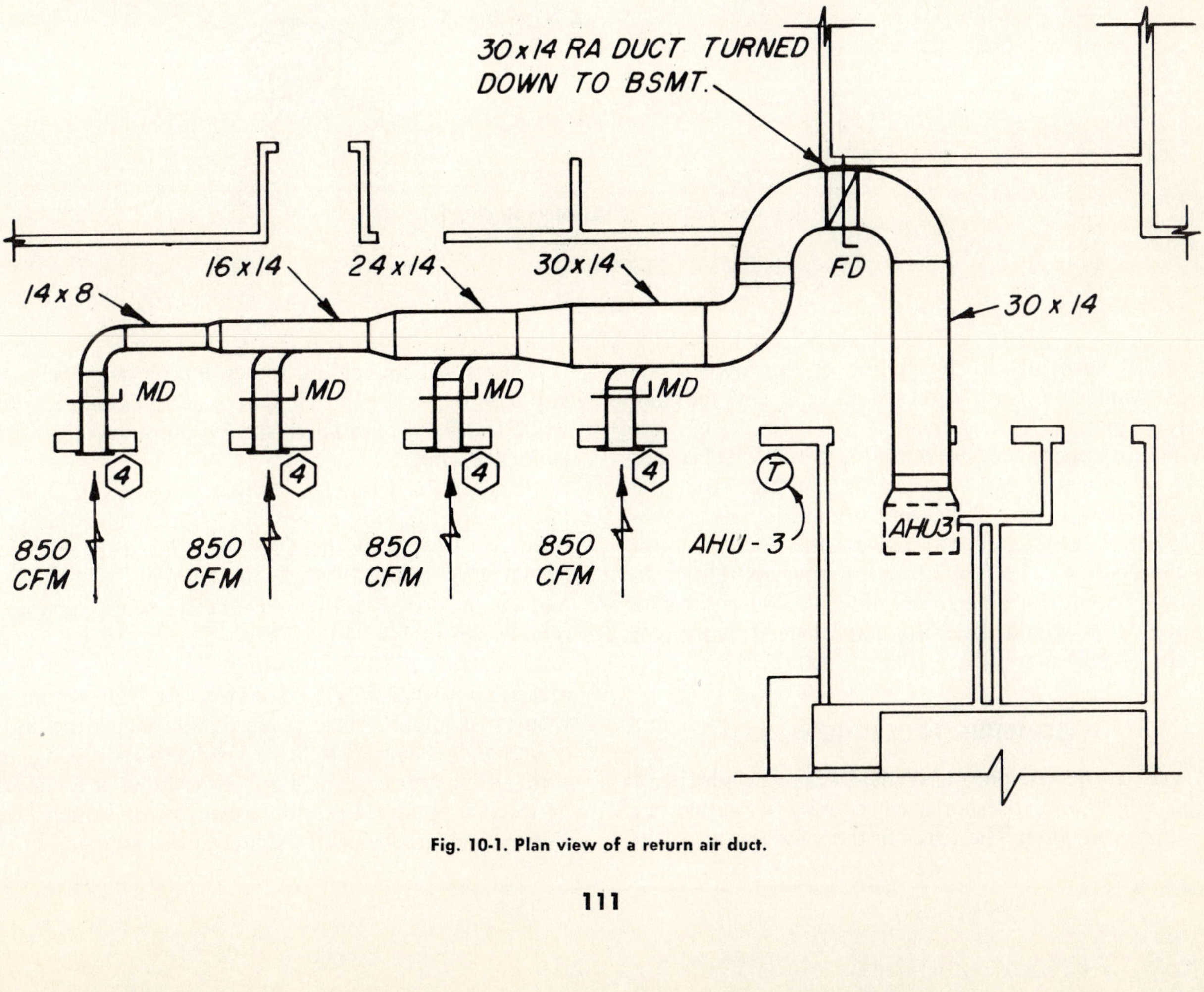

Fig. 10-1. Plan view of a return air duct.

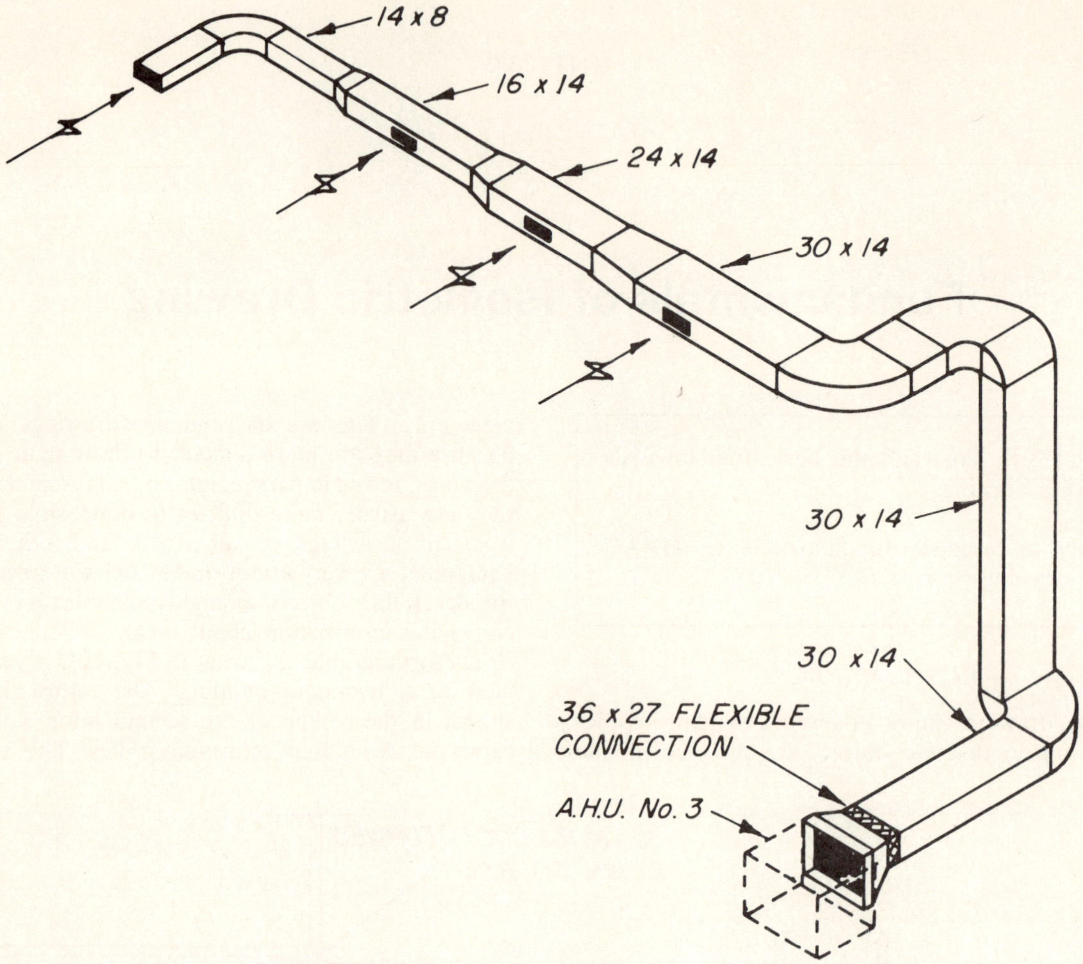

Fig. 10-2. Isometric view of the return air duct in Fig. 10-1.

then drops vertically to the ceiling of the lower floor and is connected to the return-air side of the air-handling unit.

While an experienced workman in the HVAC trades would be able to "read" the drawing in Fig. 10-1, the isometric view of the same return-air duct, as shown in Fig. 10-2, is clearer and leaves little doubt about how the duct is to be run. In fact, a new apprentice in the trade should be able to see how the duct is to be installed; this would free the experienced workman for other jobs.

ISOMETRIC PROJECTION

A view projected onto a vertical plane in which all of the edges are foreshortened equally is called an *isometric projection.* The edges in the view that are 120 degrees apart are called the *isometric axes.* The three surfaces shown in the view are called *isometric planes,* while the lines parallel to the isometric axes are called *isometric lines.*

Because the isometric axes are always 120 degrees apart and the edges are foreshortened equally, the isometric axes and the lines parallel to them can be drawn with a 30-60-90-degree triangle, and the edges can be measured with an isometric scale such as the one shown in Fig. 10-3.

Isometric HVAC drawings, however, are usually measured with a standard architect's scale rather than with an isometric scale. This results in a drawing that is somewhat larger than an isometric drawing made with an isometric scale. Fig. 10-4 shows a comparison between a condensing unit drawn as an isometric projection and drawn as an isometric drawing.

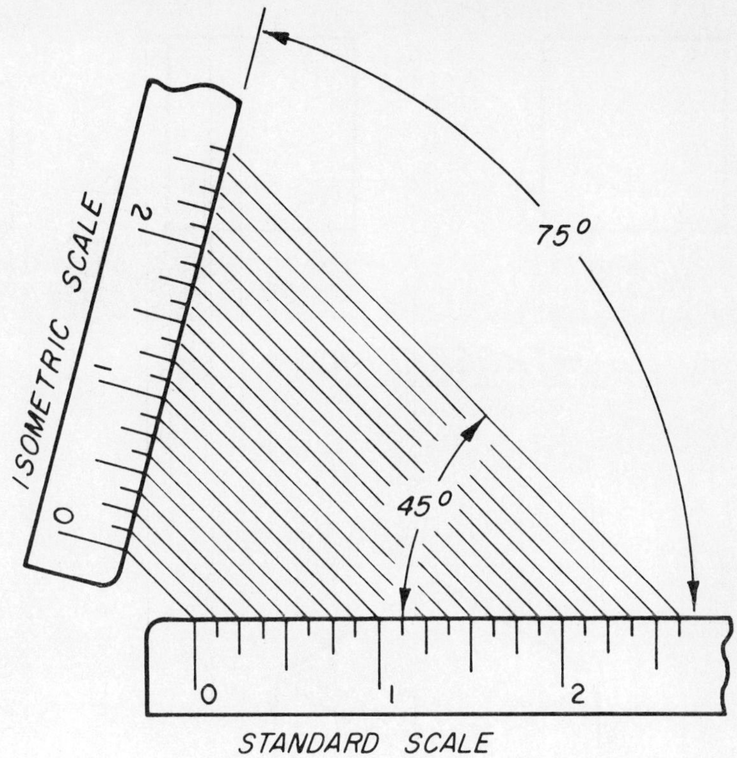

Fig. 10-3. Comparison between isometric and standard scales.

ISOMETRIC DRAWING

Isometric drawing is a method of showing a three-dimensional view on a flat surface, such as a sheet of tracing paper. This method of drawing is preferred for certain HVAC details on working drawings because it is possible to draw isometric lines to scale in the same manner as floor plans are drawn in orthographic views. The basic rules for isometric drawing are as follows:

1. Vertical lines remain the same in both orthographic and isometric views, as shown in Fig. 10-5.
2. Horizontal lines in an orthographic view are projected at an angle of 30 degrees in an isometric view, as shown in Fig. 10-6.

Fig. 10-7 shows three orthographic views of a typical residential bathroom before the plumbing fixtures have been drawn. Fig. 10-8 shows an isometric view of this same bathroom.

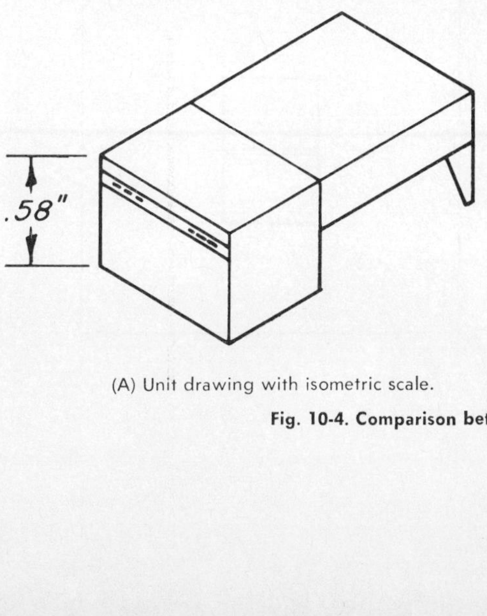

(A) Unit drawing with isometric scale.

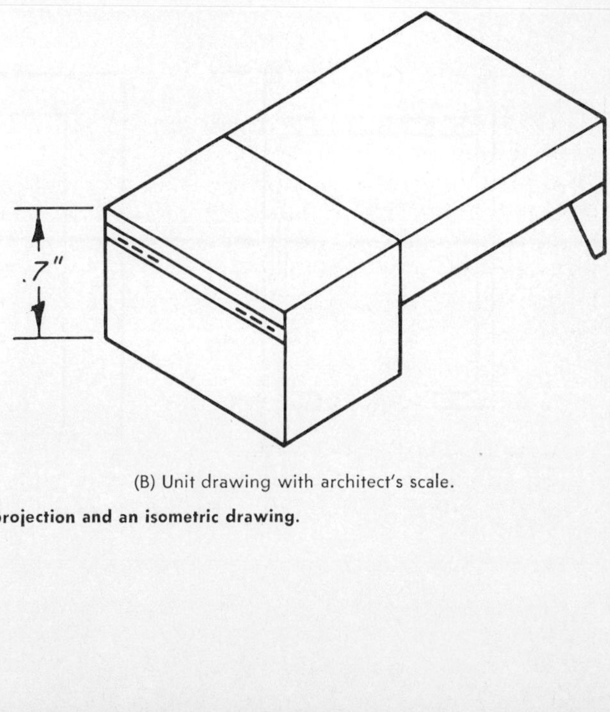

(B) Unit drawing with architect's scale.

Fig. 10-4. Comparison between an isometric projection and an isometric drawing.

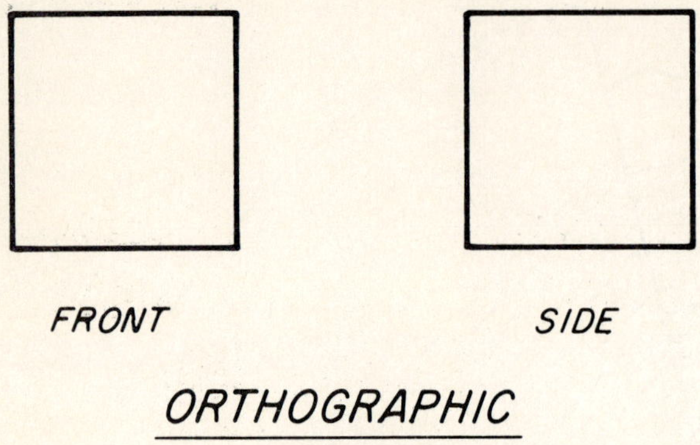

(A) Orthographic view.

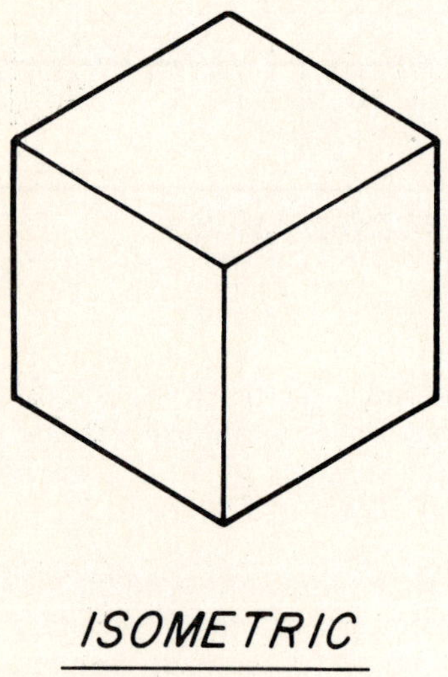

(B) Isometric view.

Fig. 10-5. Example showing that vertical lines remain the same in both orthographic and isometric views.

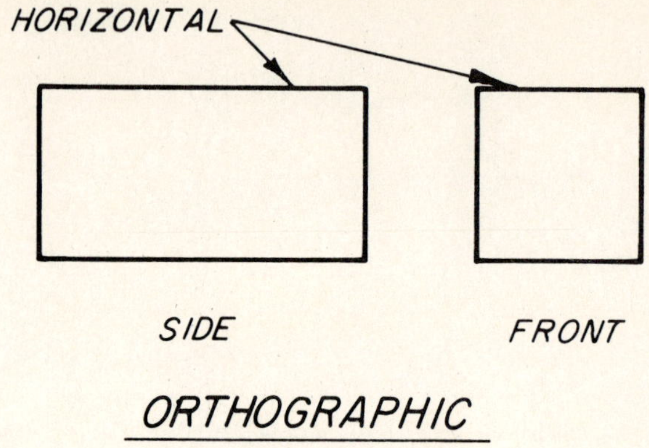

(A) Orthographic view.

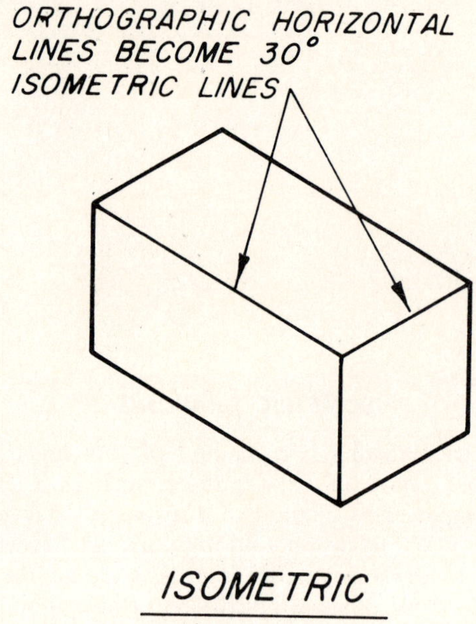

(B) Isometric view.

Fig. 10-6. Example showing how horizontal lines differ in orthographic and isometric views.

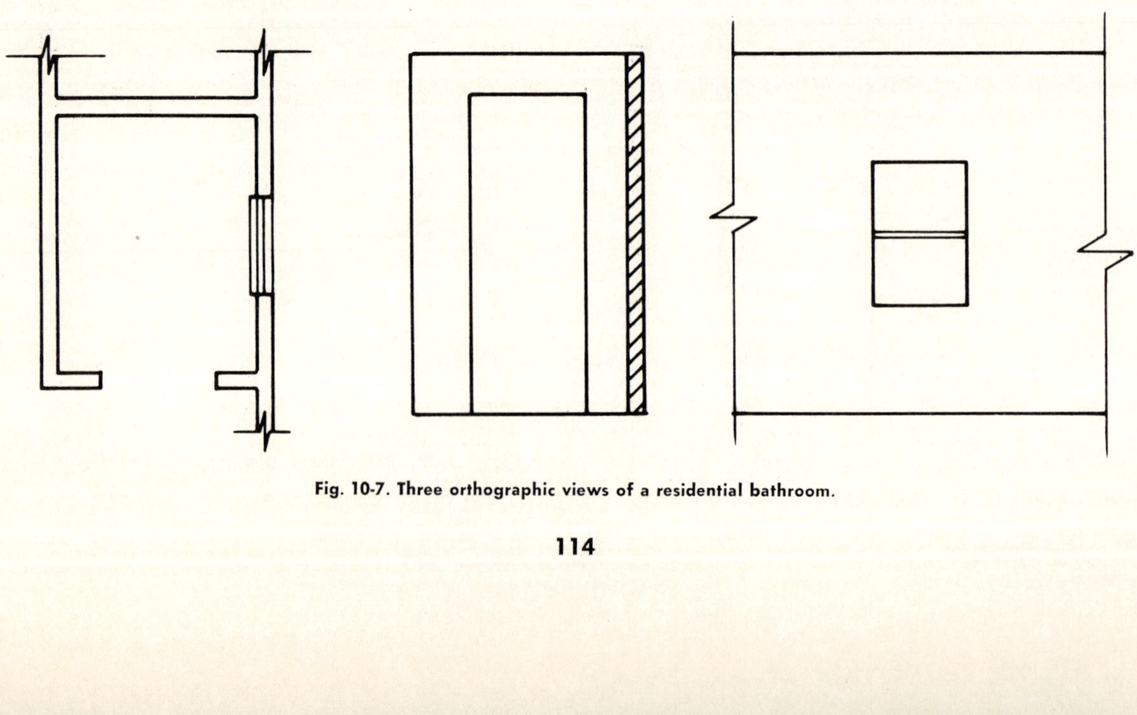

Fig. 10-7. Three orthographic views of a residential bathroom.

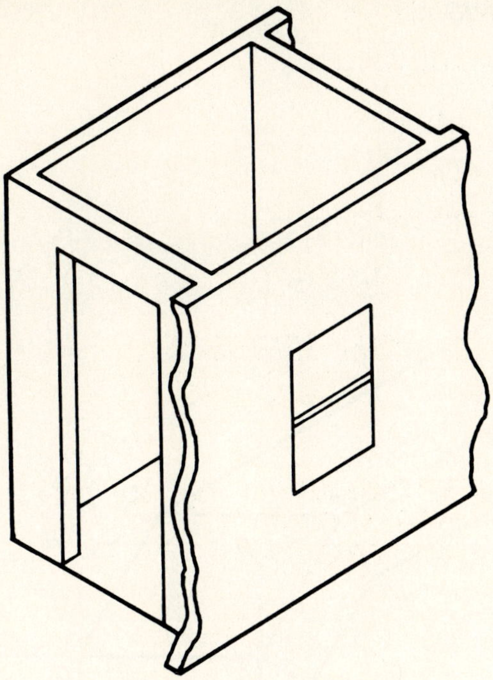

Fig. 10-8. Isometric view of the bathroom in Fig. 10-7.

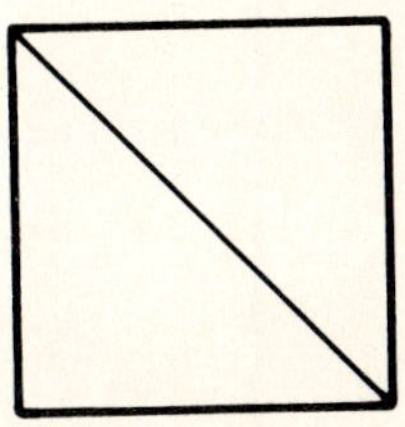

Fig. 10-9. Simple method of drawing a 45-degree angle.

Notice that the vertical lines in the orthographic view remain vertical in the isometric view, while the horizontal lines in the orthographic view project at 30-degree angles in the isometric view and are therefore no longer "horizontal lines."

The lengths of the lines—both vertical and horizontal—remain the same in both views since the same scale is used for drawing both.

ISOMETRIC ANGLES AND CURVES

Angles and curves are the most difficult parts of isometric drawing. But, if the following methods are carefully followed, the student will be able to draw both angles and curves with neatness and accuracy.

Fig. 10-9 shows a perfect square having four angles of 90 degrees each. If two of these angles are divided in half with a diagonal line, the result will be two 45-degree angles at each of the bisected corners.

A similar method is used to lay out 45-degree angles in isometric drawings. For example, in order to draw a 45-degree angle in an air-duct diagram, the following method is used.

Fig. 10-10 shows a run of rectangular ductwork starting at the air-handling unit. The ductwork lines are drawn with heavy solid lines using the 30-60-90-degree triangle to points A and A_1. The lines are extended (using a light broken line) to a predetermined distance beyond points A and A_1, to points B and B_1; a distance of ½ inch is used in Fig. 10-10. Points C and C_1 are found by measuring ½ inch vertically. This establishes two sides of a cube. Then, points A and A_1 are connected to points C and C_1 with diagonal lines which will be 60 degrees. These represent the actual 45-degree angle of the duct as it would appear in elevation.

Next, point A is connected to A_1 and point C connected to C_1, horizontal lines being used in each case. Note that if line segments AC and A_1C_1 are the same angle (60 degrees), then line segment CC_1 will set itself up as equal to line segment AA_1.

Next, line segments CE and C_1E_1 are drawn and connected at the bottom with line segment EE_1. The remaining duct is drawn in the same manner, with a 30-60-90-degree triangle. To show the depth of the duct, a 60-degree line is made from point A, intersecting the line previously drawn at point D. Next, an extending 60-degree line is drawn from point D (parallel to line segment AC).

The next step is to draw an extending 30-degree line from point C until it intersects extending line D. From this point the remaining duct can be completed in standard isometric style by using the 30-60-90-degree triangle.

Circles or curved lines are best made with special isometric templates, one of which is shown in Fig. 10-11.

A circle in an orthographic drawing becomes an ellipse in an isometric drawing. Semicircles, likewise, become part of an ellipse in an isometric drawing. Although an ellipse can be drawn by using a compass and four different centers, it is much faster and neater to use an ellipse or isometric circle template. Fig. 10-12 shows how to locate the four centers needed to draw an isometric circle with a compass.

Isometric circles are used in HVAC drawings for a number of applications; we will use an oil tank as an example. In order to draw the oil tank shown in Fig. 10-13, begin by drawing a light horizontal line to represent the center line of the tank.

Place the appropriate-size ellipse template on the center line and draw the ellipse. This represents the left end of the tank. Continue by drawing the horizontal lines AB and CD (Fig. 10-14), scaling the appropriate distance. Again, place the ellipse template on the center line and draw arc BD. Fig. 10-14 shows the three steps used to draw the isometric view of the oil tank in Fig. 10-13.

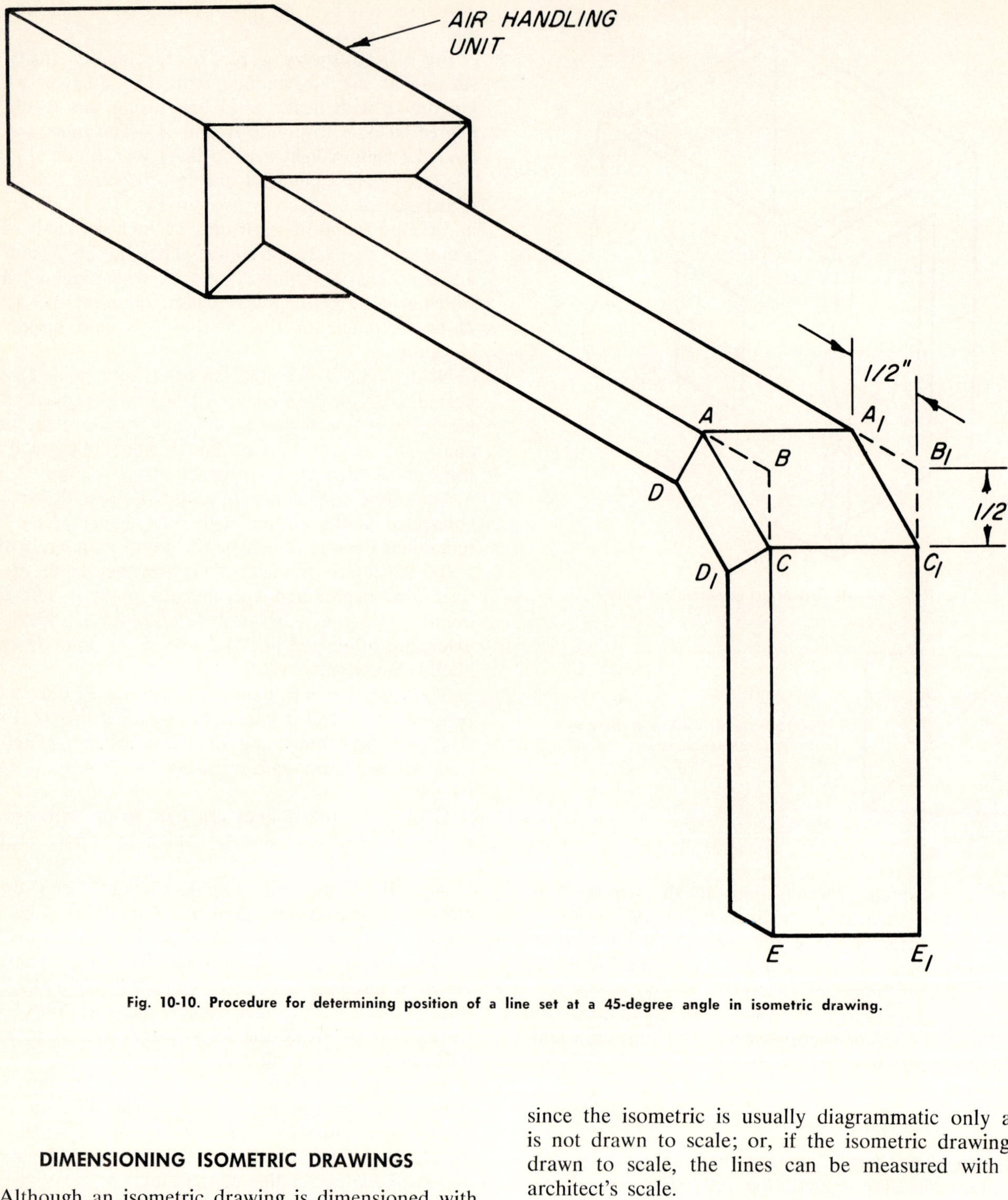

Fig. 10-10. Procedure for determining position of a line set at a 45-degree angle in isometric drawing.

DIMENSIONING ISOMETRIC DRAWINGS

Although an isometric drawing is dimensioned with extension lines in much the same manner as an orthographic drawing, isometric dimensioning is somewhat more difficult because less room is available on the single view of the isometric drawing. Fig. 10-15 shows a dimensioned isometric drawing of a condensing unit. However, in all but a few isometric drawings used in HVAC drafting, the dimension lines are usually omitted, since the isometric is usually diagrammatic only and is not drawn to scale; or, if the isometric drawing is drawn to scale, the lines can be measured with an architect's scale.

In rare instances where isometric drawings are dimensioned, the following rules should be followed:

1. Dimension lines and numerals should be kept off the view wherever possible.
2. The radius dimensions of an arc should be drawn from the center of the enclosing isometric square to the arc.

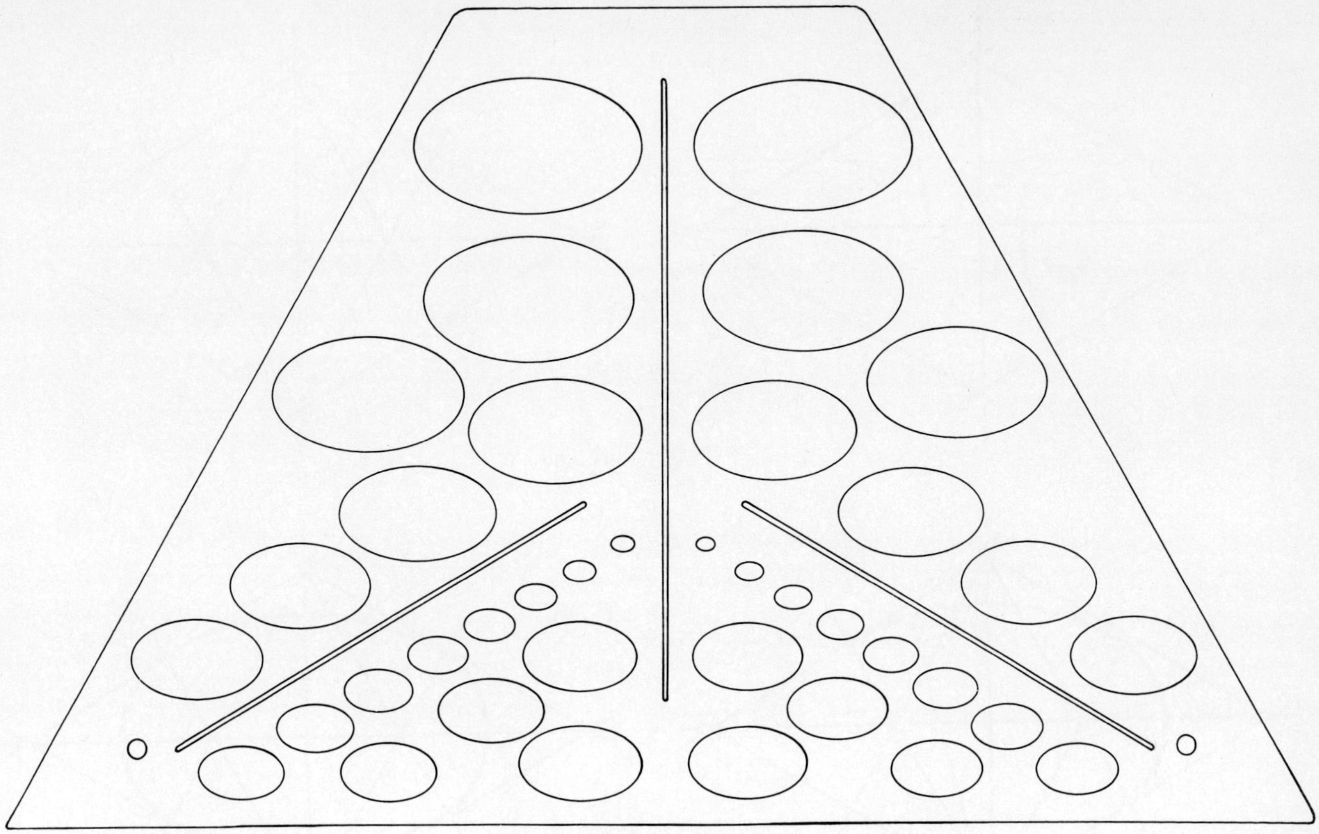

Fig. 10-11. Isometric ellipse template.

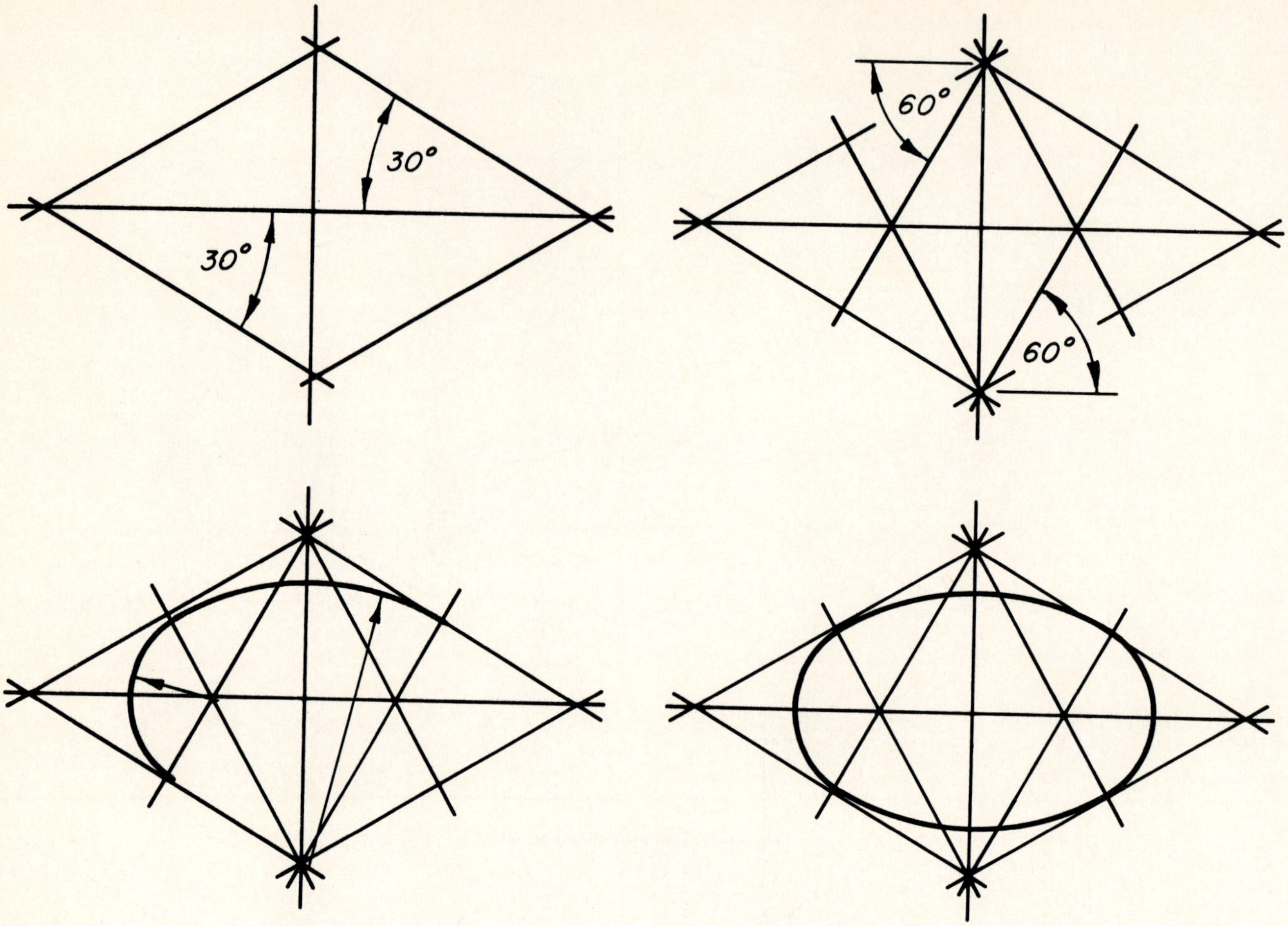

Fig. 10-12. Method used to locate the four centers needed to draw an isometric circle with a compass.

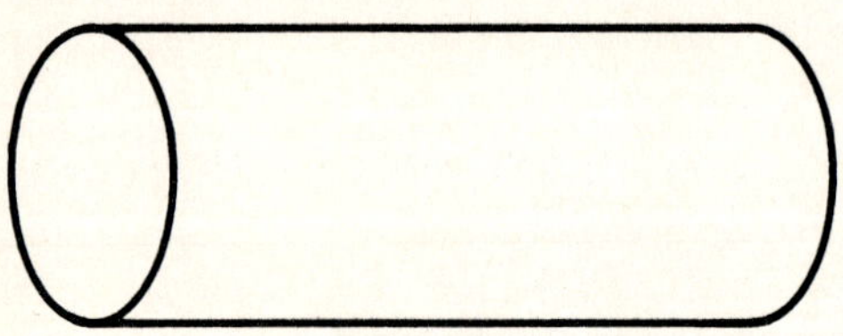

Fig. 10-13. Isometric view of an oil tank.

3. Visible surfaces should be dimensioned in preference to hidden surfaces.
4. Both extension lines for a dimension should be extended from isometric visible edges.
5. Vertical numerals should be used in dimensioning. The numerals and the arrowheads should be made in isometric.

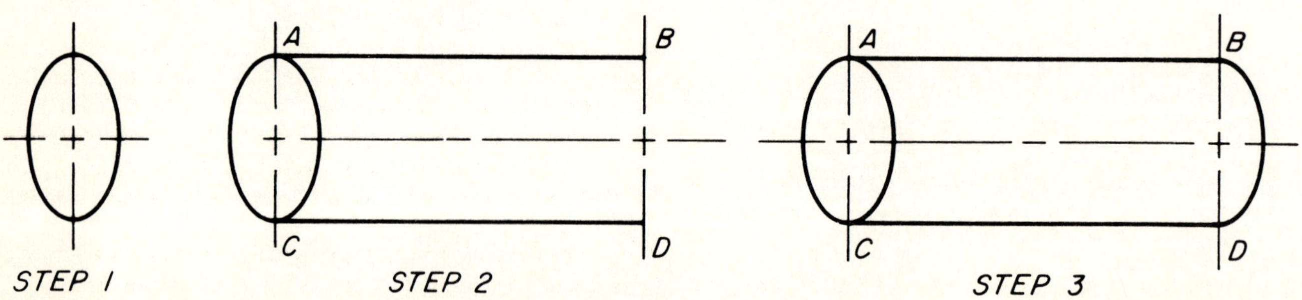

Fig. 10-14. The three steps used in drawing the oil tank in Fig. 10-13.

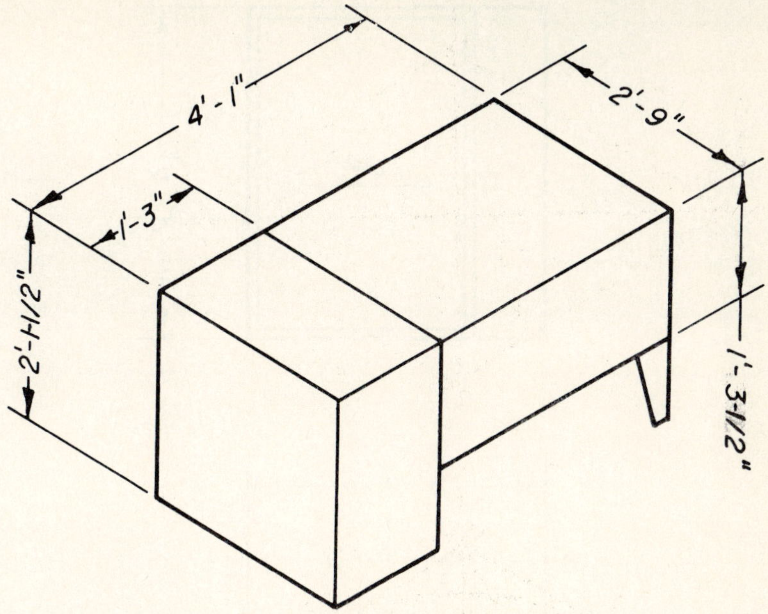

Fig. 10-15. Example of a dimension isometric drawing of a condensing unit.

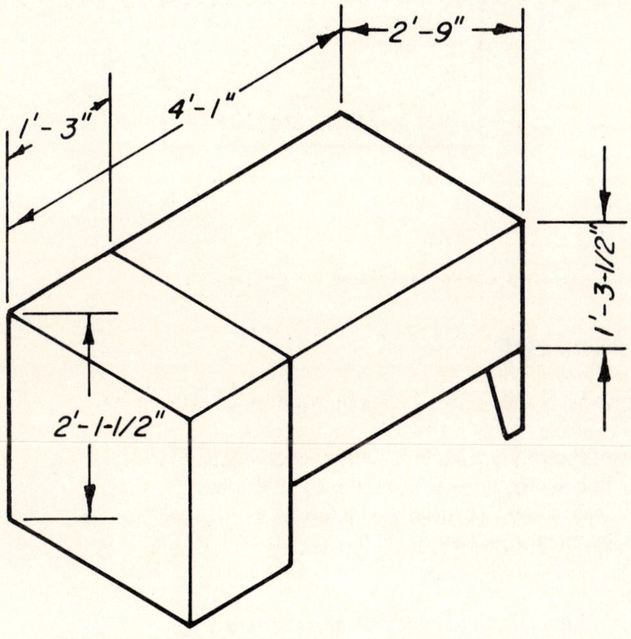

Fig. 10-16. Isometric drawing showing incorrect dimensioning.

6. Horizontal and vertical dimension lines on the orthographic views should be parallel to isometric axes in the isometric view.
7. Notes need not be lettered in isometric; rather, standard vertical single-stroke lettering may be used.

The correct application of these rules is shown in the isometric drawing of Fig. 10-15. In contrast to

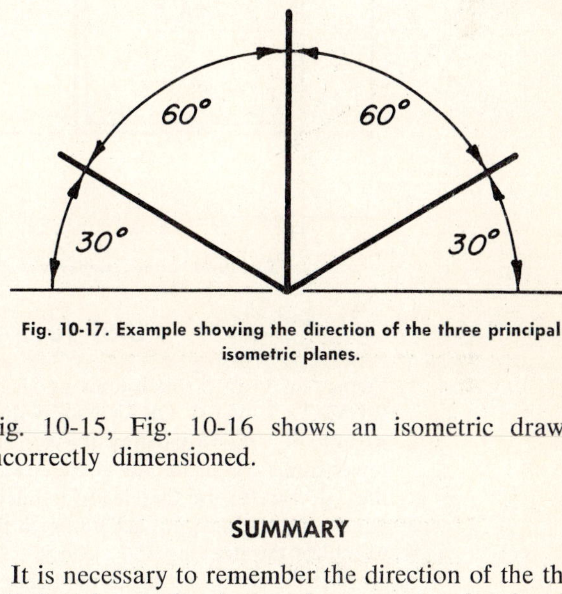

Fig. 10-17. Example showing the direction of the three principal isometric planes.

Fig. 10-15, Fig. 10-16 shows an isometric drawing incorrectly dimensioned.

SUMMARY

It is necessary to remember the direction of the three principal isometric planes when you are drawing isometric views. See Fig. 10-17.

All lines parallel to one of the three axes shown in Fig. 10-17 are isometric lines, and lines not parallel to one of these axes are nonisometric lines. All isometric lines may be drawn with the 30-60-90-degree triangle.

Nonisometric lines are drawn by first locating their end points on isometric lines and then connecting the end points with lines.

The best way to accurately draw circles and arcs on isometric drawings is to use an ellipse or isometric circle template.

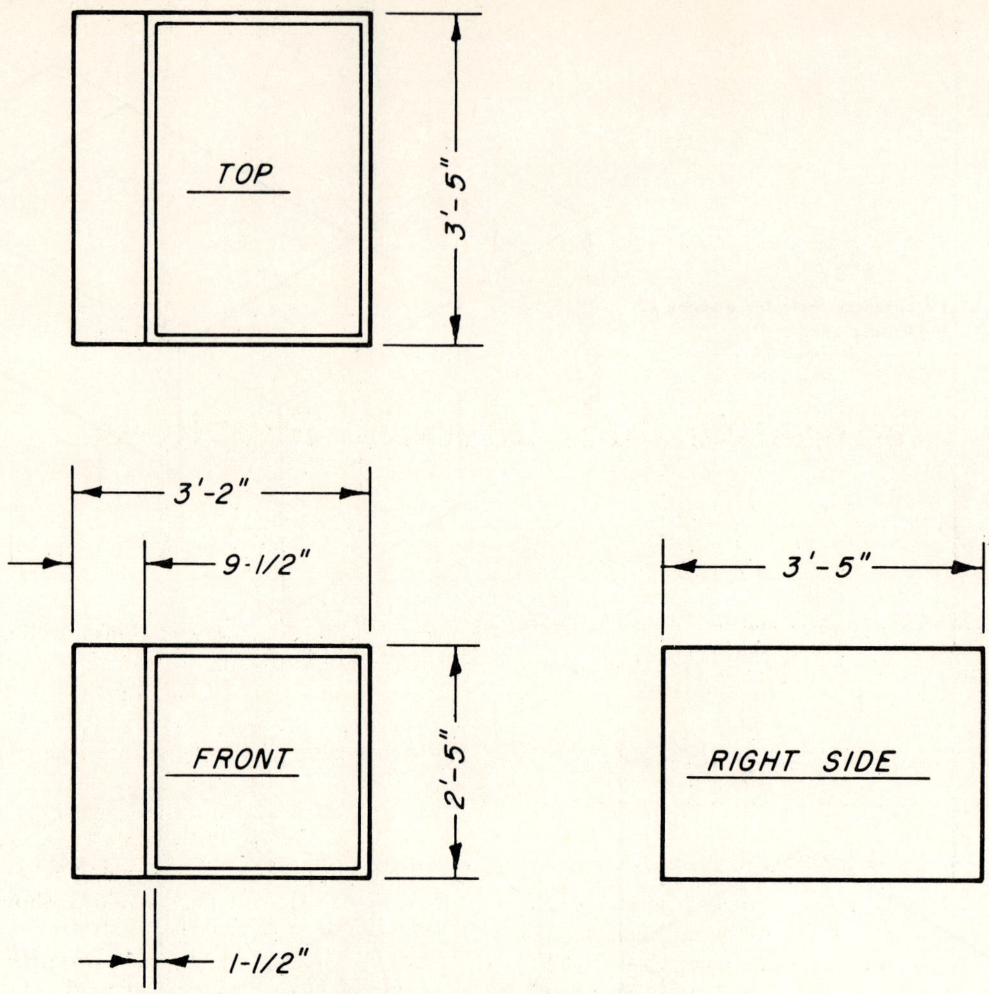

Fig. 10-18. Orthographic view of condensing unit for use in this assignment.

UNIT 10 — ASSIGNMENT

1. Prepare an 11-by-17-inch drawing sheet by drawing border lines. Letter the title block: ISOMETRIC VIEW OF CONDENSING UNIT.
2. Refer to Fig. 10–18 and draw the isometric view of the condensing unit shown. Locate the lower corner 3¼ inches above the lower border line and 6½ inches from the right border line. Be sure to use 2-H lead for linework; H lead should be used for lettering.
3. Dimension the drawing according to the rules given in this unit.
4. Complete the drawing and clean it.
5. Answer the following fill-in-the-blank questions.
   a. A view projected onto a vertical plane in which all of the edges are foreshortened equally is called a(n) __________.
   b. The three surfaces shown in a view are called __________.
   c. The edges in the view that are 120° apart are called __________.
   d. The lines parallel to the isometric axes are called __________.
   e. A(n) __________ in an orthographic drawing becomes an ellipse in an isometric drawing.

UNIT 11

Isometric Duct Layouts: Supply Air

OBJECTIVES: To draw supply-air duct diagrams for buildings

To develop ability in isometric duct drawing

GENERAL

In Unit 10 we learned that nearly all isometric drawings are made with the 30-60-90-degree triangle and that the main guidelines in all isometric drawings are either vertical, horizontal, or at angles of 30 degrees. All lines that are not 30, 60, or 90 degrees from the horizontal plane are nonisometric lines and therefore cannot be measured or drawn to scale in an isometric drawing.

ISOMETRIC PAPER

Some draftsmen use a special isometric paper to aid them in making isometric drawings. This type of paper is printed with lines that run vertically, horizontally, and at 30-degree angles from the horizontal plane. The print is usually a "nonprint" blue ink which will not reproduce on a blueline print made from the tracing.

This paper can be very helpful to both beginners and professionals, especially for making rough sketches of an object prior to preparing the working drawing. However, some draftsmen prefer to slide isometric-ruled paper under the tracing paper as a guide for the finished drawing.

RISER DIAGRAMS

In preparing working drawings for HVAC systems, the draftsman will occasionally be called upon to draw riser diagrams of various ducts within the system. Usually, such diagrams will only be required where extremely difficult routing of the ductwork may be encountered or else in cases where it is determined that a plan and sectional view of the ductwork is not sufficient.

Ductwork riser diagrams include not only the vertical ducts but also the horizontal ducts. To illustrate, a pictorial drawing of a supply air duct is shown in Fig. 11-1. A plan view, such as would be shown by a HVAC draftsman, is shown in Fig. 11-2; this drawing is usually sufficient for the experienced workman to install the system correctly, since the symbols *A* and *B* tell the workman that this is a duct section and that duct risers do exist. However, the person reading the drawings will have to carefully check the architectural and structural drawings to ascertain that no obstacles are in the way of the ducts.

If an obstacle did occur in the ductwork in Fig. 11-2 or even if there were a special routing of the ductwork, the HVAC drawings should contain a section of the ductwork (see Units 12 and 13). Fig. 11-3 shows a sectional view of the duct system shown in Figs. 11-1 and 11-2. Compare all three drawings and relate the identical items in each.

While the sectional view in Fig. 11-3 and the plan view in Fig. 11-2 give sufficient data for a correct installation of this particular system, more-complex duct systems may require even greater detail. Fig. 11-4 shows the same duct system previously discussed in this unit, but this time it is drawn in an isometric view. Notice that this view leaves little doubt as to the exact way the duct system should be installed, since all bends, offsets, connections, and obstacles are shown in one view.

The isometric duct drawing in Fig. 10-2 of Unit 10 is a two-line isometric duct drawing and, of course,

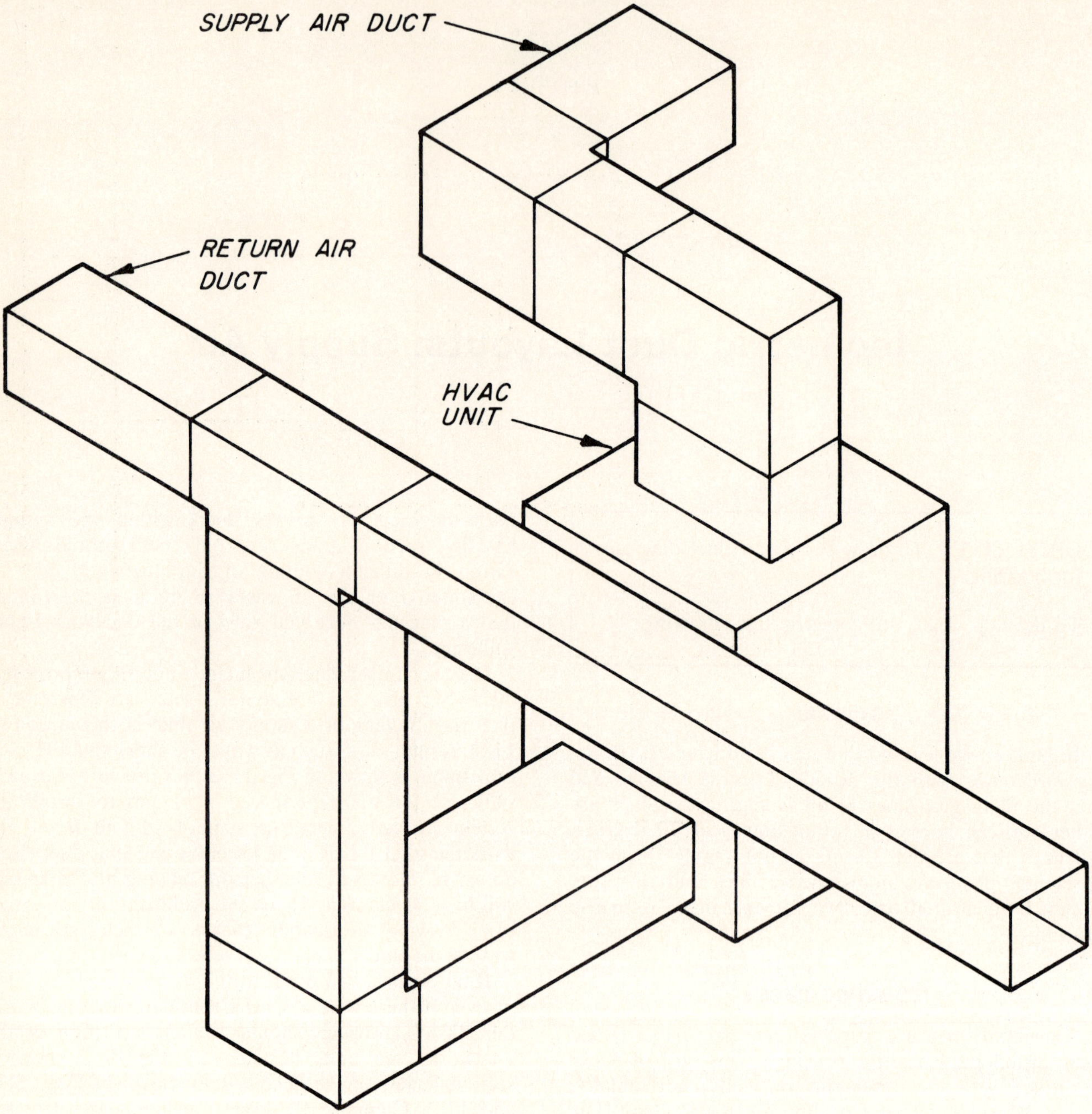

Fig. 11-1. Pictorial drawing of an HVAC unit showing supply and return air ducts.

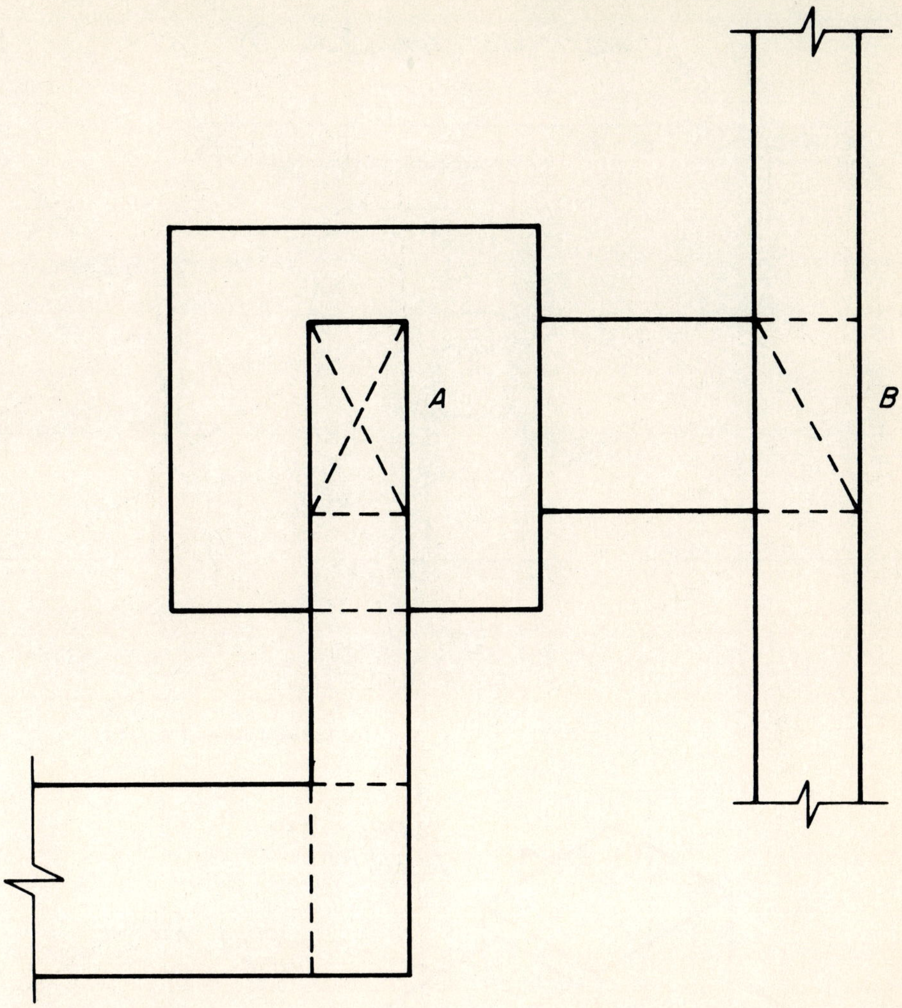

Fig. 11-2. Plan view of Fig. 11-1.

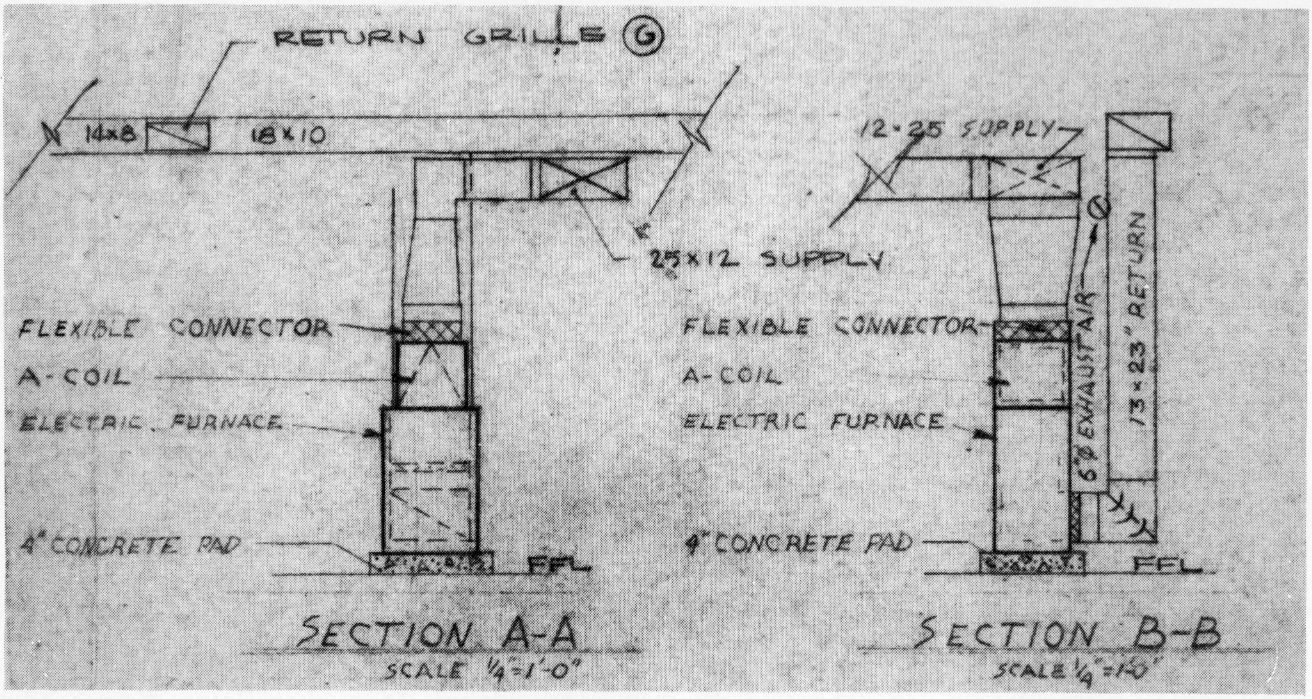

Fig. 11-3. Sectional views of Fig. 11-1.

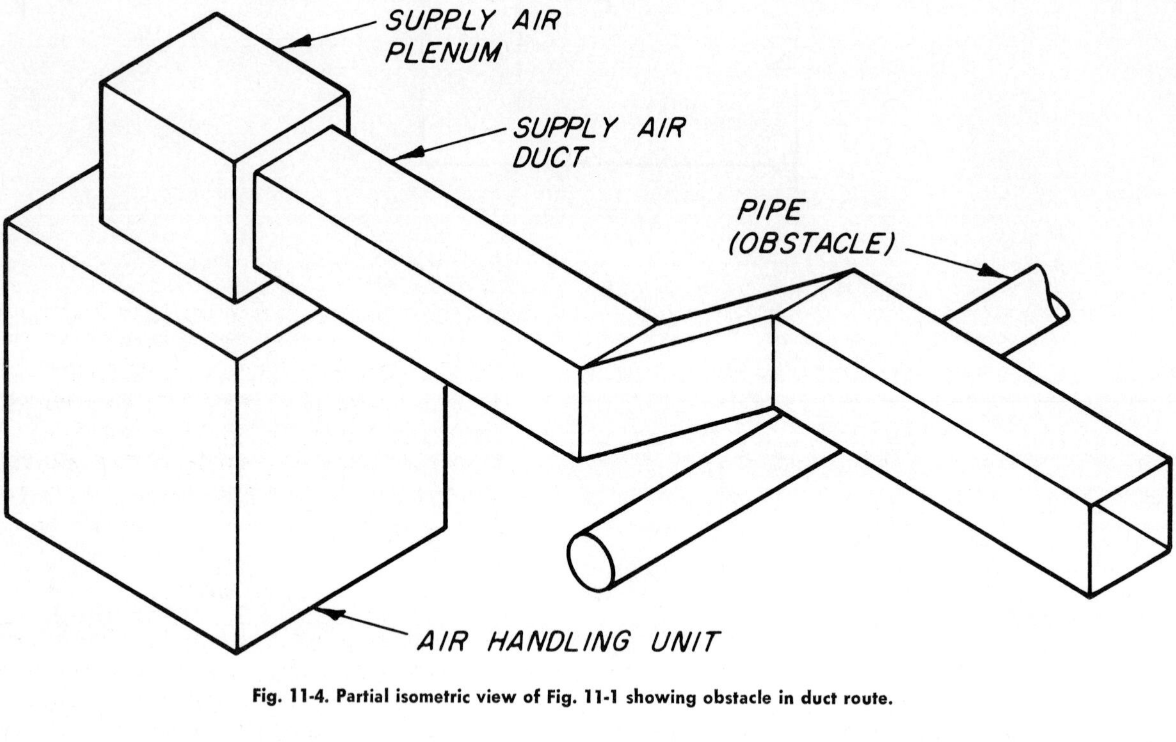

Fig. 11-4. Partial isometric view of Fig. 11-1 showing obstacle in duct route.

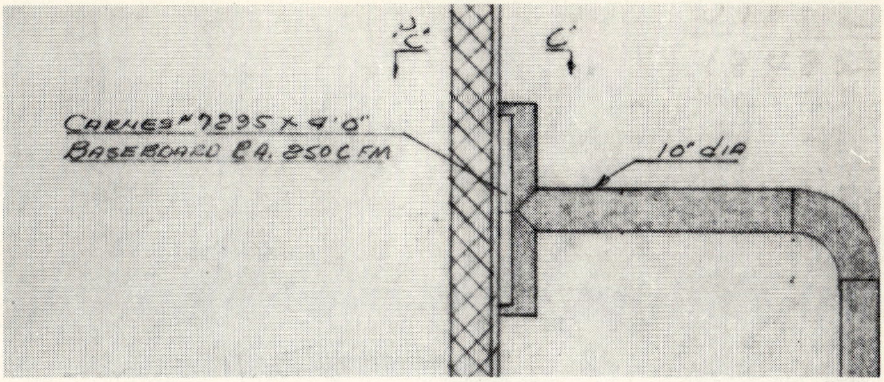

Fig. 11-5. Plan view of floor diffuser and related ductwork.

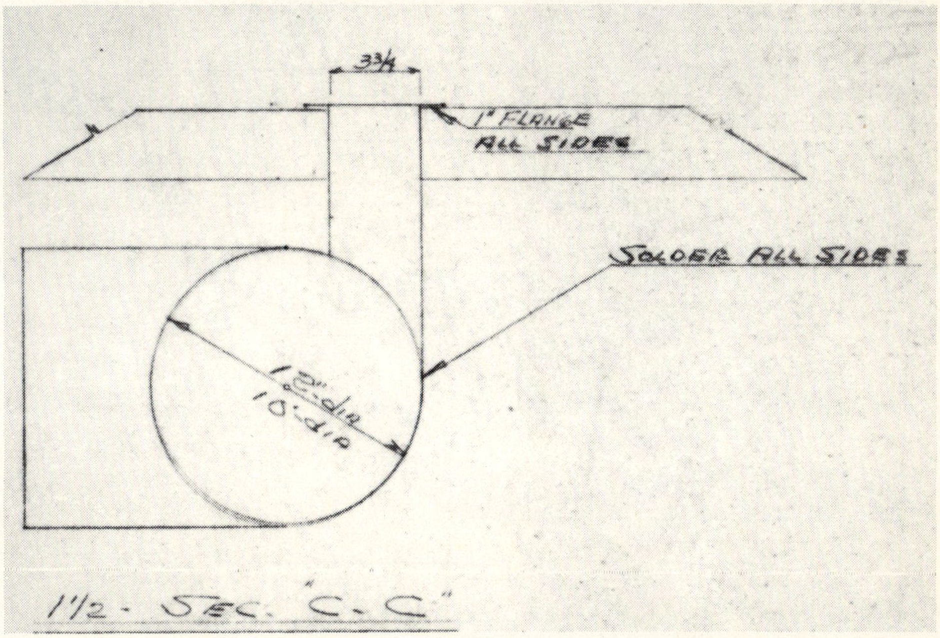

Fig. 11-6. Sectional view of Fig. 11-5.

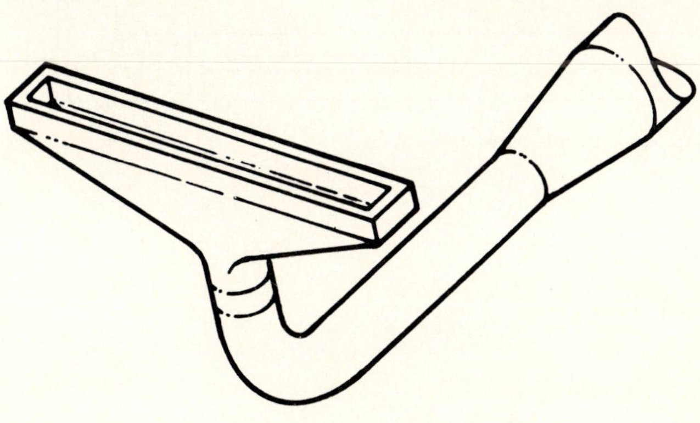

Fig. 11-7. Isometric view of Fig. 11-5.

it is very clear. However, such a drawing is time-consuming and may be more than is necessary for most applications. Most HVAC drawings will contain single-line isometric drawings rather than two-line because they are much faster to draw and are usually sufficient to indicate the routing of the ductwork.

The illustrations in Figs. 11-5 through 11-13 show examples of actual supply-air riser diagrams drawn in an isometric view. All use the single-line drawing technique. If each diagram is studied carefully, the reader will gain a good knowledge of how duct riser diagrams should appear on working drawings.

SUMMARY

Isometric diagrams are used on HVAC drawings as supplemental details, to show the workmen more clearly how the duct system is to be installed.

Single-line duct diagrams should be used for most isometric drawings because they take less time to draw than two-line diagrams and are usually sufficient for most purposes.

Some HVAC isometric views are drawn to scale, but most of the time they are not.

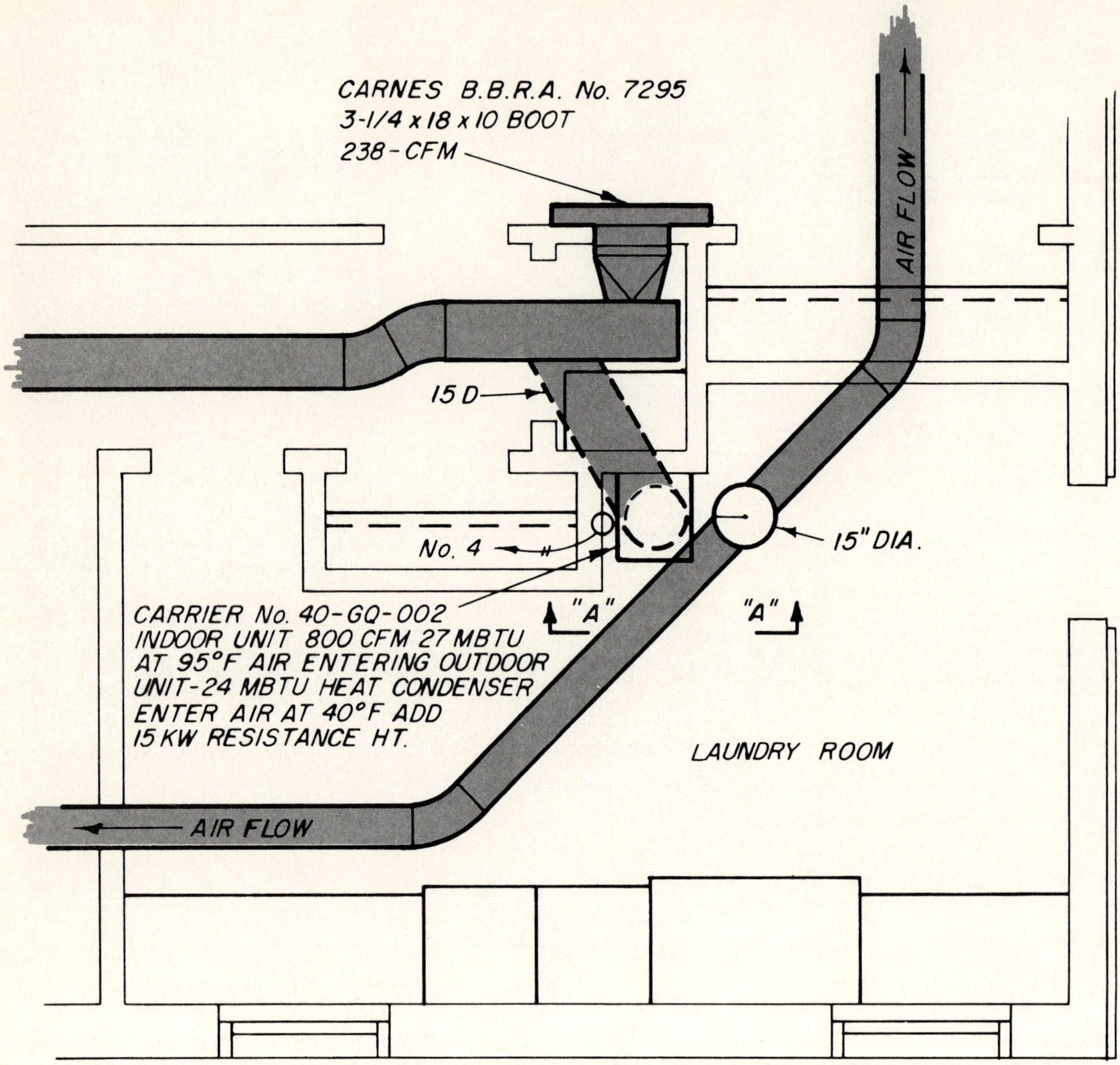

Fig. 11-8. Plan view of air-handling unit and under-floor duct.

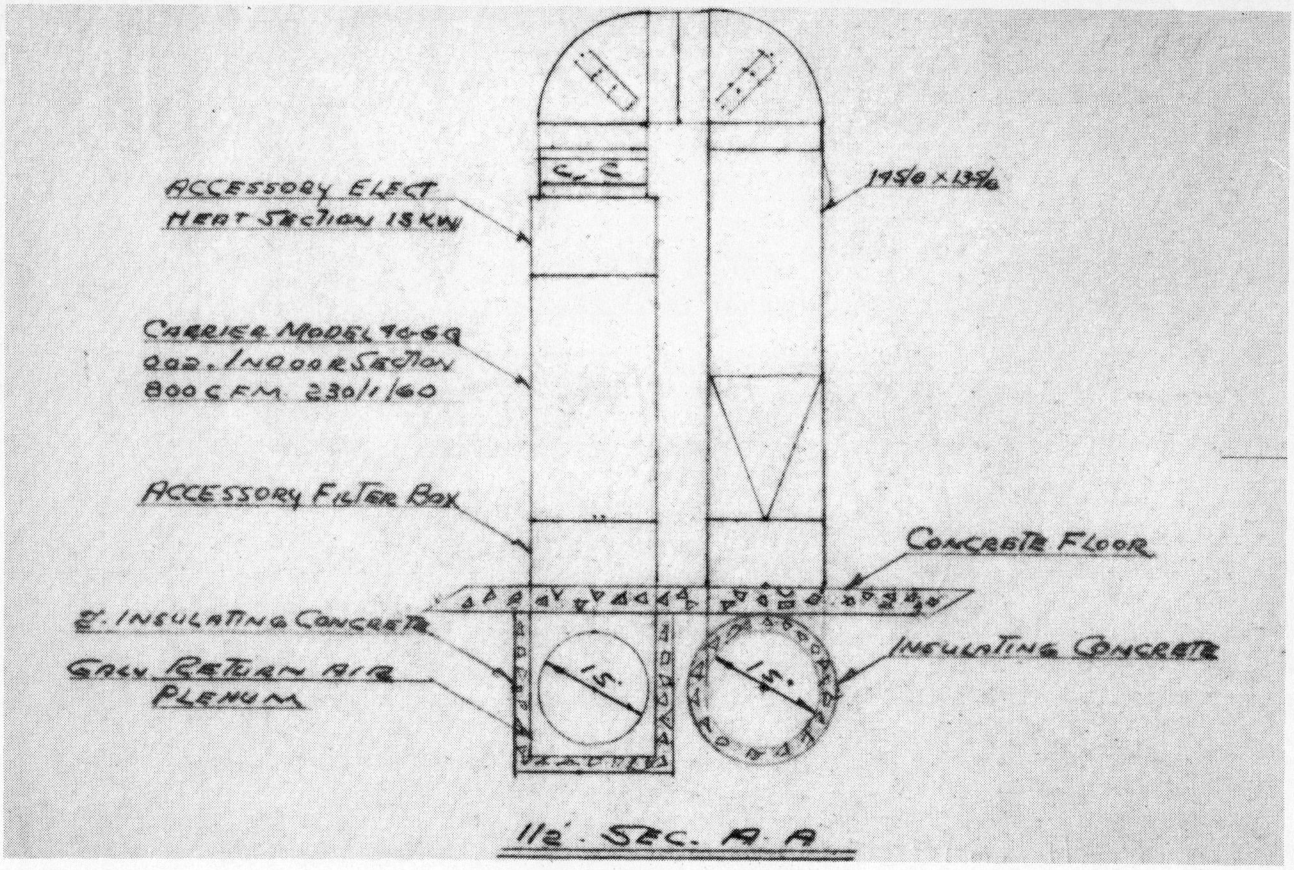

Fig. 11-9. Sectional view of Fig. 11-8.

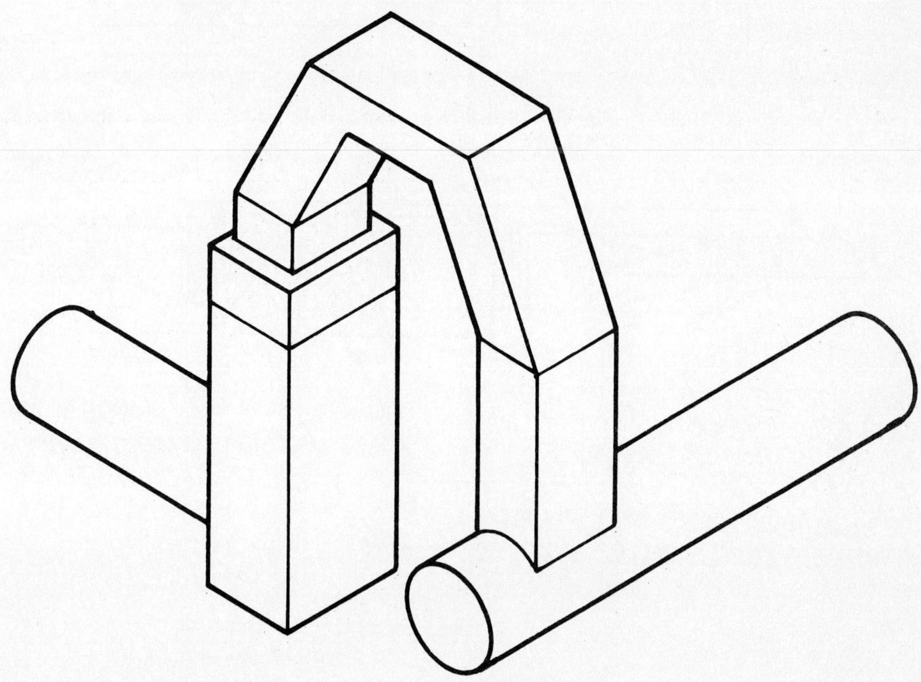

Fig. 11-10. Isometric view of Fig. 11-8.

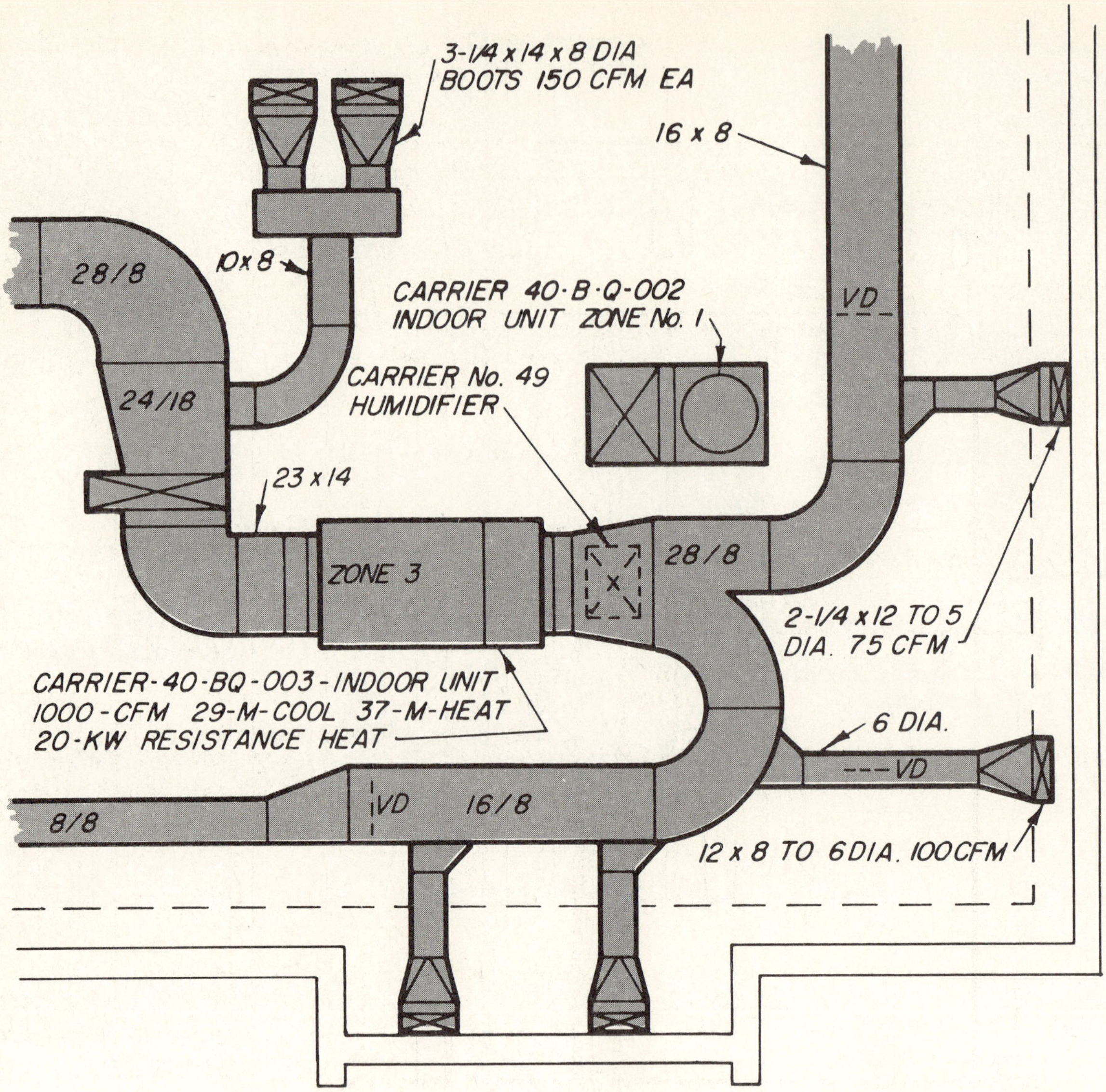

Fig. 11-11. Plan view of a horizontally mounted furnace with related ductwork.

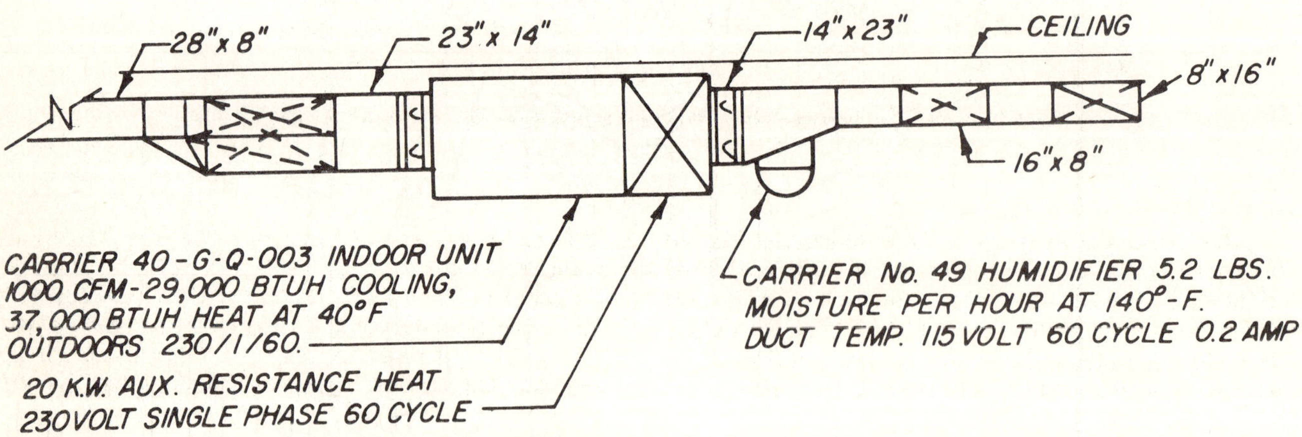

1/2" SEC. "X-X"

Fig. 11-12. Sectional view of Fig. 11-11.

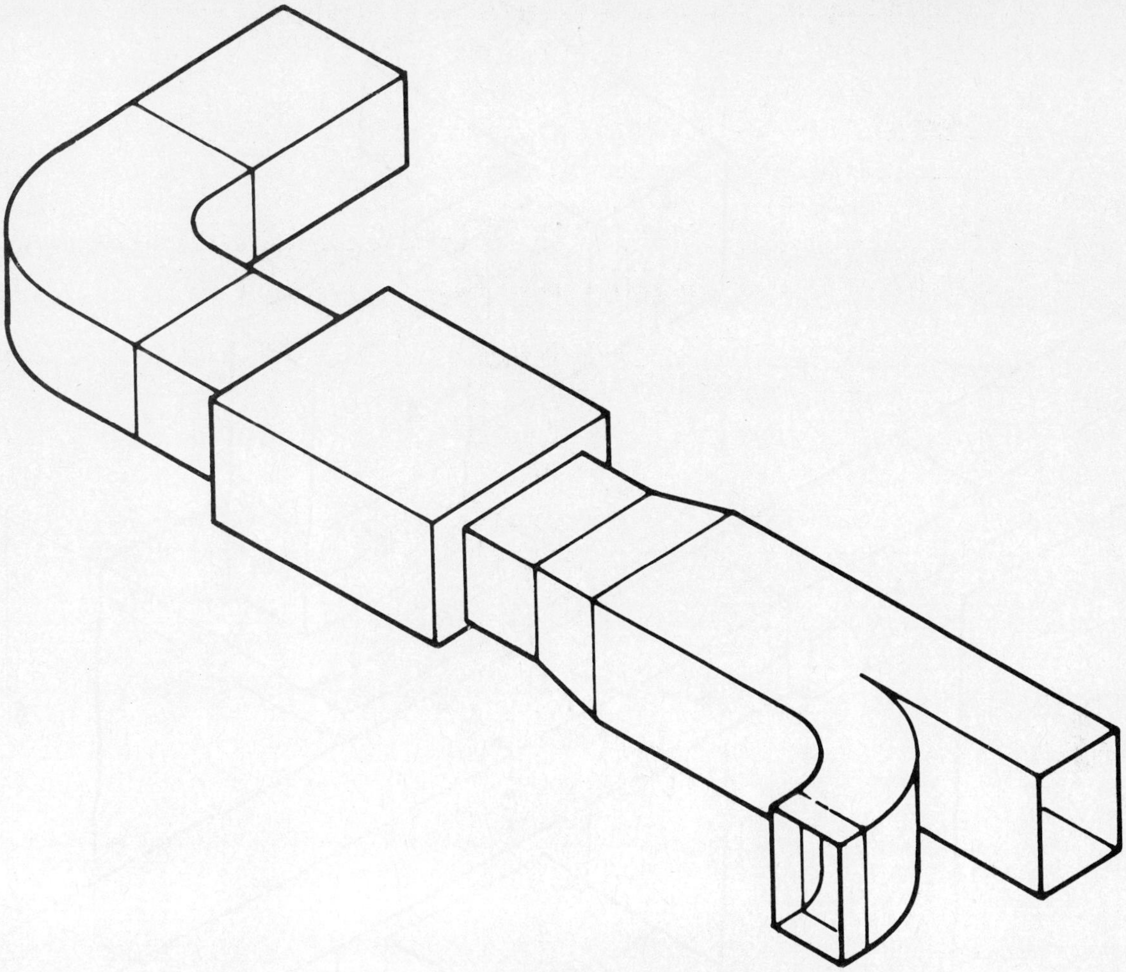

Fig. 11-13. Isometric view of Fig. 11-11.

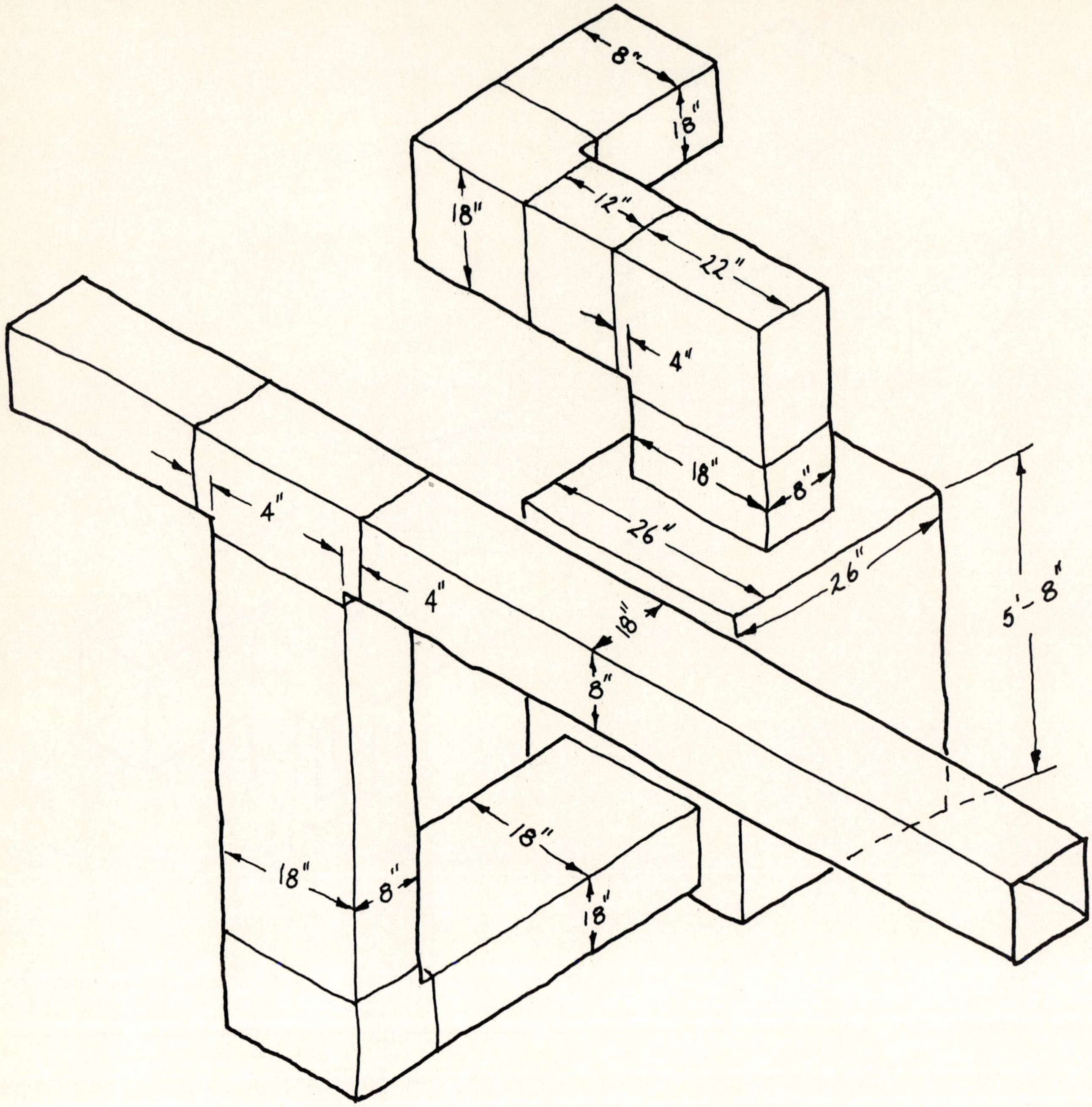

Fig. 11-14. Engineer's sketch to use for residence in this assignment.

UNIT 11 — ASSIGNMENT

1. Prepare two sheets of 11-by-17-inch tracing paper with border lines and title blocks.
2. On the first sheet, draw an isometric duct diagram for the residence shown in Appendix D of this book. Use Fig. 11–14 as a guide.
3. Using the other sheet you prepared, draw an isometric duct diagram for the commercial building in Appendix D. The sketch in Fig. 11–15 can be used for reference.

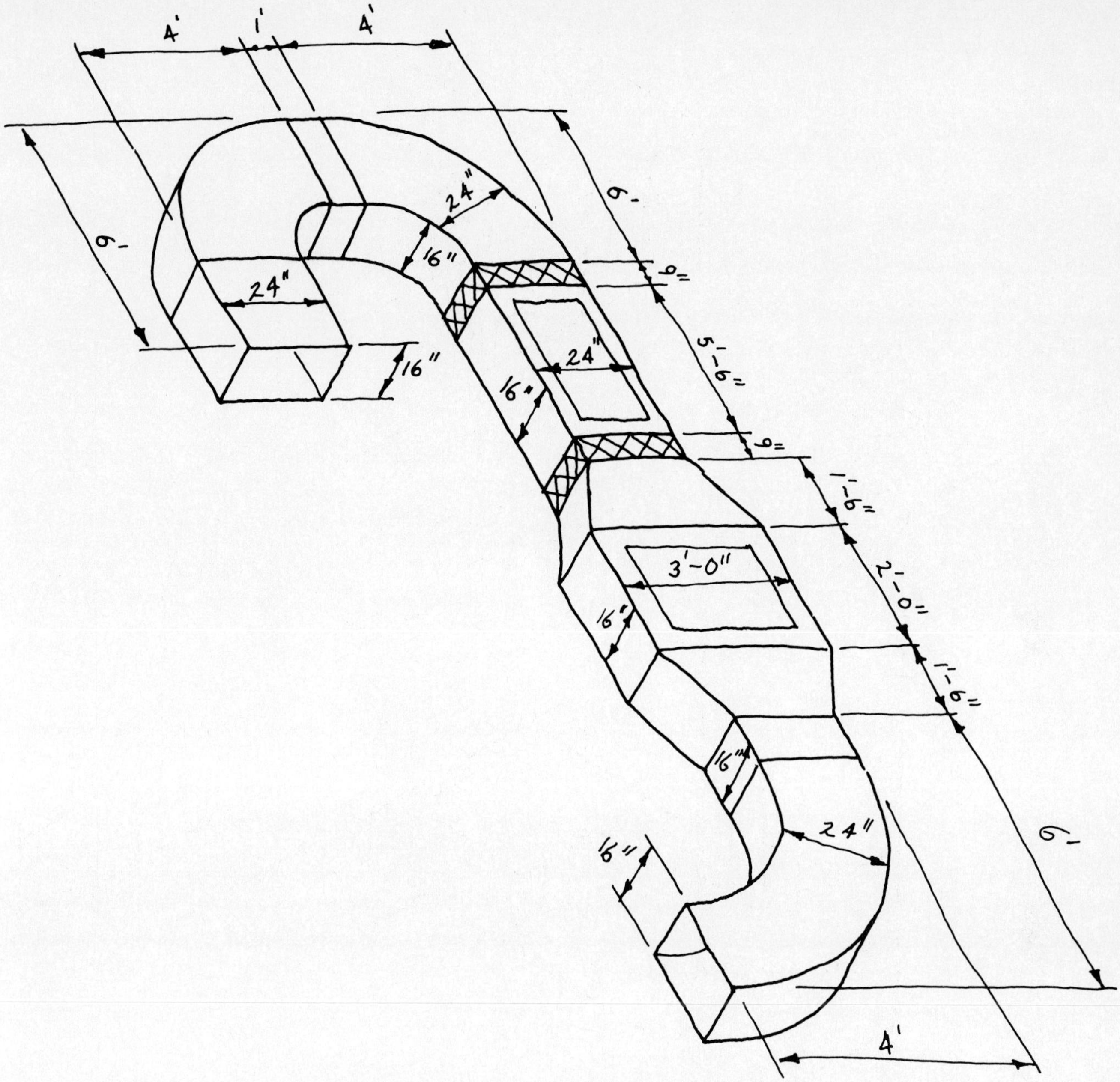

Fig. 11-15. Engineer's sketch to use for commercial building in this assignment.

4. Answer the following True or False questions.

	True	False
a. Riser diagrams of ducts within a system are always required, even for a simple job.	____	____
b. Ductwork riser diagrams include both horizontal and vertical ducts.	____	____
c. Two-line isometric duct drawings are very clear but usually aren't necessary.	____	____
d. HVAC isometric drawings are always drawn to scale.	____	____
e. Single-line duct drawings take up less time and are usually clear enough, so they are used most often.	____	____

SECTION IV

Sectional Drawing and Details

UNIT 12

Building Sections

OBJECTIVES: To obtain an overall mental picture of buildings as they are shown on architectural drawings

To point out the importance of using and understanding sectional drawings of buildings as aids in the preparation of HVAC drawings

CROSS SECTIONS

If any object, such as a duct, building, pipe, etc., were sliced or sawed into two parts, the cross section is what would be seen at the point where the section was taken. For example, everyone is familiar with the outside appearance of a golf ball, but few of us have seen the inside of one. Therefore, we are not certain how the ball is constructed. However, if we were to place a golf ball in a vice and saw it in half with a hacksaw, we would easily see how the golf ball was constructed, or at least we would have a view of the internal construction.

Sometimes the construction of a building is difficult to show in regular projection views. If too many dotted or broken lines are used to show hidden objects in the building, the views become confusing and difficult to read. To show the internal construction clearly, we imagine the building cut into sections, as if with a saw. The floor plan of the building shown in Fig. 12-1 will illustrate this point.

SLEEVES AND CHASES

Let us assume that a HVAC draftsman wishes to know the construction of the outside wall of this building in order to route the duct system correctly. Therefore, for the cross section of the outside building wall, imagine a vertical cut through the outside wall on the line *A-A,* as shown in Fig. 12-1. The arrows indicate the direction from which the section is viewed. If the portion of this wall that is nearest the viewer were taken away, the remaining portion would be seen from the direction indicated by the arrows. It would appear as shown in Fig. 12-2. The section itself is indicated as Section *A-A*.

It can be seen from the wall section shown in Fig. 12-2 that the building wall is constructed of 8-inch-thick masonry block. If the HVAC draftsman is to show a 4-inch-round duct in this wall, some modification of the structure will have to be made.

The concrete block could be cut, but it would be easier and cheaper to install a *sleeve* or provide a *chase* during construction.

Sleeves are short pieces of pipe placed in concrete or masonry walls to provide an opening for the duct. The opening avoids expensive cutting during construction and allows for duct expansion and replacement.

A *chase* is a space used to conceal a pipe, air-conditioning duct, electrical conduit, etc., in a masonry wall. If the chase is recessed, the finished wall can be flush (Fig. 12-3); if the chase cannot be recessed, the finished wall may have to be thicker to enclose the piping (Fig. 12-4). It is the HVAC draftsman's responsibility in most cases to locate the chase and determine its size for the necessary pipe risers, VTRs (vents through roof), etc.

VISUALIZATION OF A SECTION

In dealing with sections, the draftsman must use a considerable amount of visualization. Some sections

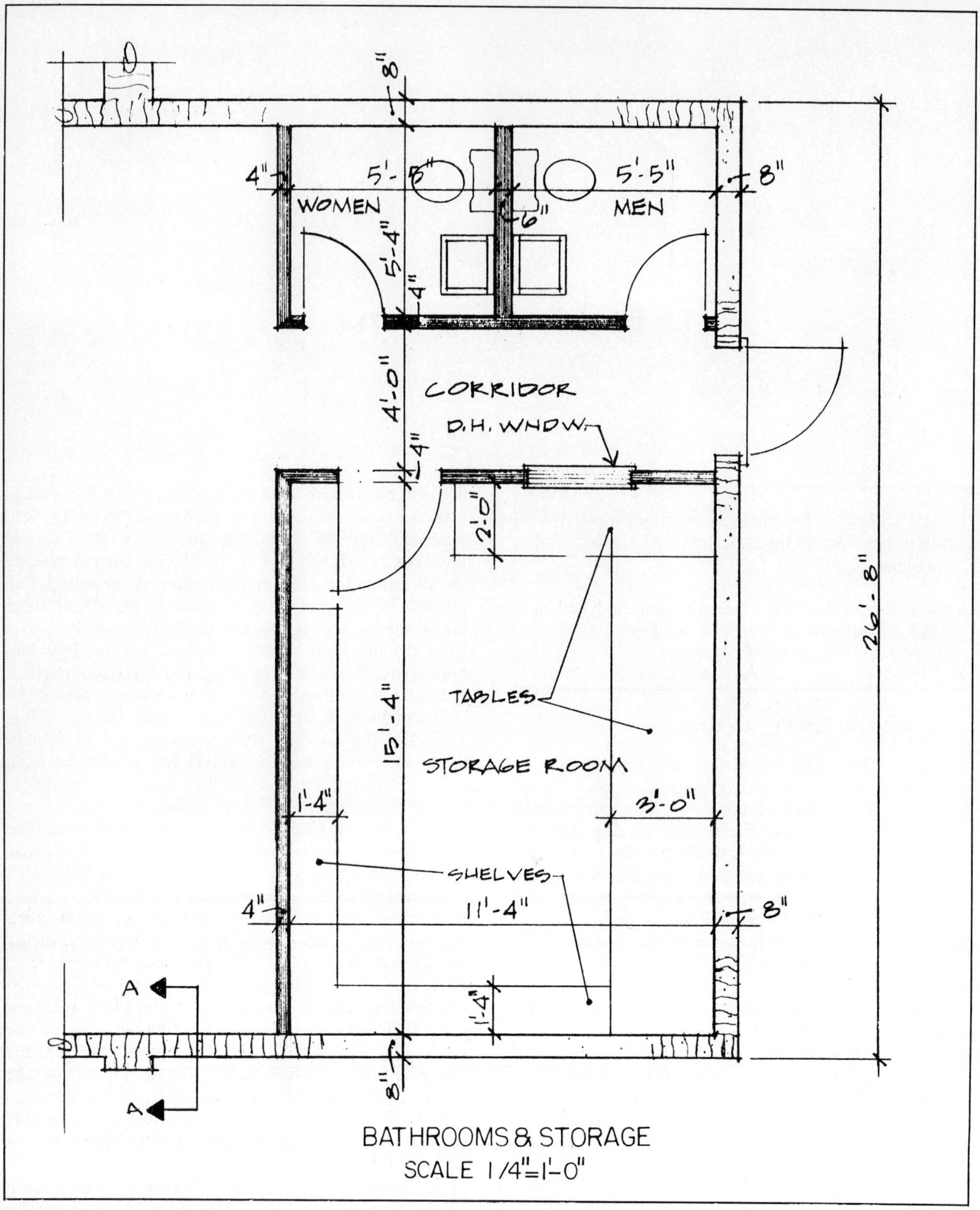

Fig. 12-1. View of a building cut in half horizontally.

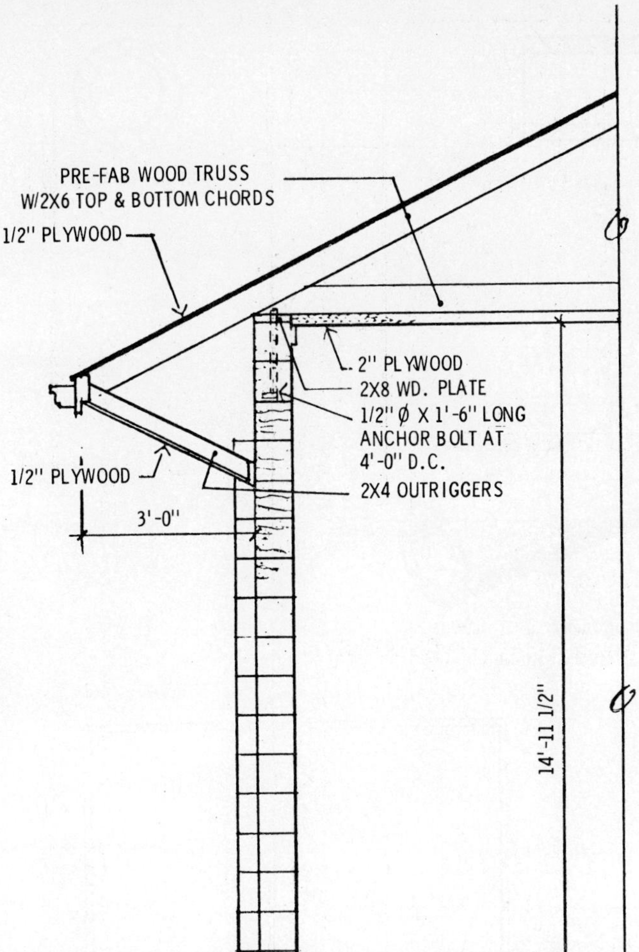

Fig. 12-2. View of a building cut in half vertically.

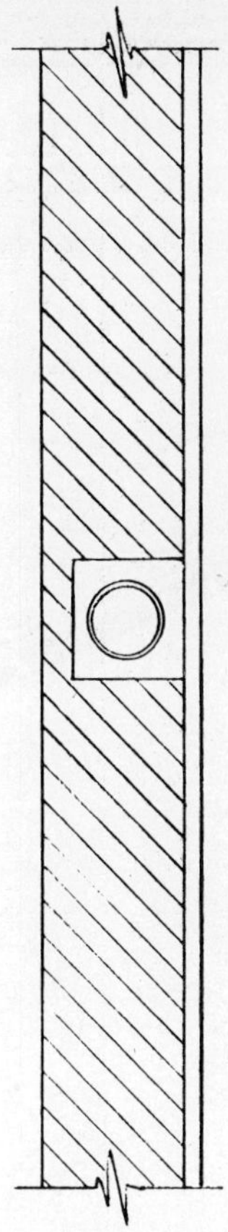

Fig. 12-3. Wall section showing round duct in a chase.

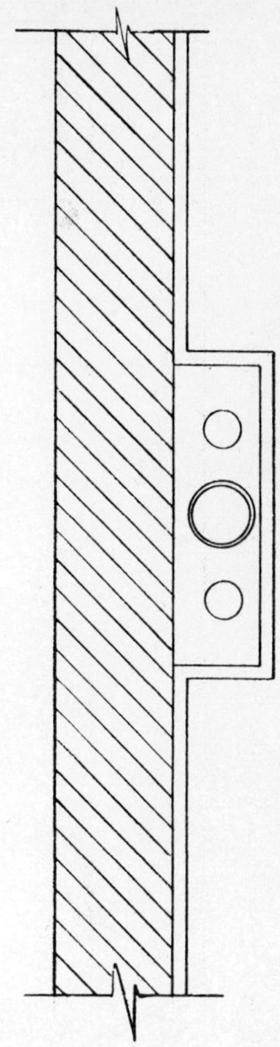

Fig. 12-4. Wall section furred out to box-in round ductwork.

are very easy to visualize, but some are extremely difficult. There are no given rules for determining what a section will look like. For example: a piece of round tile duct, cut vertically as shown in Fig. 12-5A, will be in the shape of a rectangle; cut horizontally as shown in Fig. 12-5B, it will be a circle; and cut on the slant as shown in Fig. 12-5C, it will be an ellipse.

Fig. 12-6 shows an isometric drawing of a building which, if cut vertically through the width, will produce Section *A-A* (Fig. 12-7); if cut vertically through the length, will produce Section *B-B* (Fig. 12-8); and if cut horizontally, will produce Section *C-C* (Fig. 12-9). The cross lines always indicate sectional lines or parts that are cut through.

It would be well for the student to practice drawing sections of different objects freehand in order to better understand the basic principles.

(A) Cut vertically.

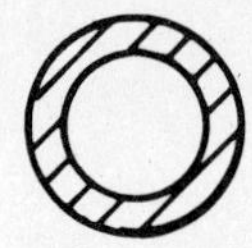

(B) Cut horizontally.

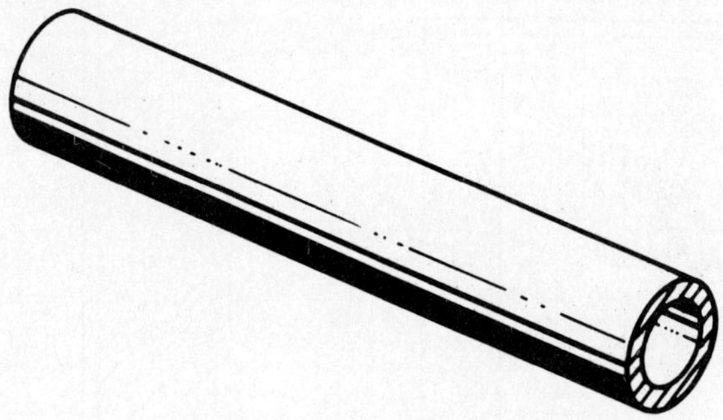

(C) Cut diagonally.

Fig. 12-5. Sectional view of tile duct.

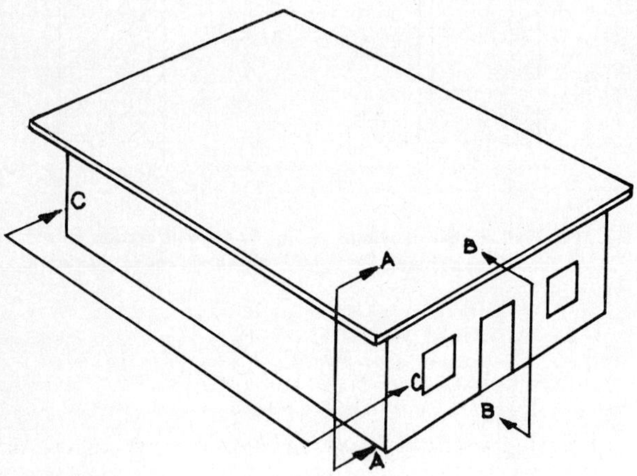

Fig. 12-6. Isometric drawing of a building showing sectional lines.

Fig. 12-9. Section C-C of the building in Fig. 12-6.

Fig. 12-7. Section A-A of the building in Fig. 12-6.

Fig. 12-8. Section B-B of the building in Fig. 12-6.

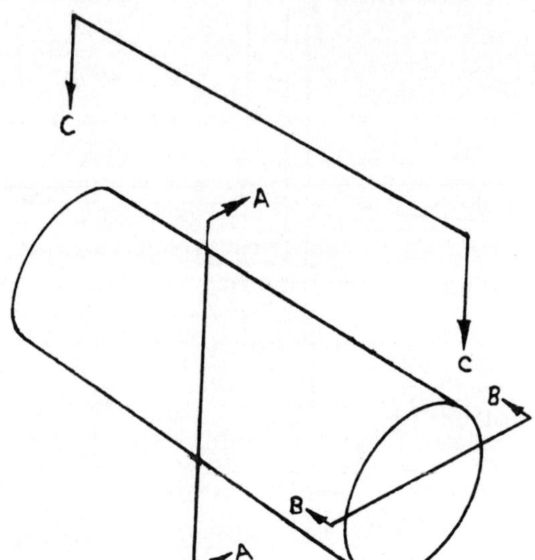

Fig. 12-10. Drawing for the assignment problem in this unit.

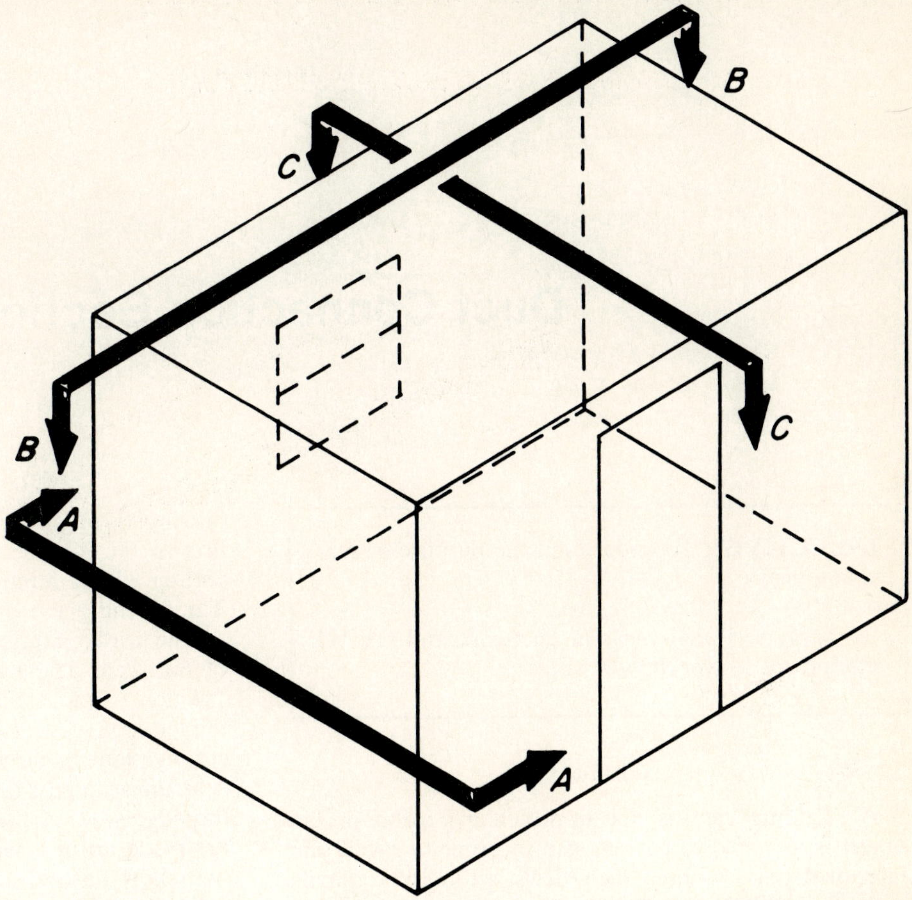

Fig. 12-11. Isometric view of a building for the assignment problem in this unit.

SUMMARY

A knowledge of sections is just as important to the HVAC draftsman as a knowledge of working drawings or graphic symbols.

If any object were cut into two pieces, the cross section is what would be seen at the point where the section was taken.

In dealing with sections, one must use a considerable amount of imagination.

UNIT 12 — ASSIGNMENT

1. Prepare an 11-by-17-inch sheet of tracing paper, drawing border lines and a title block.
2. Turn to Fig. 12–10. Draw Sections A–A, B–B, and C–C of the length of pipe. Draw in plan view, using the scale ½ inch = 1 foot.
3. Referring to Fig. 12–11, draw sections A–A, B–B, and C–C of the building. Use plan or elevation view, as required. Scale: 1/16 inch = 1 foot.
4. Answer the following fill-in-the-blank questions.
   - **a.** A(n) __________ is a short piece of pipe placed in concrete or masonry walls to provide an opening or masonry walls to provide an opening for the duct.
   - **b.** A(n) __________ is a space used to conceal a pipe, air-conditioning duct, etc., in a masonry wall.
   - **c.** In dealing with sections, the draftsman must use a considerable amount of __________.
   - **d.** If any object were cut into two pieces, the __________ is what would be seen at the point where the section was taken.
   - **e.** A finished wall may have to be thicker to enclose piping if the chase cannot be __________.

Duct Connection Sections

OBJECTIVES: To understand the purpose of sectional views

To apply sectional views of ductwork and HVAC equipment to working drawings

GENERAL

A sectional view is one in which a portion of the object is assumed to be removed in order to reveal the interior details. Without such views, all invisible details within an object must be shown in the exterior view with hidden lines. This may require a confusing number of hidden lines to show the interior details and cause the drawing to be difficult to read.

The theory of the construction of a sectional view is illustrated in Figs. 13-1, 13-2, and 13-3. Two exterior views of a rectangular duct section are shown in Fig. 13-1. In Fig. 13-2A a cutting plane is shown passing through the duct. Fig. 13-2B shows the portion of the duct section between the observer and the cutting plane removed to reveal the interior details. The cutting plane is removed in Fig. 13-2C.

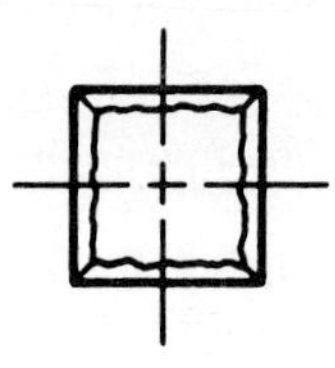

(A) End view.

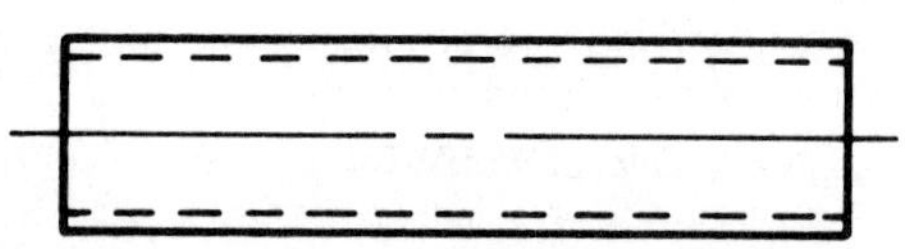

(B) Cross-sectional view.

Fig. 13-1. Two exterior views of a rectangular duct section.

The last illustration (Fig 13-3) again shows two views of the duct. The front view is a sectional view called "section." Notice that a cutting plane line is drawn in the left-side view of the duct, as shown in Fig. 13-3. The cutting plane line is used to show where the cutting plane is assumed to pass through an object. The line represents the edge view of the cutting plane. Cutting plane lines are made up of a series of long dashes, alternated with two short dashes, each of which is drawn as a heavy solid line. Fig. 13-4 gives an example of a cutting plane line.

Arrowheads on the ends of the cutting plane line in Fig. 13-4 show the direction in which the section is viewed. Letters are sometimes used with cutting plane lines to identify the cutting plane and the corresponding sectional views.

SECTION LINING

In the sectional view (Fig. 13-3), *section lining* or *cross-hatching* is drawn across those surfaces that are in contact with the cutting plane. Section lining is used in sectional views to indicate the various materials of construction. For the common metals, the section lining is made with fine lines, usually drawn at angles of 45 degrees. The lines in Fig. 13-3, however, are used to represent rigid insulation which is lining the duct. A few common section-lining symbols are given in Fig. 13-5.

The draftsman should take care to draw the section lining with uniform spacing. For small parts, the lines should be $\frac{1}{32}$ inch apart; on large parts, they should be $\frac{3}{16}$ inch apart. These distances should be measured by the student until he has acquired enough skill to space the lines uniformly by eye.

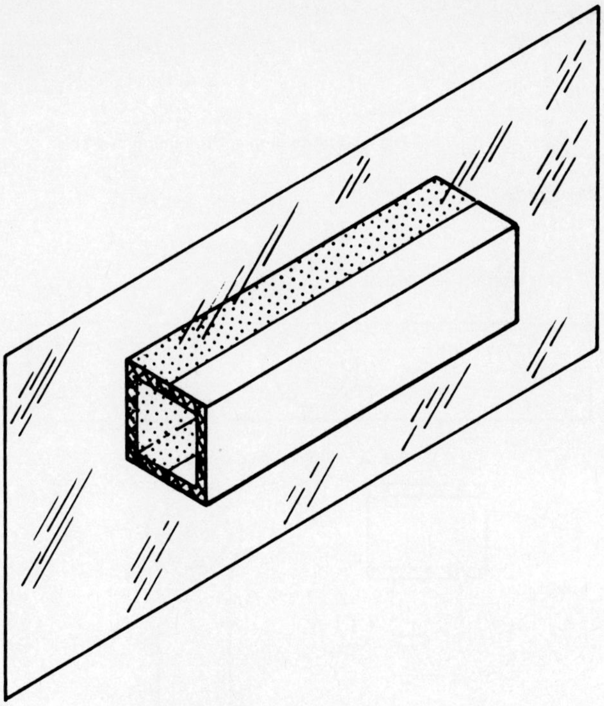

(A) Cutting plane passing through the duct.

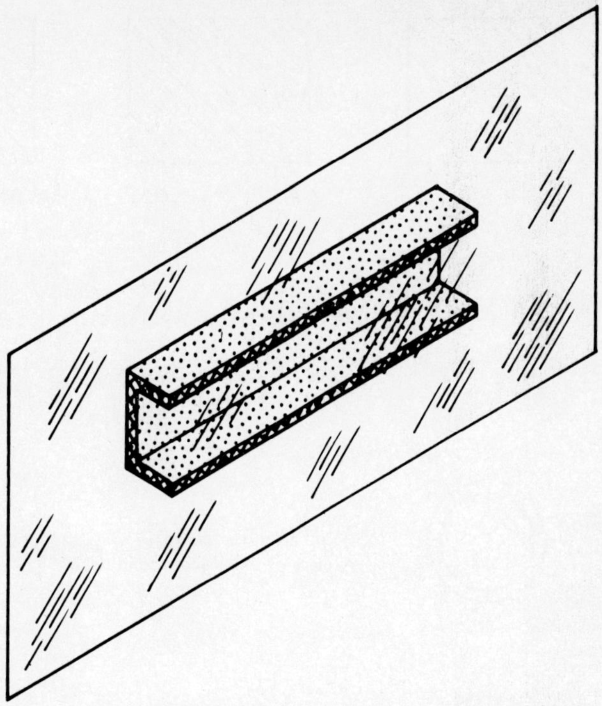

(B) Portion of duct between cutting plane and the viewer is removed.

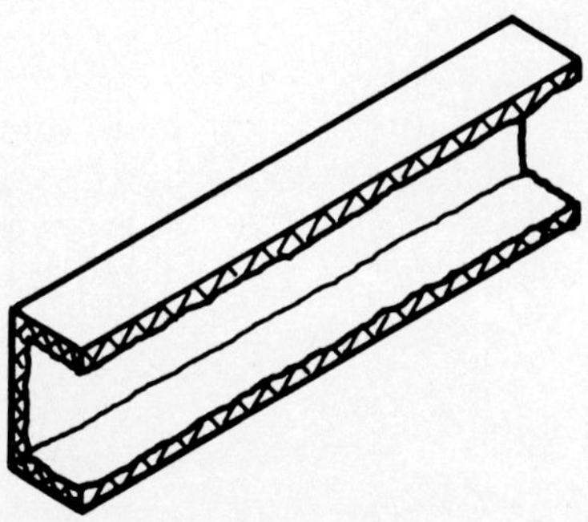

(C) View after cutting plane is removed.

Fig. 13-2. Dimensional views of a rectangular duct section.

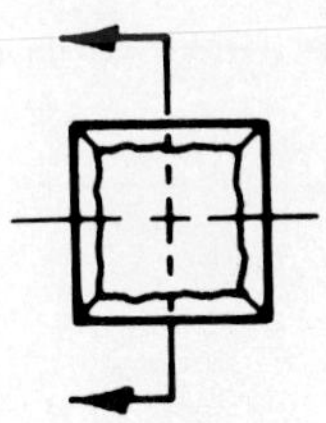

(A) End view.

(B) Cross-sectional view.

Fig. 13-3. Two more views of the rectangular duct.

Fig. 13-4. Cutting-plane line.

Some employers prefer that their draftsmen use a series of equally spaced sloping full lines for section lining of all materials. However, notes must then be used to indicate which materials are used.

In sectional views, all necessary visible details behind the cutting plane must be shown. Hidden lines, while usually not shown on sectional views, may be drawn when it is necessary to show a hidden part in the section, or to aid in dimensioning the view.

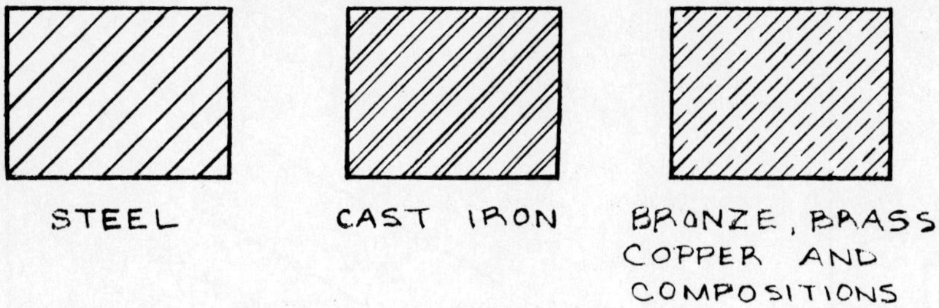

Fig. 13-5. Common section-lining symbols.

Fig. 13-8. A revolved-section view.

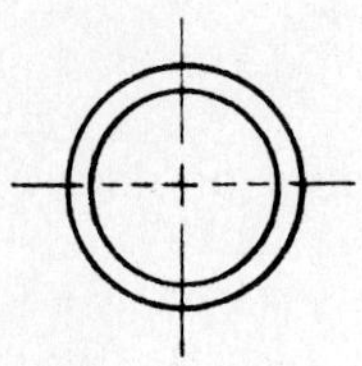

(A) Cutting plane passes halfway through a piece of underfloor round duct.

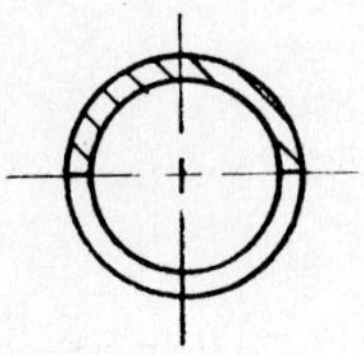

(B) Portion in front of cutting plane is removed.

Fig. 13-6. An example of a half-section view.

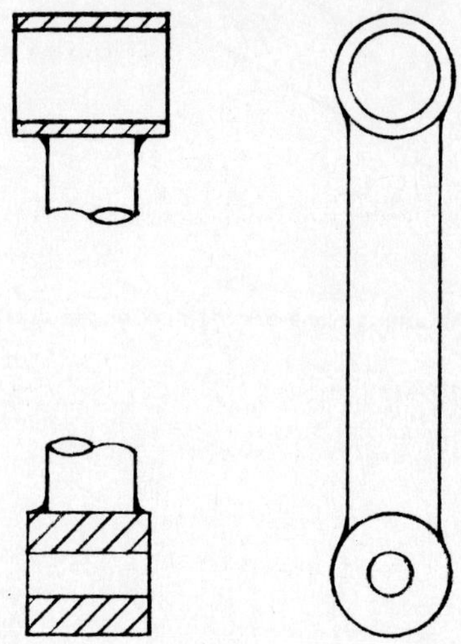

Fig. 13-9. A removed-section view.

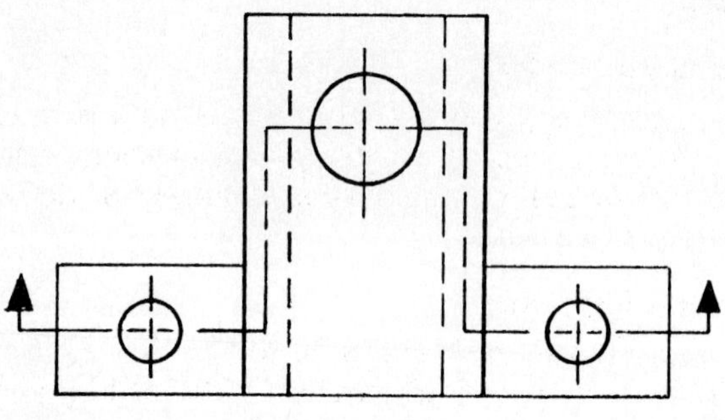

(A) Exterior view.

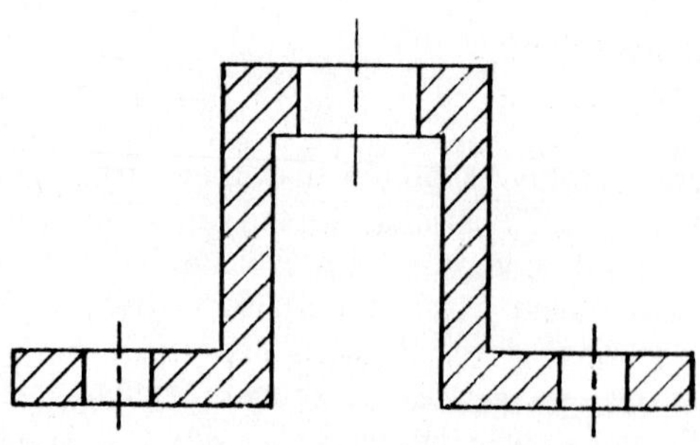

(B) Sectional view.

Fig. 13-7. An offset section view.

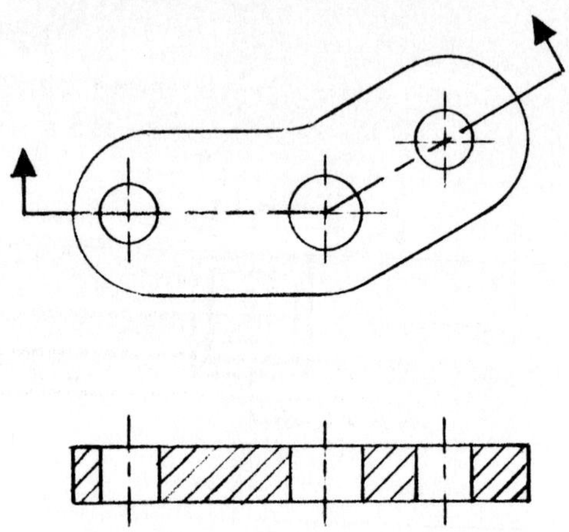

Fig. 13-10. An aligned-section view.

TYPES OF SECTIONAL VIEWS

Sectional views of air ducts, air-handling units, insulation, and other HVAC equipment or materials are classified as:

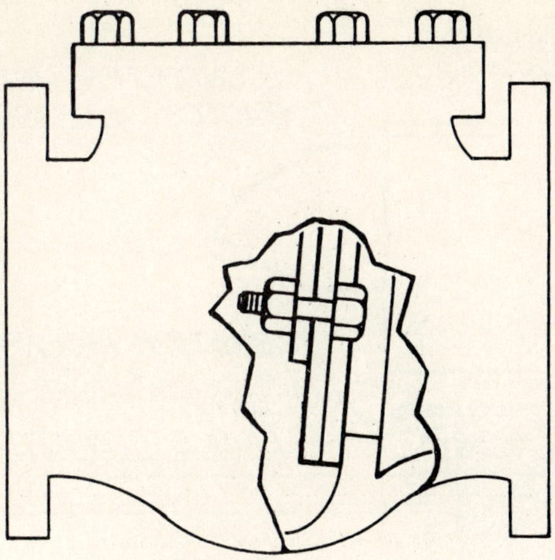

Fig. 13-11. A broken-out section view.

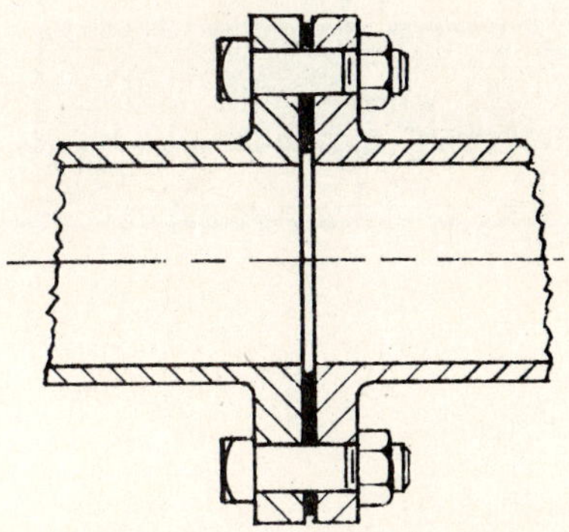

Fig. 13-12. View of an assembly section.

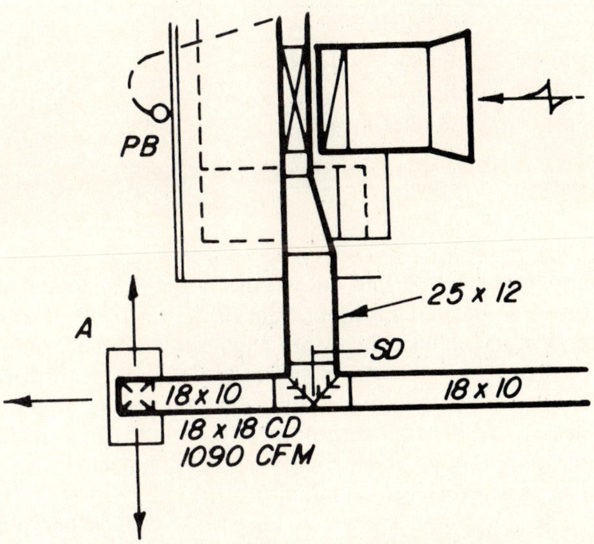

Fig. 13-13. Plan view of a partial HVAC system.

1. Full section
2. Half section
3. Offset section
4. Revolved section
5. Removed section
6. Aligned section
7. Broken-out section
8. Assembly section

The various types of sections will be discussed in the following paragraphs.

Full Section

A full section is a sectional view in which the cutting plane is assumed to pass entirely through the object. The sectional view in Fig. 13-3 is called a full section.

Half Section

A half section is a sectional view in which the cutting plane passes halfway through the object. One half of the view is shown in section, while the other half is shown from the exterior. Fig. 13-6A shows a cutting plane passing halfway thorugh a piece of underfloor round duct which is constructed of tile. Fig. 13-6B shows a portion in front of the cutting plane removed, and this portion is indicated by the cross-hatched lines.

Offset Section

An offset section is a sectional view in which the cutting plane is bent, or offset, as shown in Fig. 13-7.

Revolved Section

A revolved section is a cross section that has been revolved through 90 degrees. It is used to show the true shape of the cross section of bars and other elongated parts. Fig. 13-8 shows an example of such a section.

Removed Section

A removed, or detail, section is a cross section that has been removed from its original position to a convenient space near the principal view, such as the section shown in Fig. 13-9.

Aligned Sections

An aligned section is a sectional view in which a sloping part is rotated parallel to the cutting plane to show its true shape, such as the section shown in Fig. 13-10.

Broken-out Sections

A broken-out section, as shown in Fig. 13-11, is used when less than a half section is sufficient to show some interior detail. An irregular break line separates the section from the exterior view.

Assembly Section

Assembly drawings often are drawn in section to show how the interior parts are fitted together. However, parts that lie in the path of the cutting plane, such as bolts, screws, etc., are not drawn in section. See Fig. 13-12.

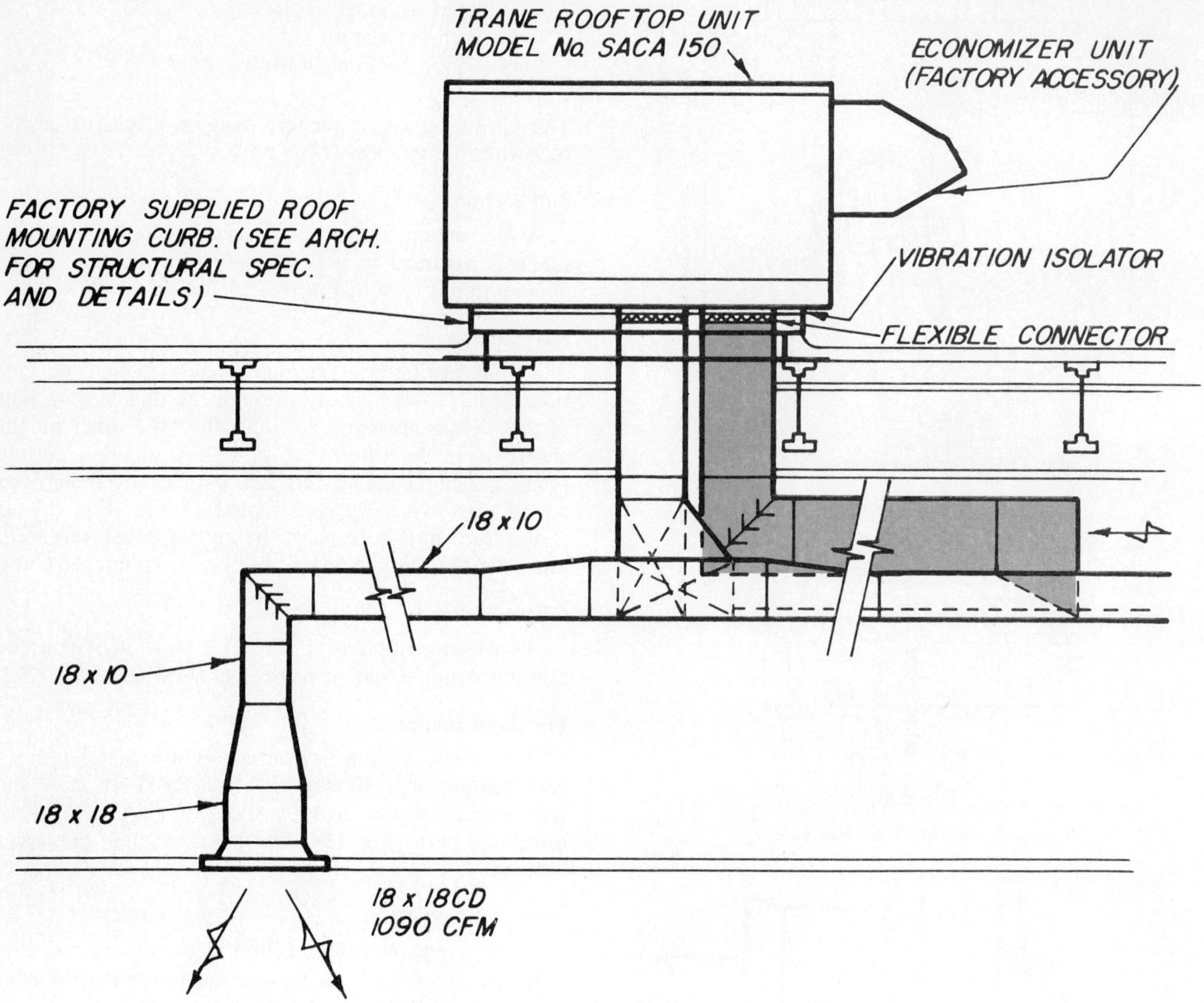

Fig. 13-14. A section view of a partial HVAC system.

PRACTICAL APPLICATION

The views shown in Figs. 13-13 through 13-17 are typical sectional views used on HVAC drawings. They should be carefully studied, and all details of each should be noted.

Fig. 13-13 shows a floor plan view of part of a HVAC system used in a department store. While this view is necessary to show the mechanics on the job the duct routing, further details of construction are still necessary. Therefore, a section detail was used to convey this additional information, and it is shown in Fig. 13-14.

In the sectional view of Fig. 13-14, the rooftop HVAC unit and the mounting details are clearly described. The unit is shown on a factory-supplied roof-mounting curb with a vibration isolator installed. The supply and return air ducts (the latter is shaded) are shown connected to the bottom of the unit with flexible connectors. These ducts are then shown to be run in the space between the roof structure and the finished ceiling. The supply duct terminates in a ceiling-mounted diffuser, while the return air duct terminates within the space above the ceiling, since this space is being used as a return-air plenum.

The floor plan shown in Fig. 13-15 is another plan view of the department store HVAC system. Sectional views were also needed to more clearly describe the construction details of this area.

Fig. 13-16 shows a sectional view of a power roof ventilator. This ventilator is used to ventilate an area within the building.

Fig. 13-17 shows a sectional view of two rooftop units with related ductwork.

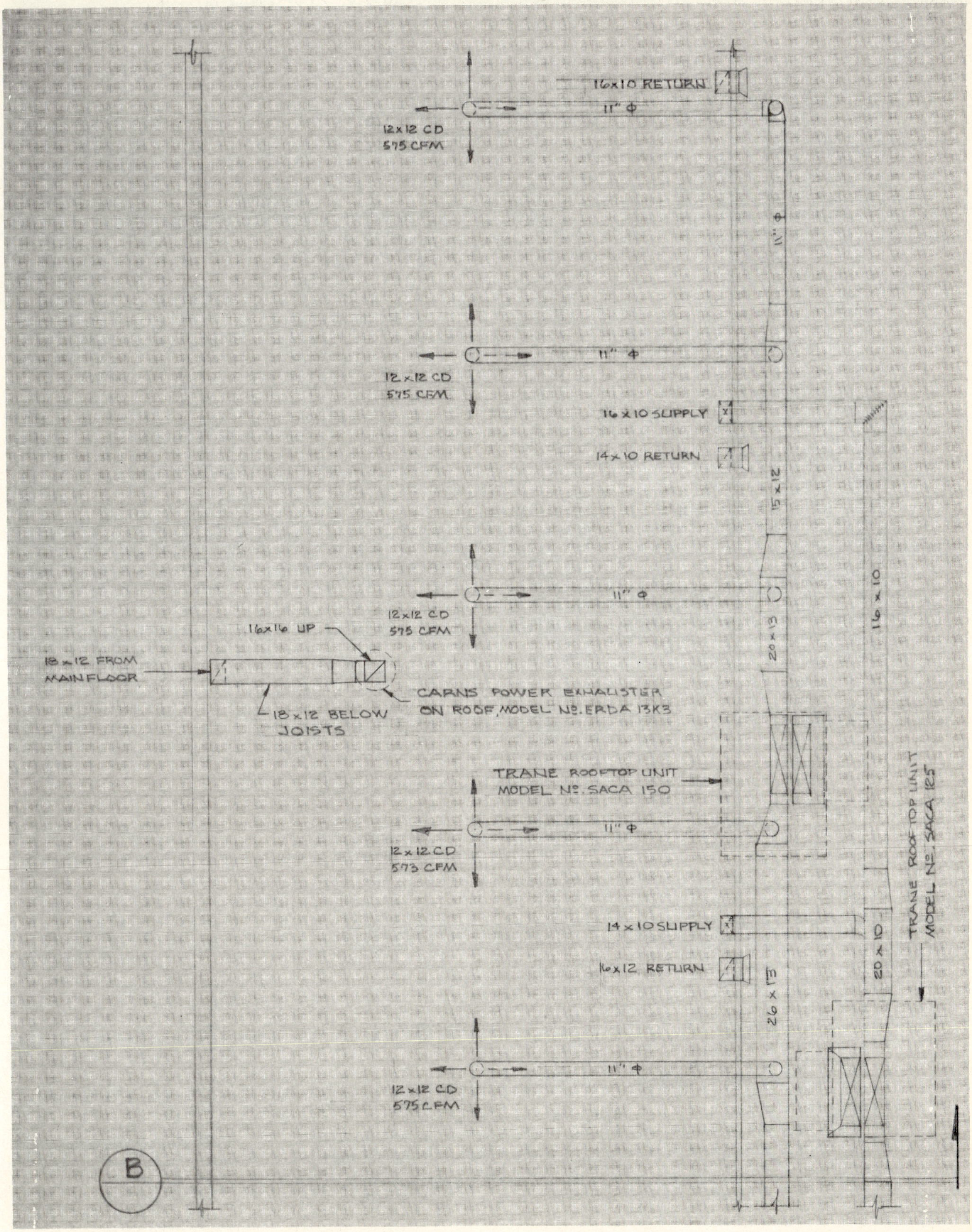

Fig. 13-15. Plan view of a portion of an HVAC system.

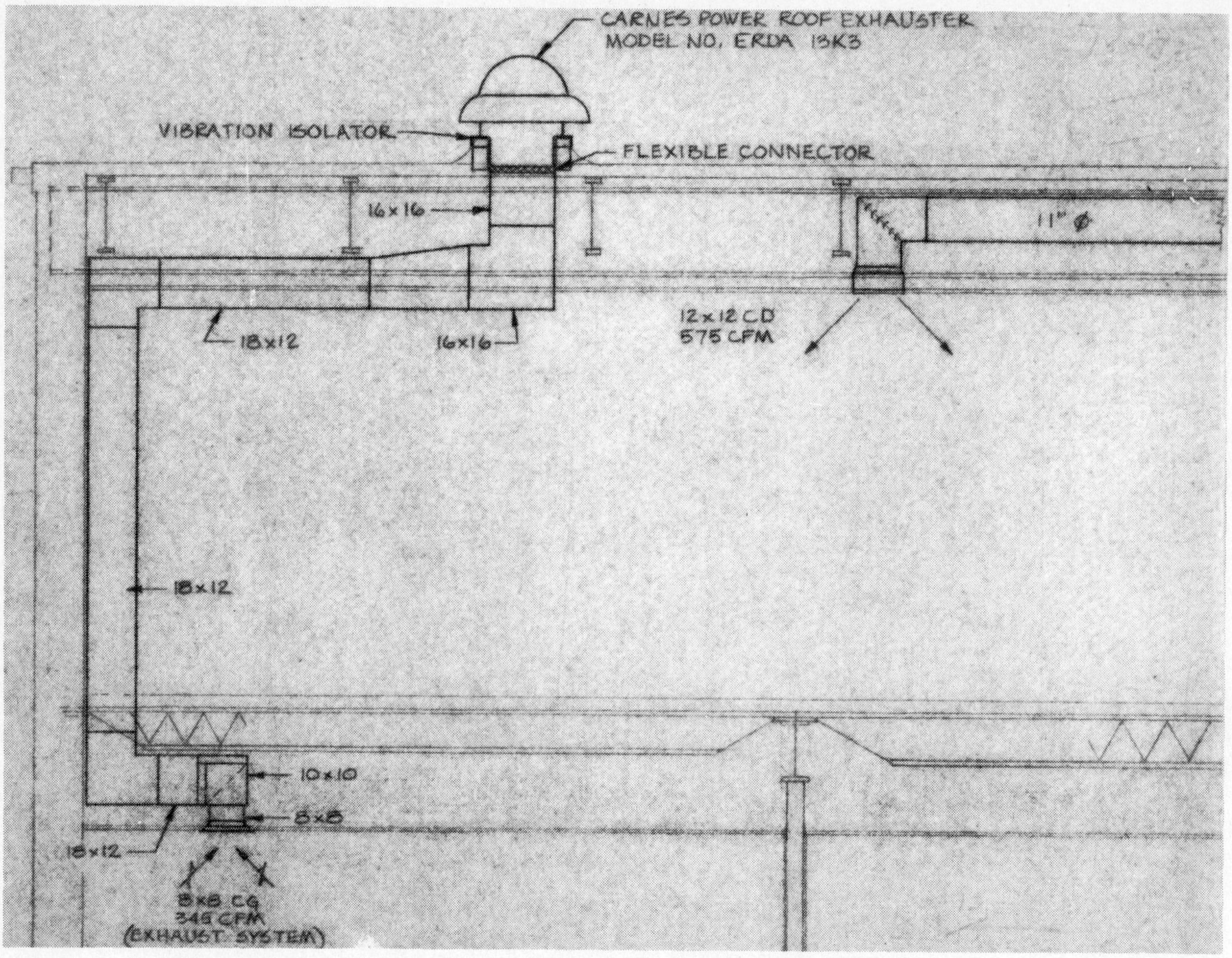

Fig. 13-16. Sectional view of the power roof ventilator in Fig. 13-15.

SUMMARY

A sectional view is one in which a portion of the object is assumed to be removed to reveal the interior detail.

The cutting plane line is used to show where the cutting plane is assumed to pass through an object.

Sectional views of various HVAC materials are classified as follows:

1. Full section
2. Half section
3. Offset section
4. Revolved section
5. Removed section
6. Aligned section
7. Broken-out section
8. Assembly section

Hidden lines, while usually not shown on sectional views, may be drawn when it is necessary to show a hidden part in the section, or to aid in dimensioning the view.

UNIT 13 — ASSIGNMENT

1. Prepare a sheet of 11-by-17-inch tracing paper with border lines and title block.
2. Turn to Fig. 13–18 to use as a guide in this assignment. Draw a sectional view, using 2-H lead to produce a heavy solid line for all equipment details.

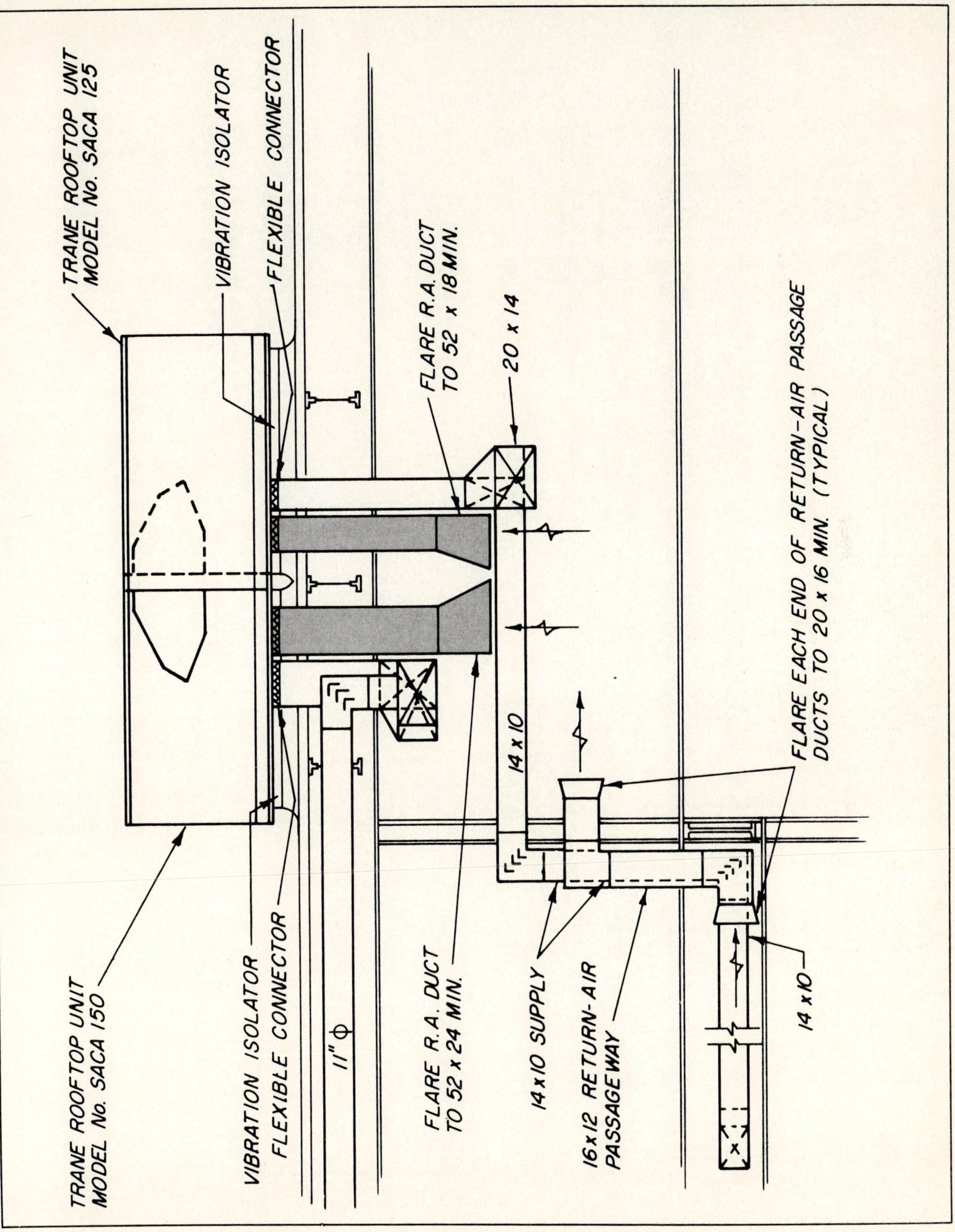

Fig. 13-17. Sectional view of the rooftop units and related ductwork in Fig. 13-15.

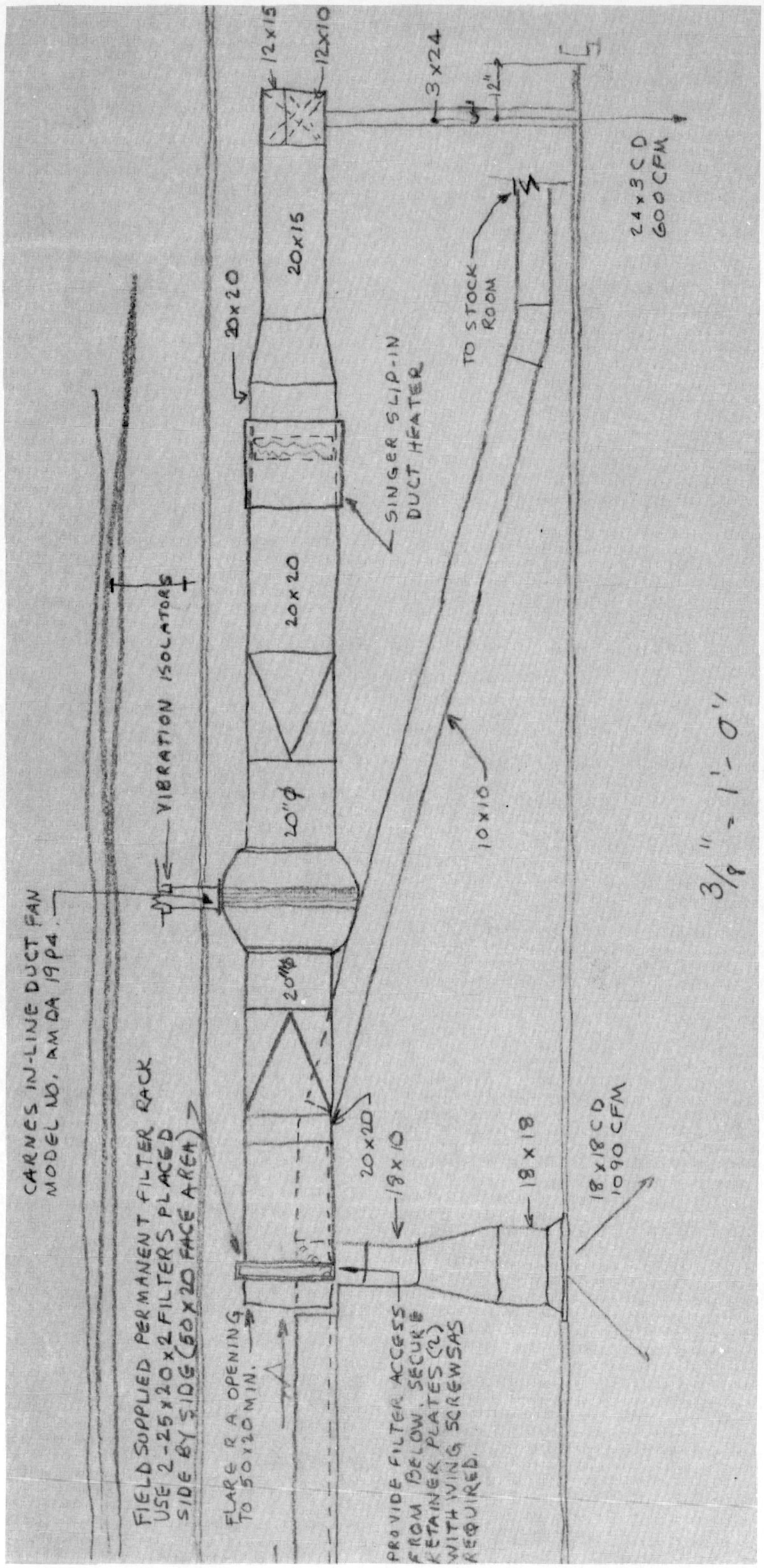

Fig. 13-18. Engineer's sketch for the assignment problem in this unit.

UNIT 14

Equipment, Duct, and Installation Details

OBJECTIVES: To understand the need for HVAC details

To relate HVAC details to working drawings

INTRODUCTION

The definition of HVAC drafting given in Unit 1 of this book stated that HVAC drafting is a language using lines, symbols, dimensions, and notations in the form of working drawings to accurately describe how a heating, ventilating, and air-conditioning system is to be installed in a building.

WORKING DRAWINGS

Lines and symbols are used to show the building walls and partitions and to show the location of equipment, ductwork, diffusers, grilles, registers, etc. The lines and symbols usually describe the exact form or shape of a structure or piece of equipment.

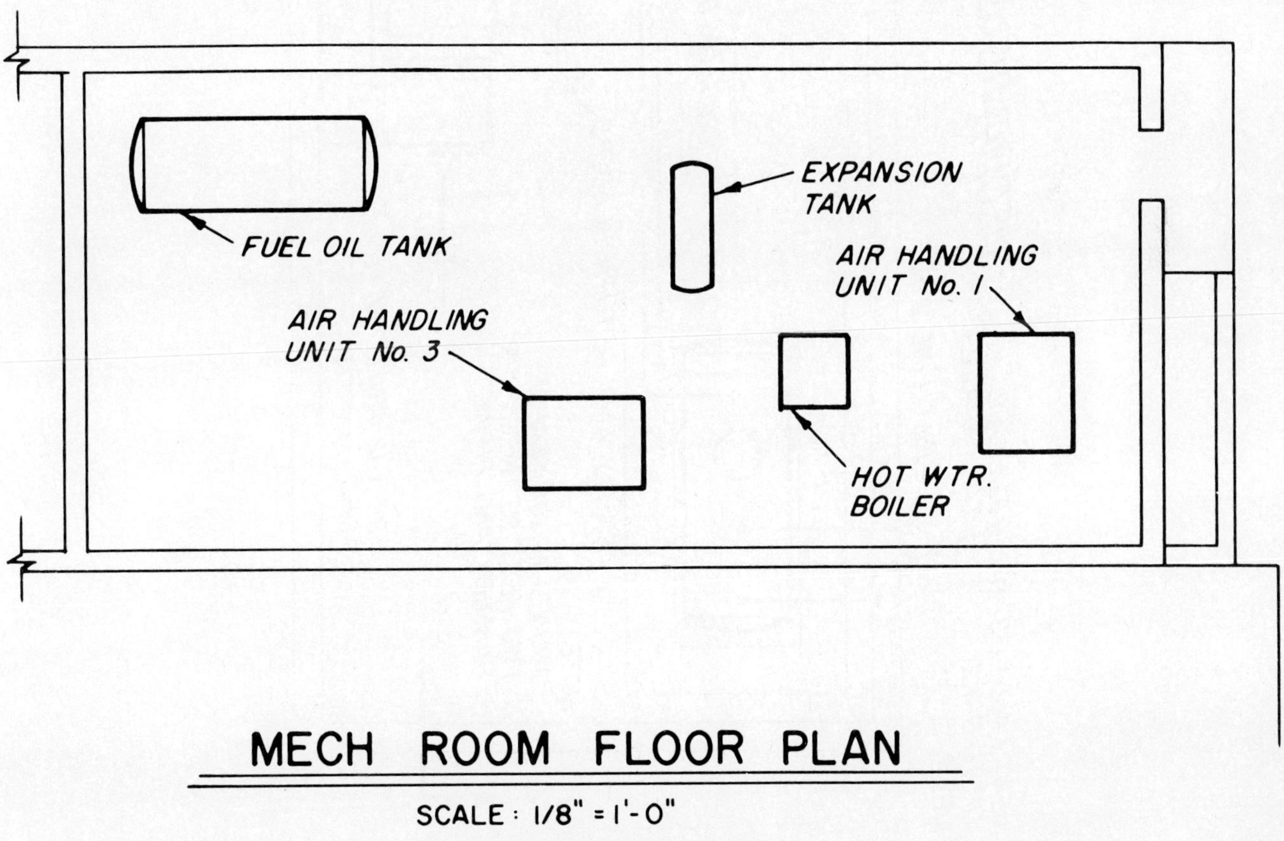

Fig. 14-1. Mechanical room floor plan.

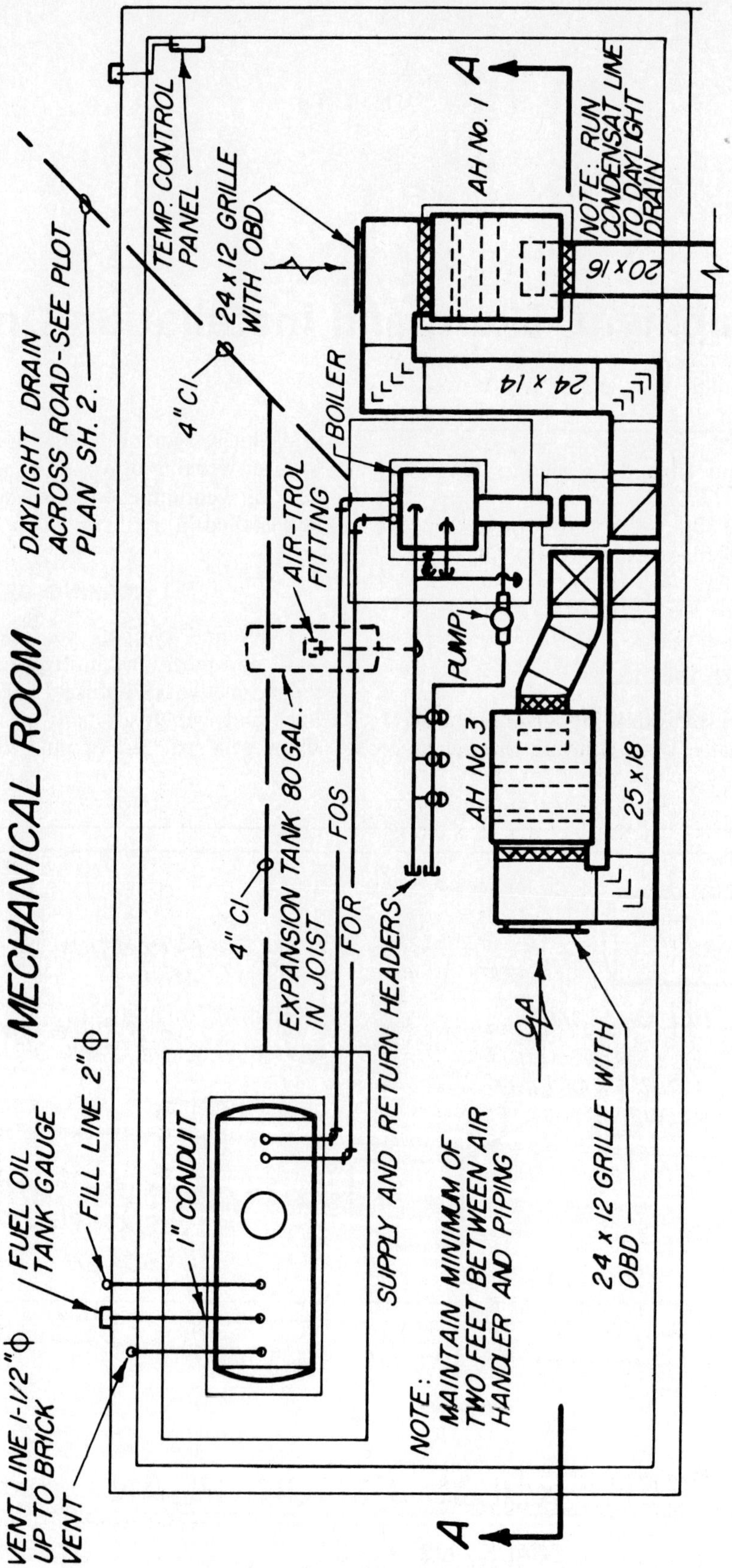

Fig. 14-2. An enlarged floor plan of Fig. 14-1.

Dimensions and notations are used to give the exact size for ducts, diffusers, and similar items or to describe a piece of equipment more clearly. Sometimes the working drawings specify the kind of material or equipment to be used on the project, but usually these descriptions are covered in the specifications, such as the HVAC Specifications in Appendix A of this book.

In general, a complete set of construction documents includes the written specifications and the working drawings, with adequate drawings and details to describe exactly how an HVAC system is to be installed.

We have covered the drawing of floor plans, sections, and other necessary items in previous units. The drawing of HVAC details will be the primary concern of this unit.

DETAIL DRAWINGS

A detail drawing is the drawing of a separate item or portion of a building or building system, giving a complete and exact description of its use and all details necessary to show the workman exactly what is required. The floor plan in Fig. 14-1 shows a good example of a HVAC drawing where an extra, detailed drawing would be desirable.

It is obvious that the mechanical room shown in Fig. 14-1 contains a hot-water boiler, air-handling unit, and other heating and cooling equipment. However, the duct connections to the air-handling unit are not clear in this drawing; thus, a detail of these connections is required as shown in Fig. 14-2.

A set of HVAC drawings, in addition to the floor plans, will sometimes require sectional drawings of parts that are not indicated with sufficient clearness on the plan drawings. The mechanical room shown in Fig. 14-1 will serve again to illustrate the need for a sectional drawing. The same room, drawn in a sectional view, is illustrated in Fig. 14-3. Notice that the oil tank, the boiler, the air-handling unit, and other mechanical items are shown very clearly. This helps the workman making the installation locate the equipment more accurately.

HVAC DETAILS

The following HVAC details were taken from actual working drawings and will be very useful to HVAC draftsmen, especially to those who are employed with contracting and consulting engineering firms. The use of each detail is fully described.

Roof Curb

The roof-curb detail illustrated in Fig. 14-4 was used on a HVAC drawing to show the contractor how the roof curbs were to be built wherever ductwork passed through the roof from the rooftop cooling units into the conditioned area within the building.

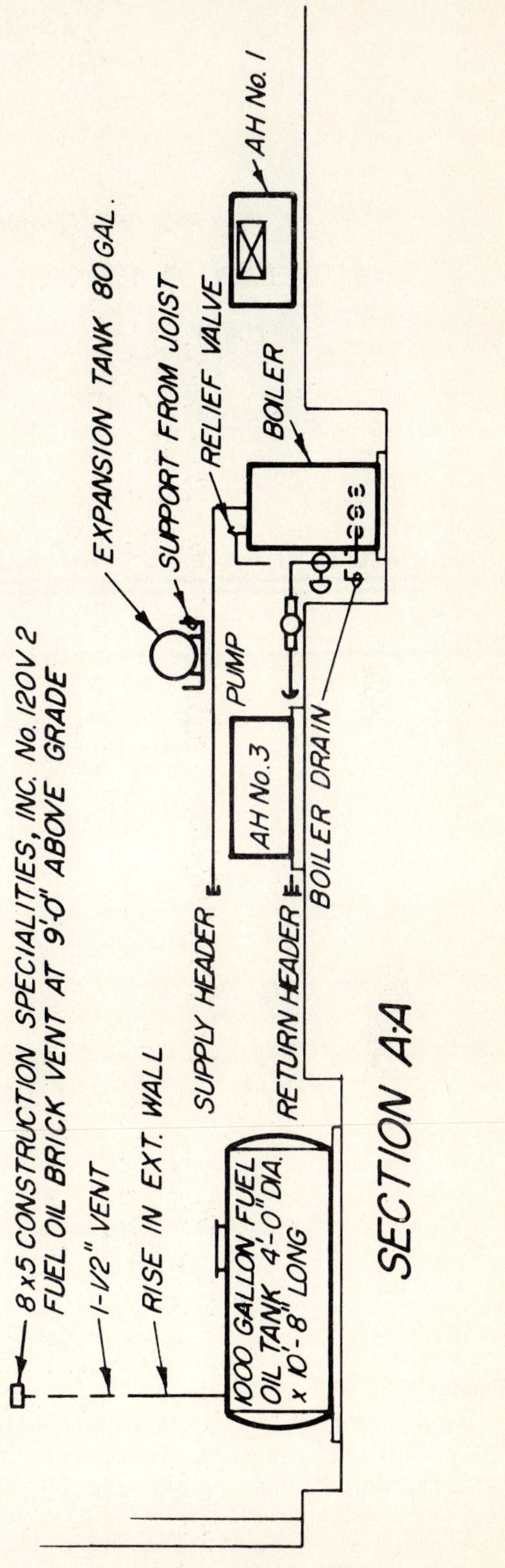

Fig. 14-3. Section of mechanical room in Fig. 14-1.

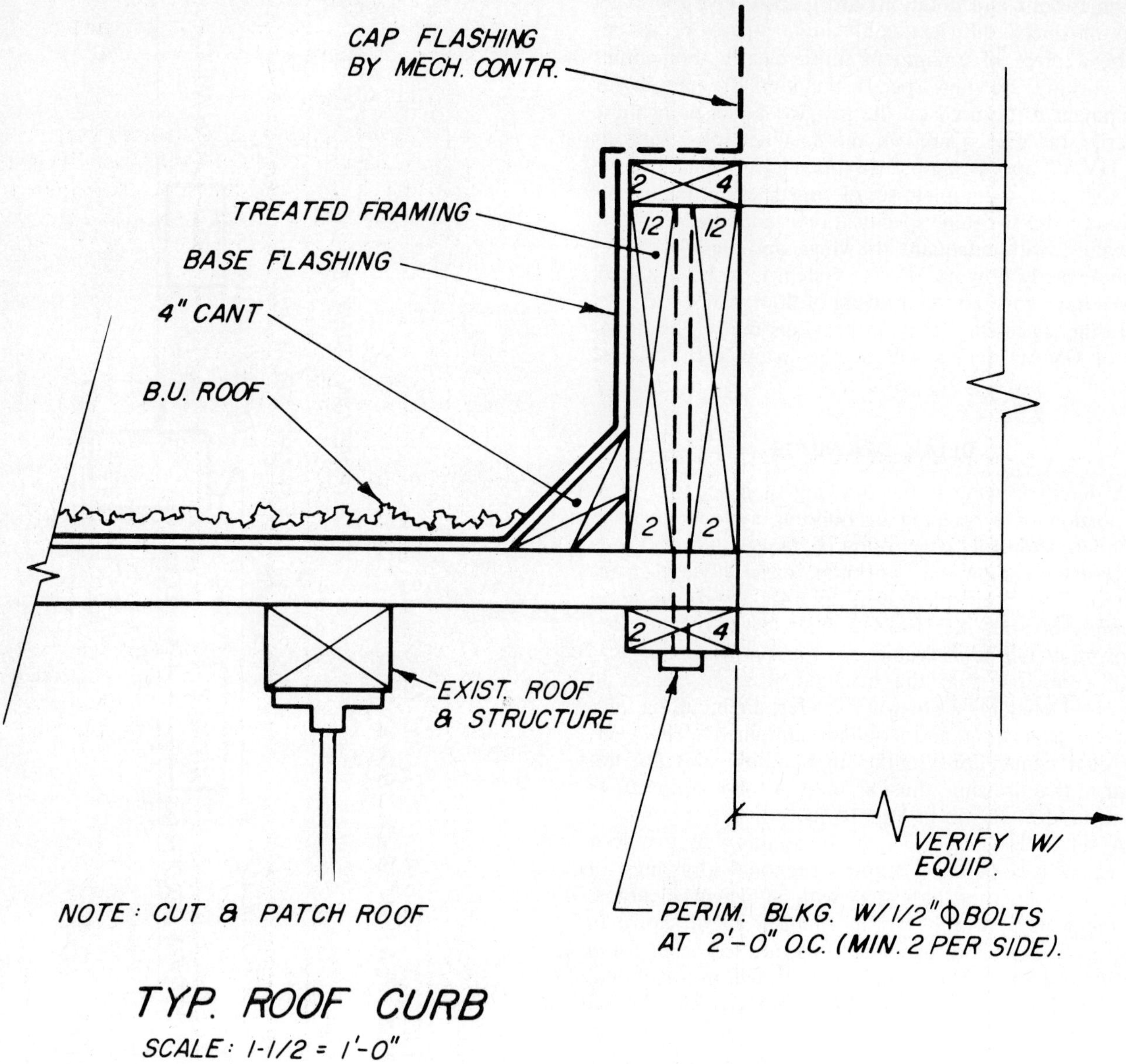

Fig. 14-4. Detailed drawing of a roof curb for HVAC equipment.

Frames

Because of the weight, location, and other characteristics of HVAC equipment, custom-built frames are often necessary on many projects. Fig. 14-5 shows how one such detail appeared on a working drawing.

A plan view of another condensing unit frame is illustrated in Fig. 14-6.

Fig. 14-7 shows Section *A-A* of the floor plan appearing in Fig. 14-6. The section shows an elevation of the frame, with the condensing unit mounted on top.

The curb and unit support plan illustrated in Fig. 14-8 shows mounting details of the rooftop heating and cooling unit which was specified on this particular project. This frame was designed to support the entire piece of equipment as well as to allow space for both the supply and return air ducts.

Transite Duct

Fig. 14-9 shows a detail of a 10-inch-round transite duct that was installed in the ground under a concrete floor. The connector of the takeoff and the floor dif-

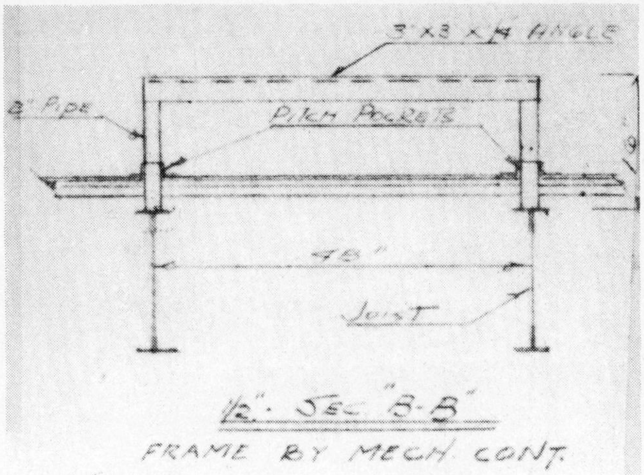

Fig. 14-5. Custom-built iron frame for condensing unit.

fuser are also shown. A support for the ductwork is shown in Fig. 14-10. The purpose of the support is to hold the duct in position during construction or during settling of the building, thus preventing a possible duct separation.

Power Roof Ventilators

Power roof ventilators are used to exhaust air from the inside of the building. Details of such ventilators are often included on HVAC drawings to show the contractor the necessary details for construction. Two such details are shown in Figs. 14-11 and 14-12.

Electric Cabinet Heaters

Electric cabinet heaters are sometimes used in the counter space under schoolroom windows. When this is done, supply and return ductwork are required to deliver the conditioned air to the space within the room. Fig. 14-13 shows an installation of an electric cabinet heater.

Underfloor Duct

Fig. 14-14 shows another detail of an underfloor duct embedded in the concrete slab. Floor registers, ducts, duct connections, and perimeter insulation details are clearly shown.

Air-Conditioning Units

Air-conditioning units are sometimes mounted on isolators, as shown in Fig. 14-15, to cut down on vibration.

Diffuser

Fig. 14-16 shows a diffuser detail for a multipurpose room in a public school building. The roof, ceiling, and beams are shown in relation to the placement of the duct. The duct size, diffuser collar, and type of diffuser are indicated.

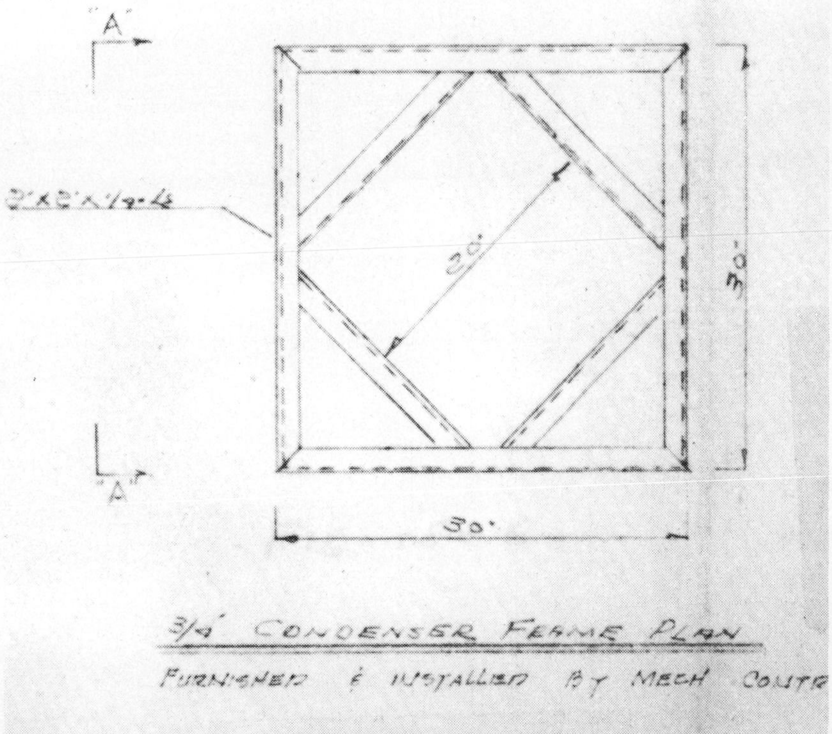

Fig. 14-6. Plan view of a condensing unit frame.

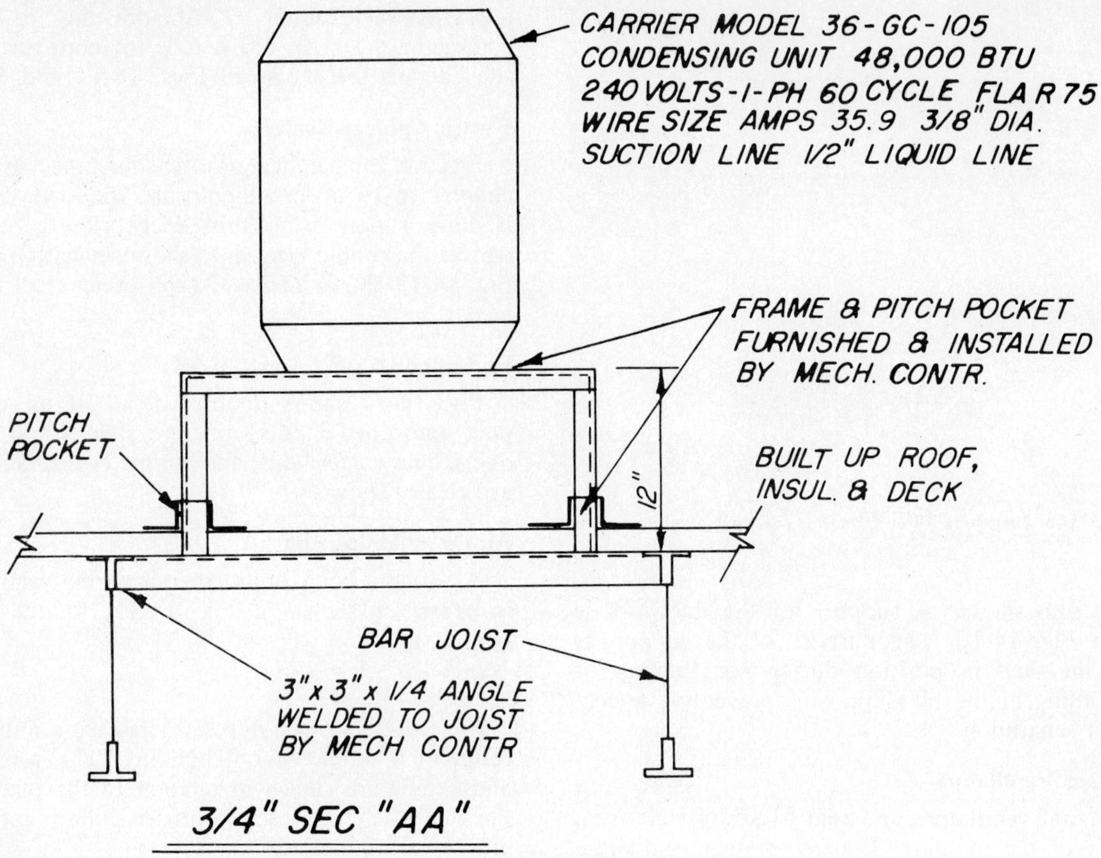

Fig. 14-7. Section A-A of the floor plan appearing in Fig. 14-6.

ROOF CURB ACCESSORY (SRC-I)
FOR SA48G AND SA61G WITH PLENUM

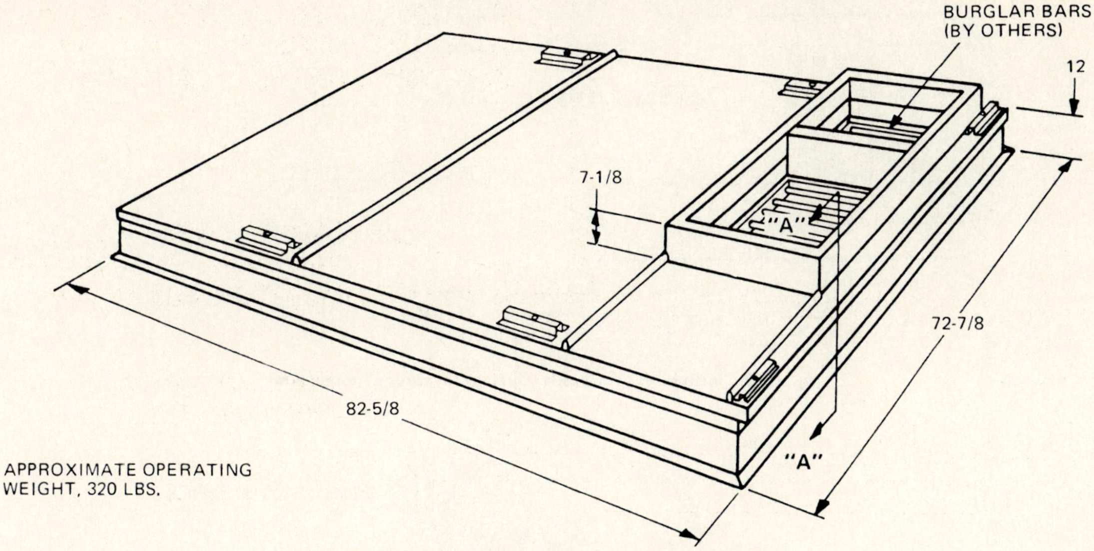

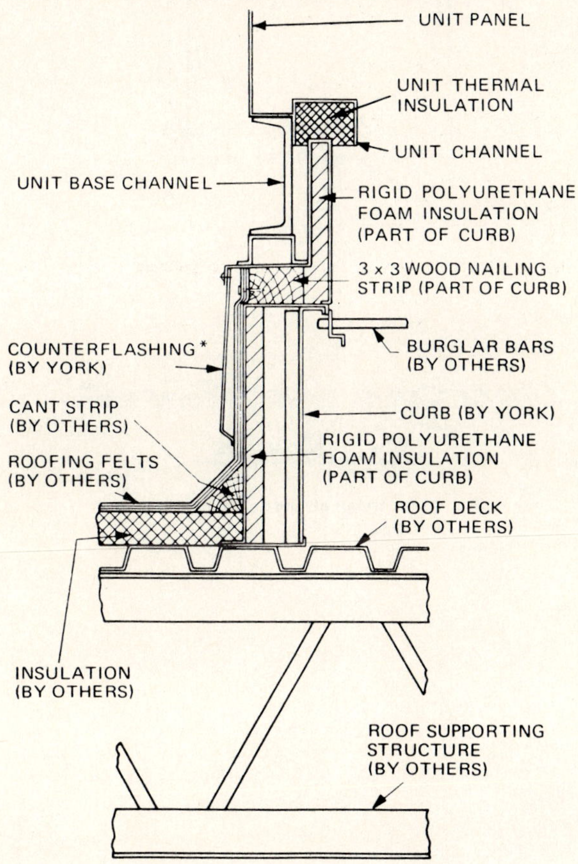

SECTION "A–A"

* Field-installed after the roofing material is nailed in place to provide positive weather seal and protect edges of roofing felts.

Courtesy York, Division Borg-Warner

Fig. 14-8. Curb and condensing unit support plan.

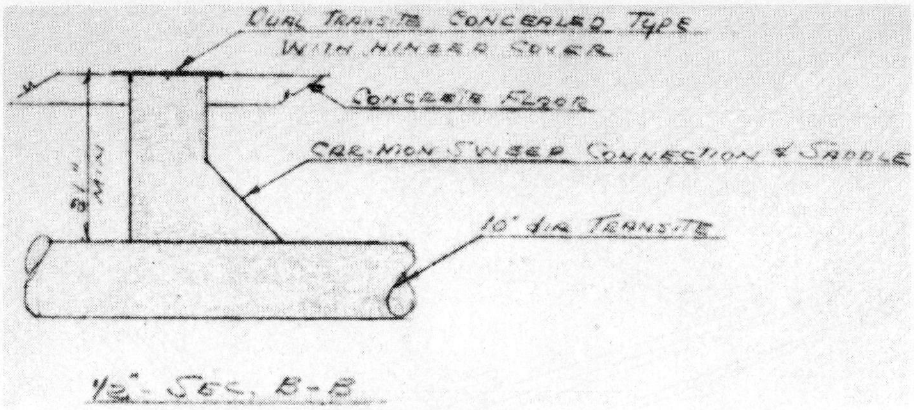

Fig. 14-9. Section of a transite duct with takeoff connection.

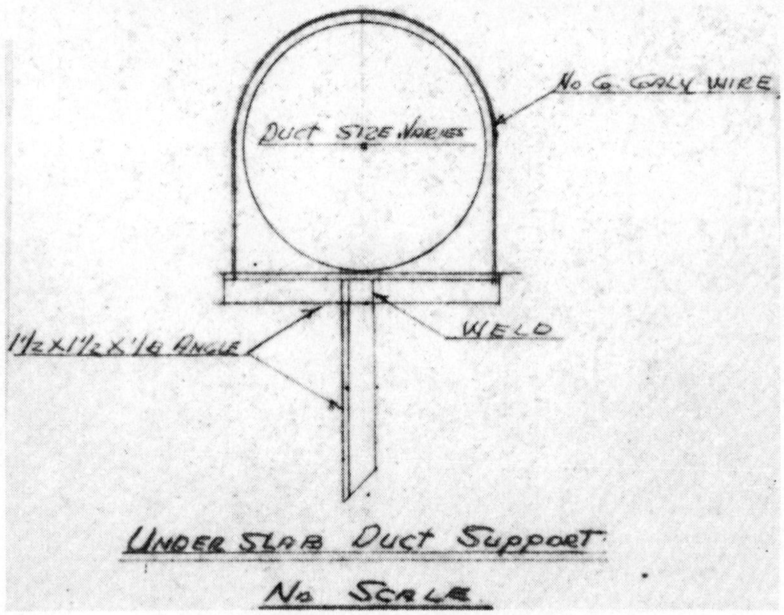

Fig. 14-10. Detail of underslab duct support.

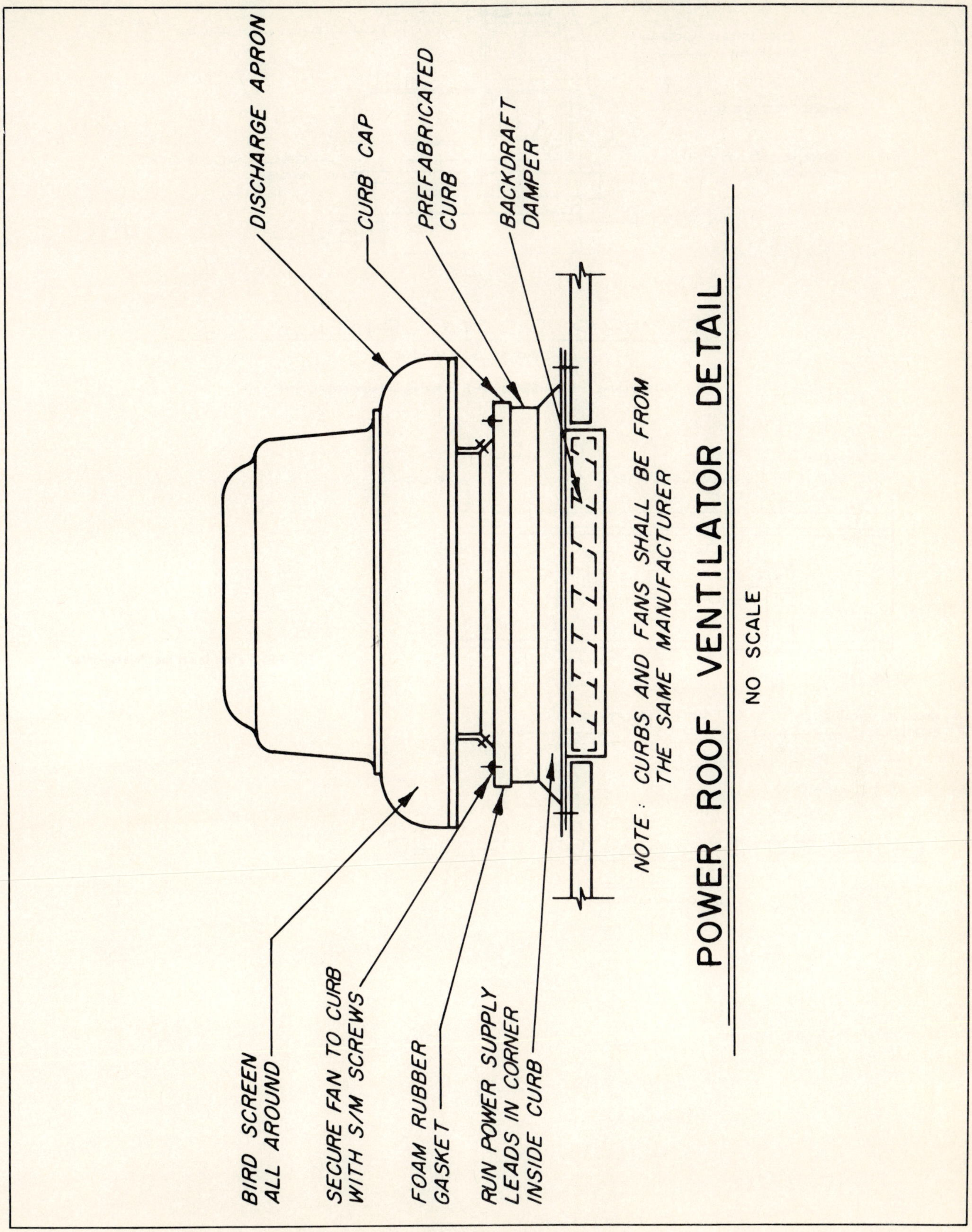

Fig. 14-11. Power roof ventilator detail.

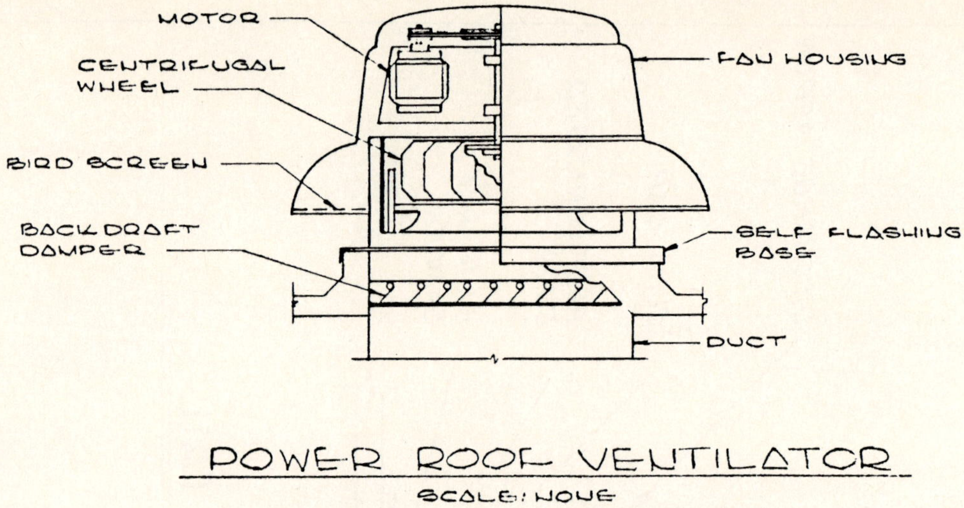

Fig. 14-12. Power roof ventilator detail showing interior arrangement.

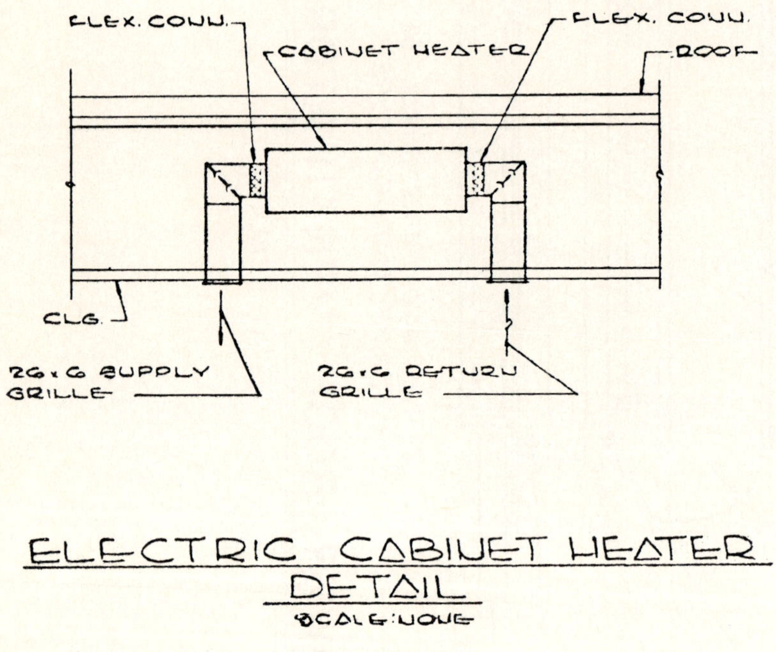

Fig. 14-13. Electric cabinet heater detail.

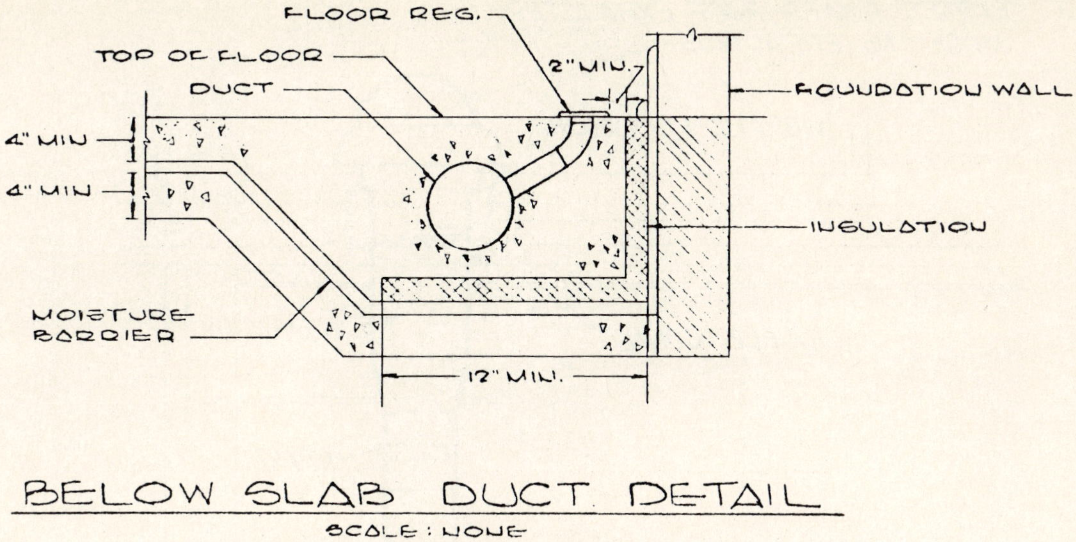

Fig. 14-14. Below slab duct detail.

Fig. 14-15. Air-conditioner support detail.

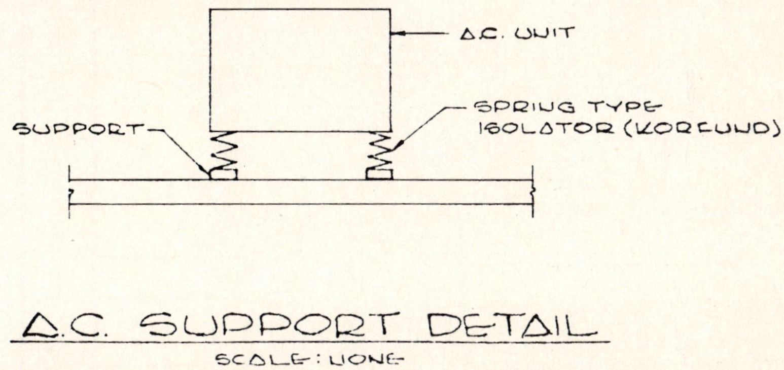

Fig. 14-16. Diffuser detail for a multipurpose room.

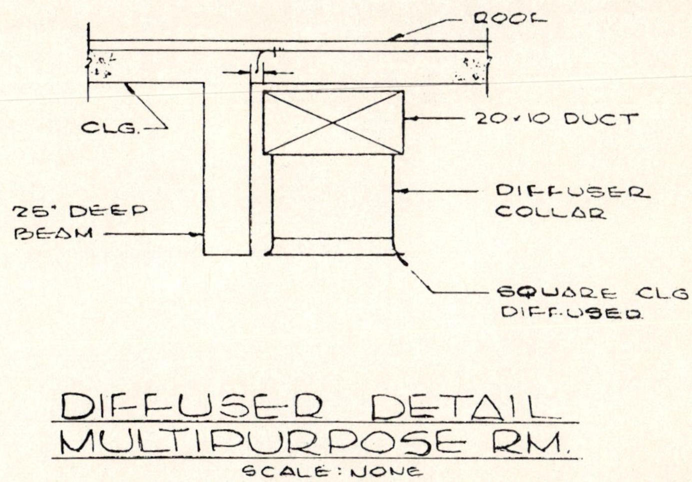

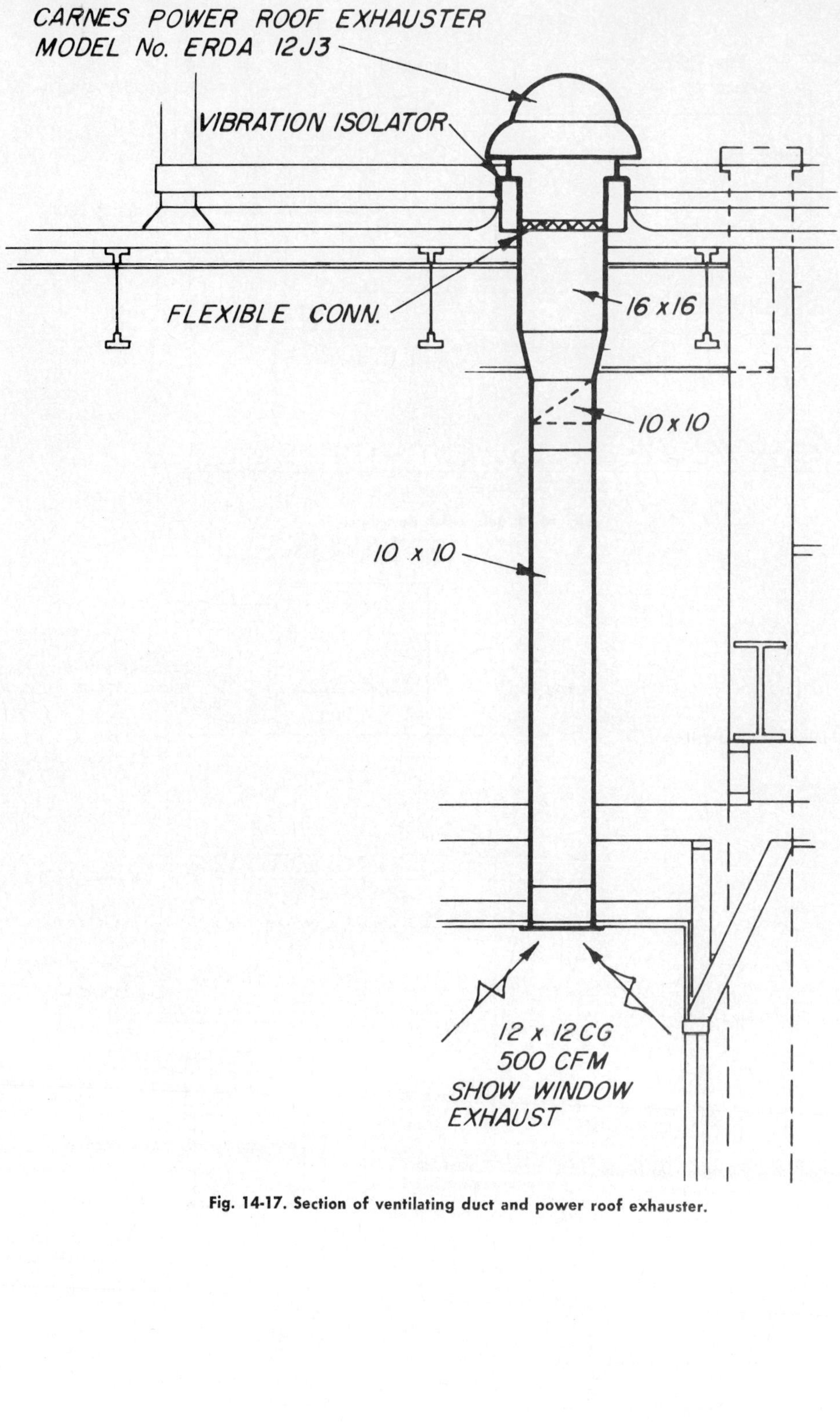

Fig. 14-17. Section of ventilating duct and power roof exhauster.

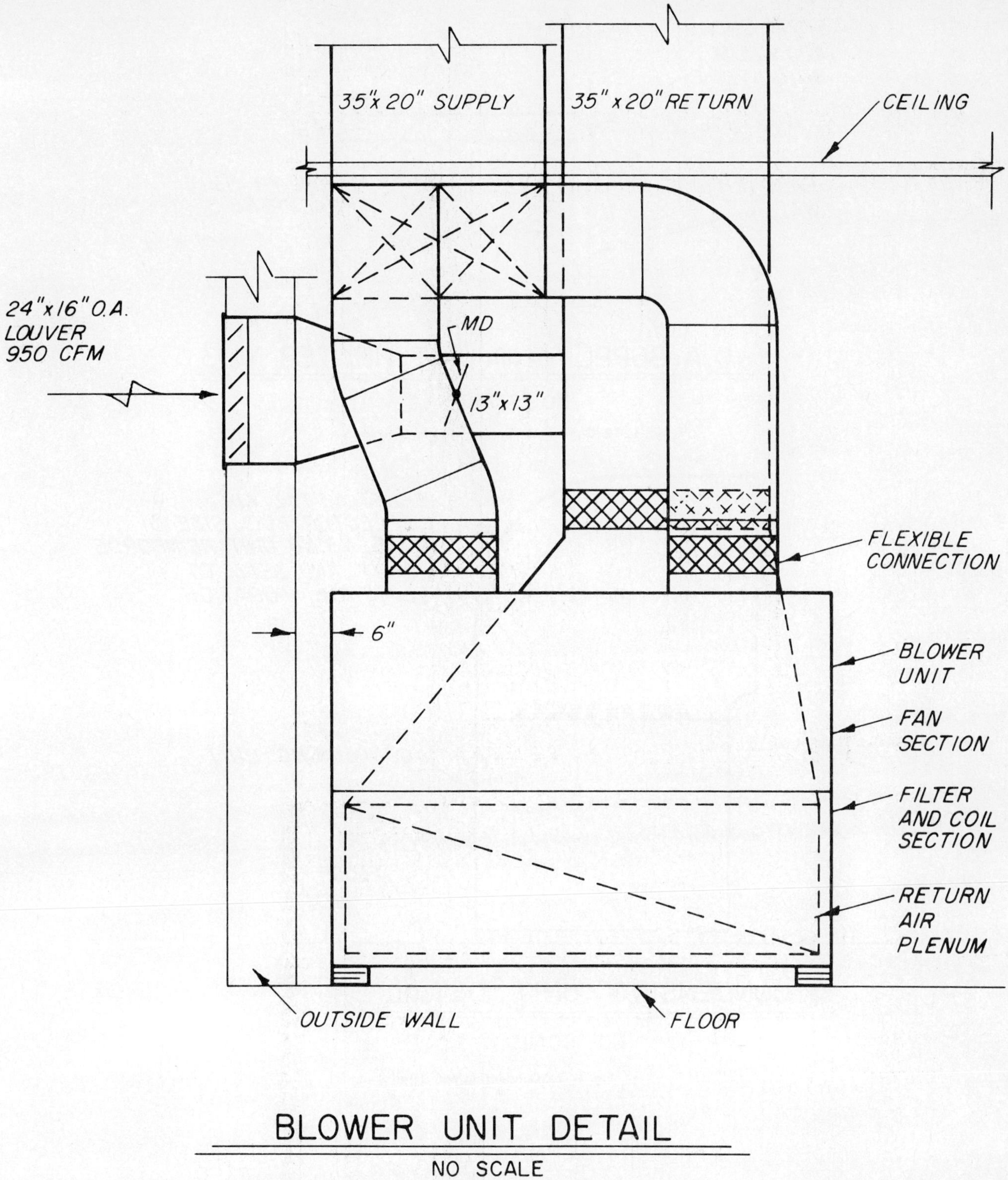

Fig. 14-18. Blower unit detail.

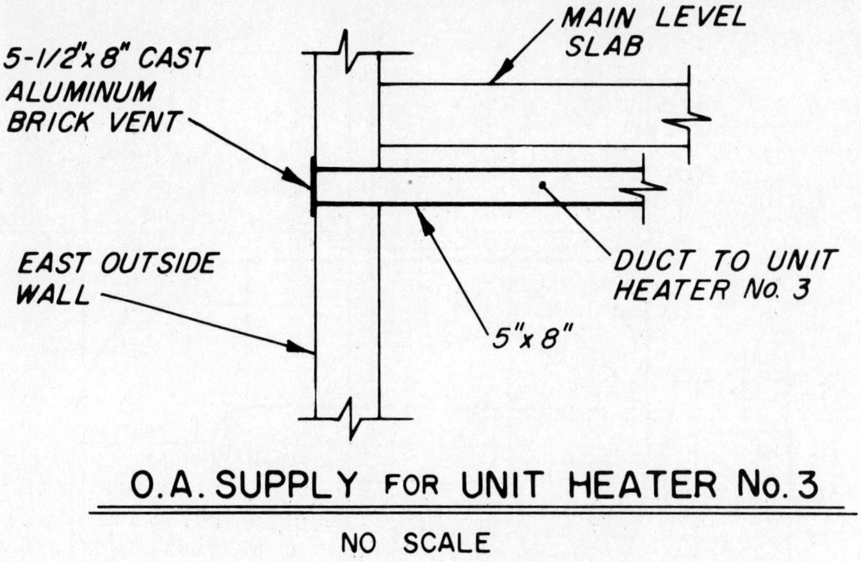

Fig. 14-19. Outside air supply detail for unit heater.

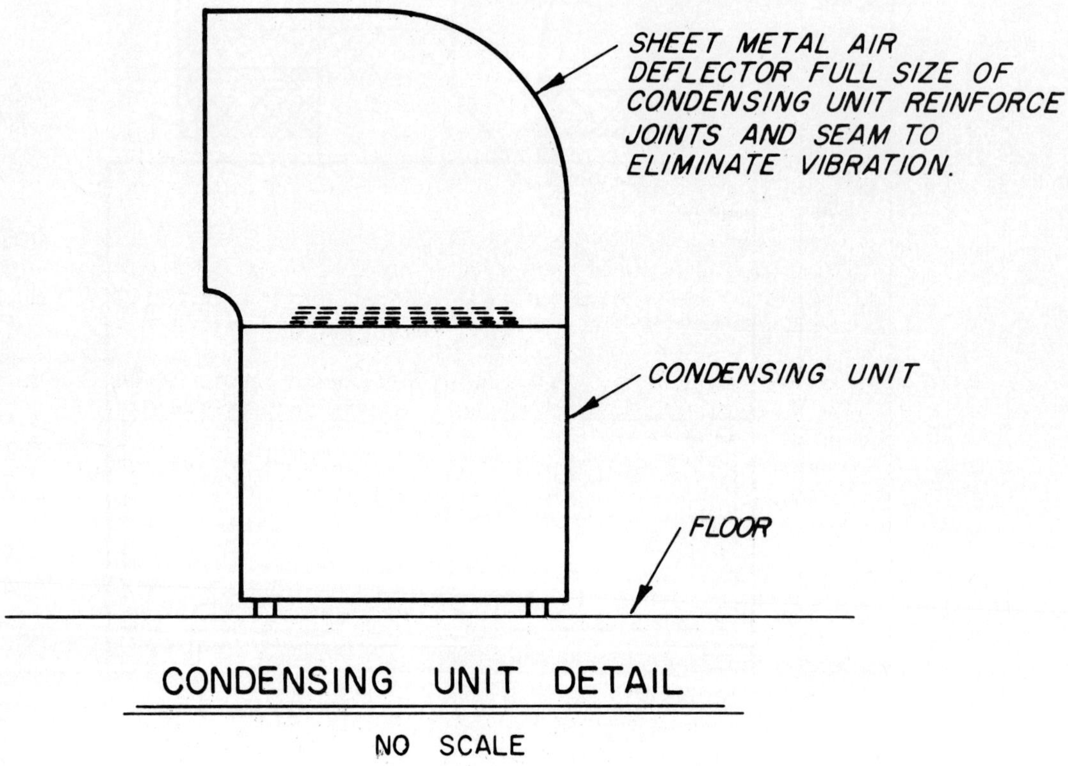

Fig. 14-20. Condensing unit detail.

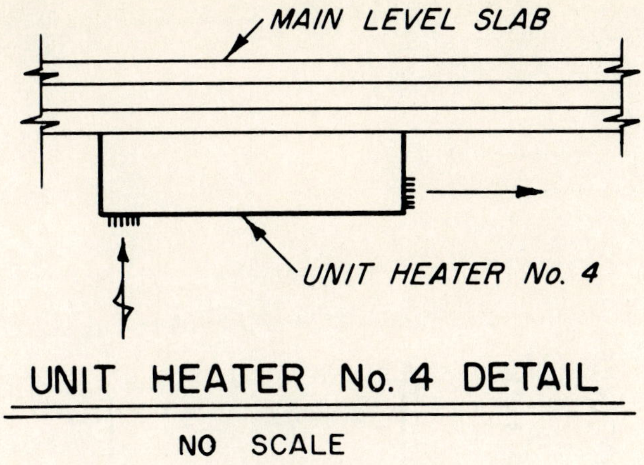

Fig. 14-21. Unit heater detail.

Department Store

The detail shown in Fig. 14-17 was used on the HVAC working drawings for a department store. Details of ventilation for the store windows are clearly shown.

Bank Building

Figs. 14-18 through 14-23 were used on the HVAC drawings for a bank building. They gave additional information to the contractor so that the HVAC system could be installed more easily and according to the specifications.

SUMMARY

A complete set of construction documents includes the written specifications and the working drawings, with adequate drawings and details to describe exactly how an HVAC system is to be installed.

A detail drawing is the drawing of a separate item or portion of a building system, giving a complete and exact description of its use and all necessary details of its construction.

UNIT 14 — ASSIGNMENT

1. Prepare an 11-by-17-inch sheet of tracing paper with border lines and title block.
2. Turn to Fig. 14–7. Using a scale of ¾ inch = 1 foot, draw the detail of the figure. Be sure to center it on the sheet.
3. Answer the following True or False questions.

	True	False
a. The lines and symbols used to show building walls and partitions and to show the location of equipment can only give a rough idea of the form or shape.	____	____
b. Custom-built frames are often necessary due to the weight of HVAC equipment.	____	____
c. Power roof ventilators are used to exhaust air from the inside of a building.	____	____
d. When electric cabinet heaters are used under windows, only supply ductwork is necessary.	____	____
e. The vibrations from air-conditioning units can be reduced by mounting them on isolators.	____	____

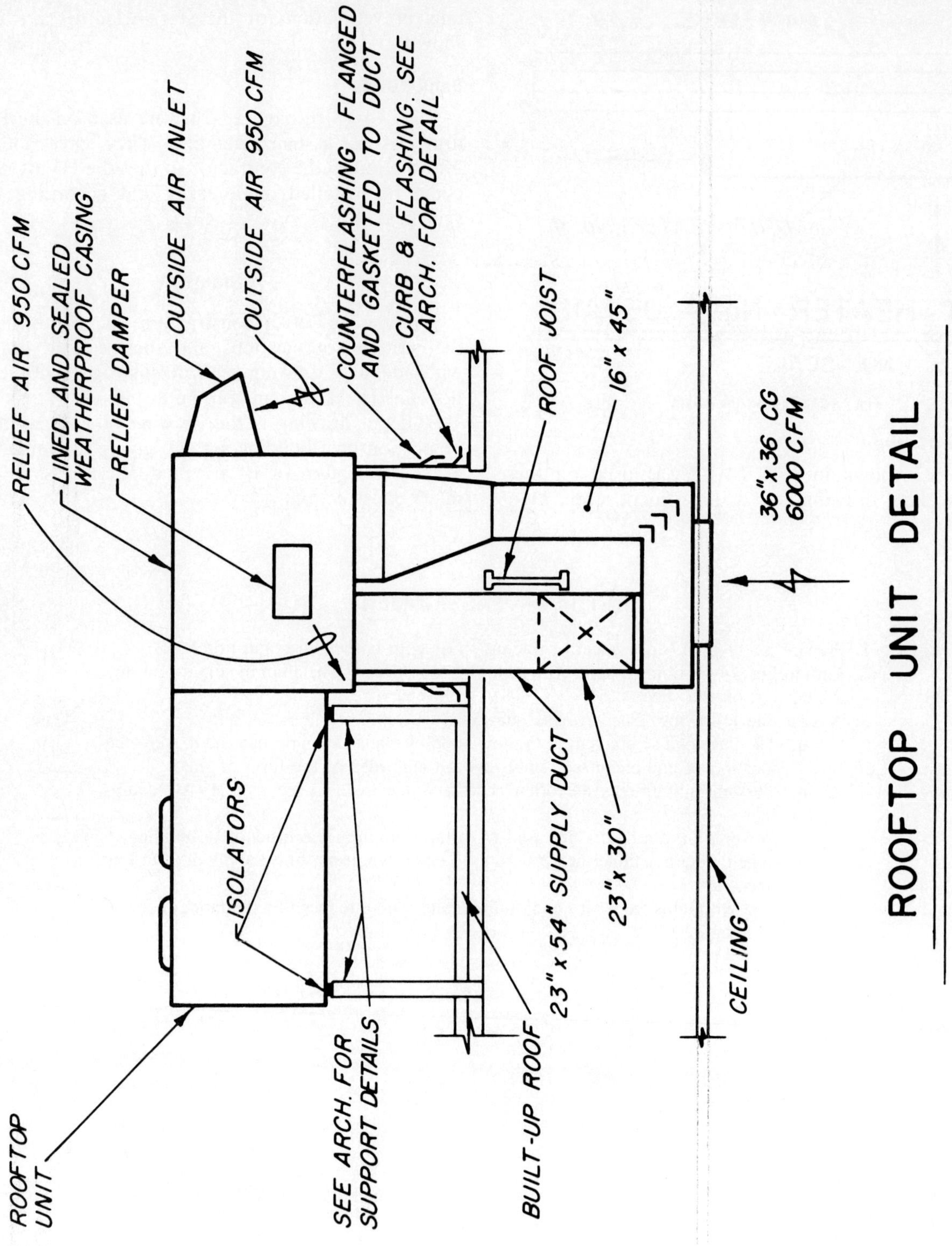

Fig. 14-22. Rooftop unit detail.

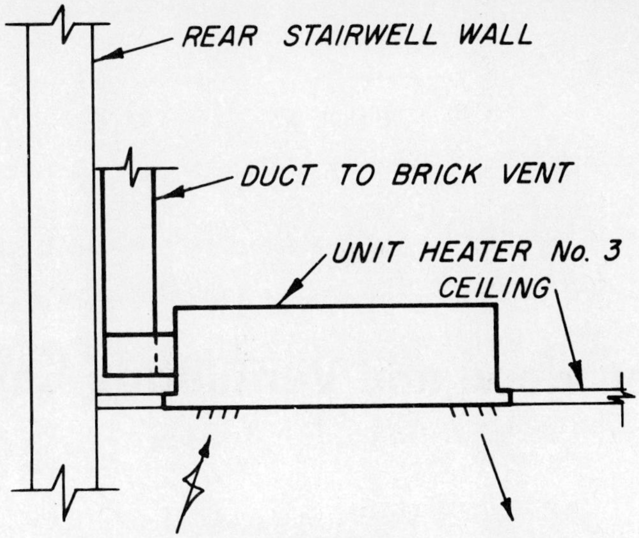

UNIT HEATER No. 3 DETAIL

NO SCALE

Fig. 14-23. Unit heater No. 3 detail.

UNIT 15

Kitchen Hood and Ventilating Layouts

OBJECTIVES: To discuss proper ventilation of restaurants, drive-ins, and kitchens of schools, hotels, and institutions

To learn how ventilation equipment should appear on HVAC working drawings

GENERAL

Good ventilation is essential in commercial kitchens to maintain efficient and sanitary operations. Improper ventilation will allow heat and odors to spread throughout the kitchen area and into adjacent rooms. It will also cause grease vapors to coat walls, ceilings, and fixtures. This often creates an undesirable dining atmosphere and results in serious cleaning problems and expensive redecorating costs.

Massive overventilation, on the other hand, imposes needless expense in equipment and operating costs. This unit will suggest solutions for solving commercial kitchen ventilating problems and will aid the reader in selecting the most efficient and economical exhaust equipment.

The diagrams in Fig. 15-1, show how air flows to the openings for different types of exhaust canopies. In designs of these types, it is essential to provide suffi-

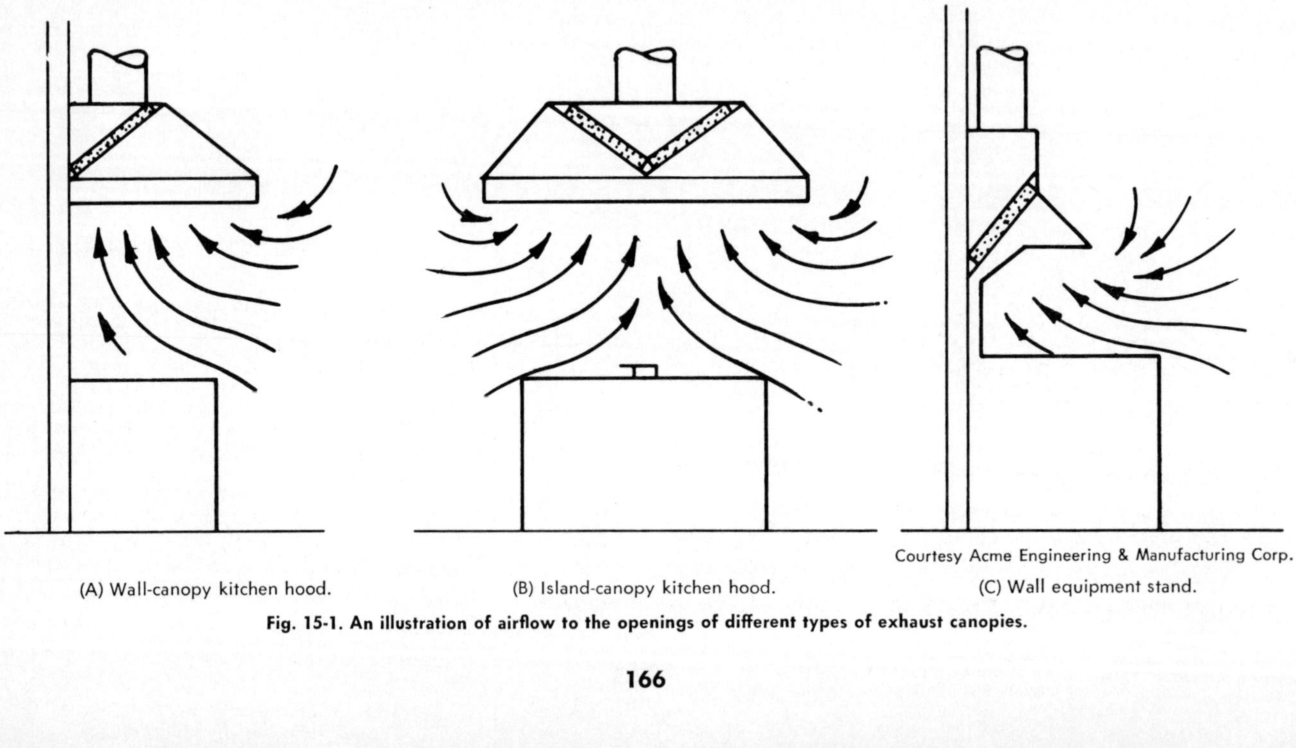

Courtesy Acme Engineering & Manufacturing Corp.

(A) Wall-canopy kitchen hood. (B) Island-canopy kitchen hood. (C) Wall equipment stand.

Fig. 15-1. An illustration of airflow to the openings of different types of exhaust canopies.

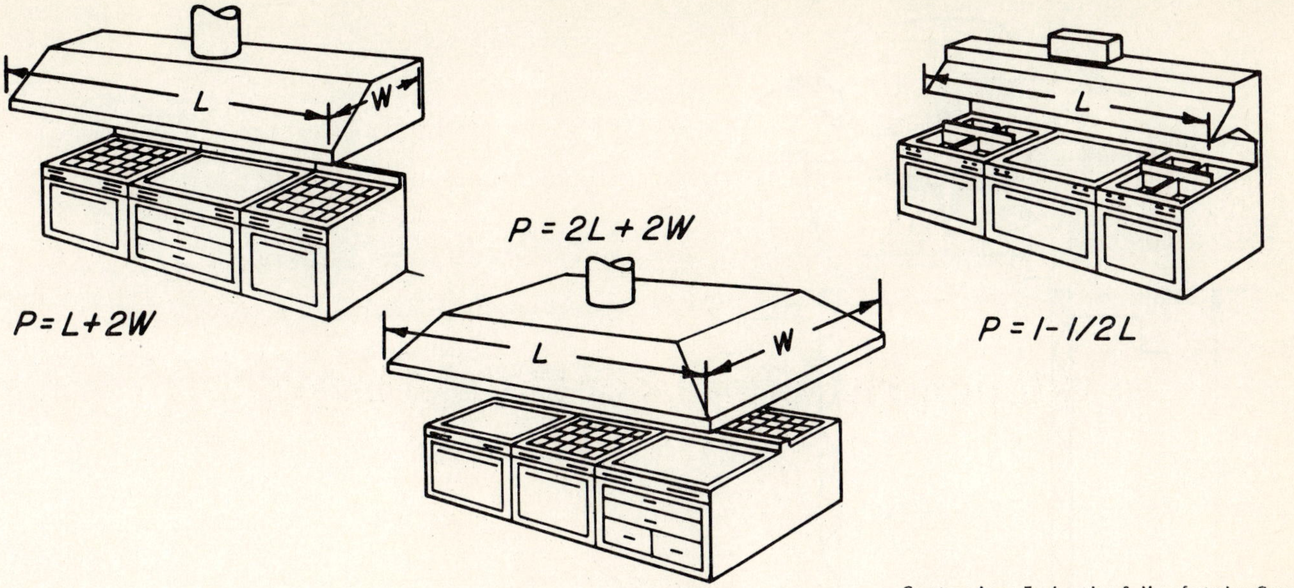

Courtesy Acme Engineering & Manufacturing Corp.

(A) Wall canopy. (B) Island canopy. (C) Wall equipment stand.

Fig. 15-2. Formulas for determining exposed perimeters of various types of kitchen hoods.

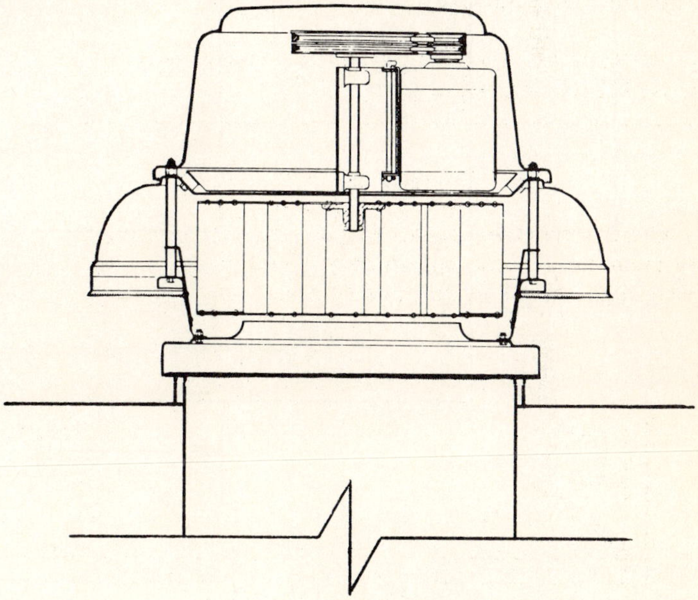

Fig. 15-3. Sectional view of an exhaust system with no turns in the ductwork.

Courtesy Acme Engineering & Manufacturing Corp.

cient air speed in the open area surrounding the exposed perimeter of the canopy to prevent the escape of heat, odors, and vapors into the kitchen. This is often referred to as "capture velocity."

SELECTION OF VENTILATING EQUIPMENT

First, determine the exposed perimeter for the appropriate type canopy—wall, island, or wall equipment stand. The exposed perimeter (P) is the total distance in feet around the open side of the canopy. Refer to the formula shown with each type of canopy in Fig. 15-2. For example, let us assume that a wall canopy is 8 feet long, by 4 feet wide. Thus, $P = 8 + (2 \times 4) = 16$ feet.

Second, determine the applicable static pressure from Table 15-2. For example, the exhaust system shown in Fig. 15-3 has no turns but does use a filter; therefore, the static pressure will be ¼ inch. The exhaust system shown in Fig. 15-4 has one turn (elbow) and also uses a filter. From Table 15-1, the static pressure will be ⅜ inch.

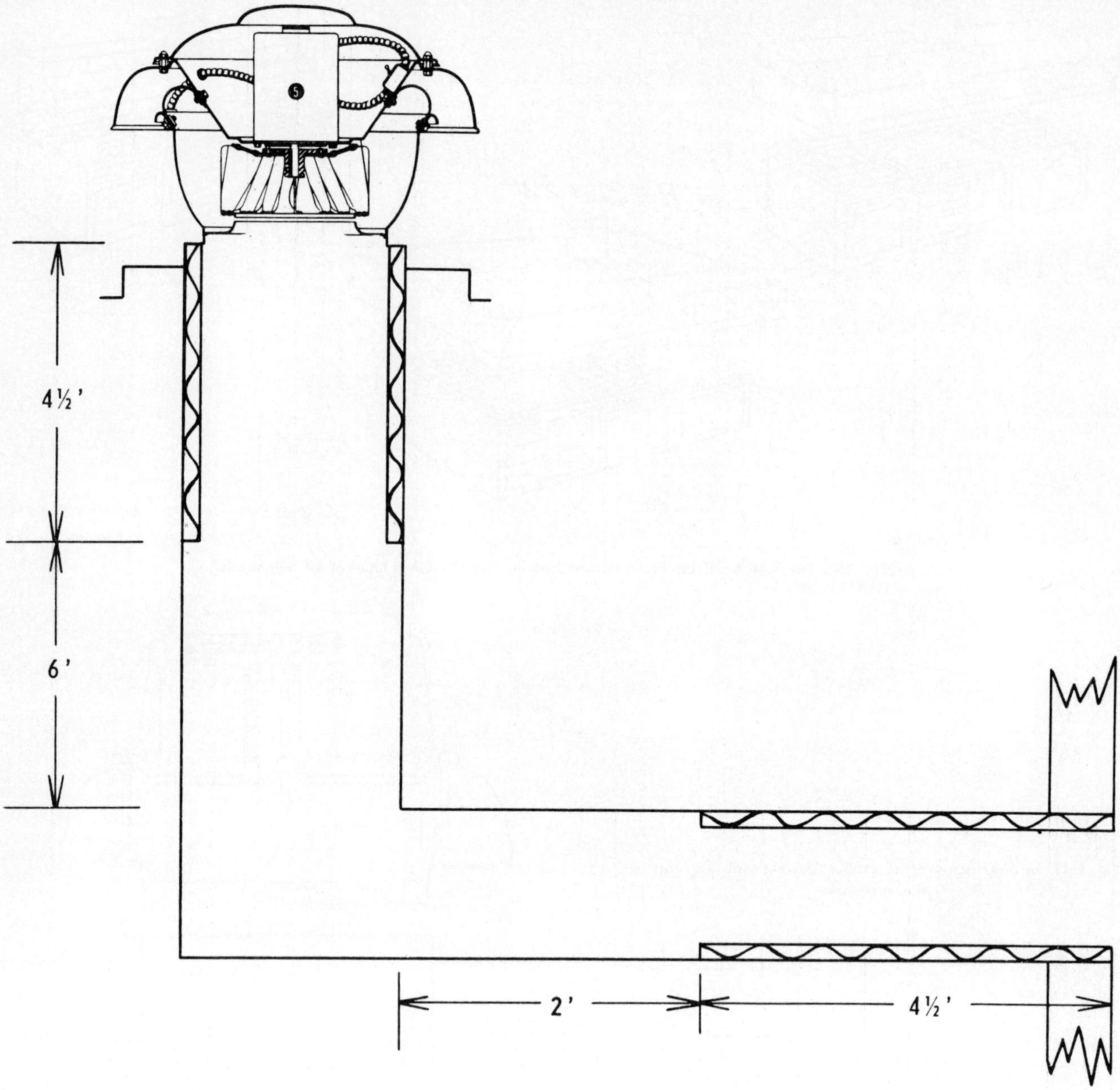

Courtesy Acme Engineering & Manufacturing Corp.

Fig. 15-4. Example of an exhaust system with one turn in the ductwork.

Table 15-1. The Static Pressures Corresponding to the Number of Turns in the Exhaust Duct

Number of Turns (elbows)	Static Pressure (inch)	
	With Filter	Without Filter
0	¼	⅛
1	⅜	¼
2	½	⅜

Third, referring to Table 15-2, select the exhauster model that is recommended for the specific requirement. For example, assume a wall canpoy has a calculated exposed perimeter of 16 feet. The exhaust duct will run directly out of the top of the canopy and through the roof, and will contain no filter. The static pressure is therefore ⅛ inch. An air-conditioning condensing unit will be mounted on the roof near the point where the canopy exhaust duct will penetrate. So the designer

Table 15-2. Recommendations for Centrimaster Exhauster Models and Total Volumes of Exhaust Air Required for Each Application*

Exposed Perimeter Open to Air Flow (ft)	CFM	Centrimaster Up-Blast Exhauster Direct Driven — Static Pressure (inch)				Centrimaster Roof Exhauster Direct Driven — Static Pressure (inch)				Centrimaster Wall Exhauster Direct Driven — Static Pressure (inch)				Centrimaster Up-Blast & Roof Exhauster Belt Driven — Static Pressure (inch)				Recommended Duct Size Diameter (inches)
		1/8	1/4	3/8	1/2	1/8	1/4	3/8	1/2	1/8	1/4	3/8	1/2	1/8	1/4	3/8	1/2	
3	525	PRU11A8	PRU11A8	PRU11C6	PRU11C6	PR76	PR76	PR82	PR82	PW76	PW76	PW82	PW82	PU or P10D3	PU or P10D3	PU or P10D3	PU or P10D3	10
4	700	PRU11A8	PRU11C6	PRU11C6	PRU11C6	PR82	PR82	PR82	PR9D4	PW82	PW82	PW82	PW9D4	PU or P10D3	PU or P10D3	PU or P10D3	PU or P10D3	10
5	875	PRU11A8	PRU11C6	PRU11C6	PRU11F4	PR82	PR82	PR9D4	PR9D4	PW82	PW11C6	PW9D4	PW9D4	PU or P10D3	PU or P10D3	PU or P10D3	PU or P10D3	12
6	1050	PRU11C6	PRU11C6	PRU11F4	PRU11F4	PR10B6	PR11C6	PR9D4	PR9D4	PW11C6	PW11C6	PW9D4	PW10E4	PU or P10D3	PU or P10D3	PU or P10D3	PU or P12F	12
7	1225	PRU11C6	PRU12D6	PRU11F4	PRU11F4	PR11C6	PR11C6	PR9D4	PR10E4	PW12D6	PW12D6	PW10E4	PW10E4	PU or P10D3	PU or P10D3	PU or P12F	PU or P12F	14
8	1400	PRU12D6	PRU12D6	PRU11F4	PRU11F4	PR11C6	PR12D6	PR10E6	PR10E4	PW12D6	PW12D6	PW10E4	PW11F4	PU or P10D3	PU or P12F	PU or P12F	PU or P12F	14
9	1575	PRU12D6	PRU12D6	PRU11F4	PRU11F4	PR12D6	PR12D6	PR10E6	PR11F4	PW12D6	PW13F6	PW11F4	PW11F4	PU or P12F	PU or P12F	PU or P12F	PU or P12F	14
10	1750	PRU13D8	PRU 14D8	PRU11F4	PRU11F4	PR13D8	PR14D8	PR11F4	PR11F4	PW13F6	PW13F6	PW11F4	PW13F6	PU or P12F	PU or P12F	PU or P12F	PU or P12F	16
11	1925	PRU14D8	PRU11F4	PRU13F6	PRU14F6	PR14D8	PR11F4	PR11F4	PR11F4	PW13F6	PW13F6	PW13F6	PW14F6	PU or P12F	PU or P12F	PU or P12F	PU or P14F	16
12	2100	PRU14D8	PRU13F6	PRU14F6	PRU14F6	PR14D8	PR11F4	PR13F6	PR14F6	PW13F6	PW13F6	PW146	PW14F6	PU or P12F	PU or P12F	PU or P14F	PU or P16G	16
13	2275	PRU13F6	PRU13F6	PRU14F6	PRU14F6	PR13F6	PR13F6	PR14F6	PR14F6	PW13F6	PW14F6	PW14F6	PW16H6	PU or P12F	PU or P16G	PU or P16G	PU or P16G	18
14	2450	PRU13F6	PRU14F6	PRU14F6		PR13F6	PR14F6	PR14F6		PW14F6	PW14F6	PW16H6	PW16H6	PU or P16G	PU or P16G	PU or P16G	PU or P16G	18
15	2625	PRU14F6	PRU14F6			PR14F6	PR14F6			PW14F6	PW16H6	PW16H6	PW16H6	PU or P16G	PU or P16G	PU or 16G	PU or P20G	20
16	2800	PRU14F6				PR14F6				PW16H6	PW16H6	PW16H6	PW16H6	PU or P16G	PU or P16G	PU or 16G	PU or P20G	20
17	2975	PRU14F6				PR14F6				PW16H6	PW16H6	PW16H6	PW16H6	PU or P16G	PU or P16G	PU or P20G	PU or P22G	20
18	3150									PW16H6	PW16H6	PW16H6	PW16H6	PU or P16G	PU or P20G	PU or P22G	PU or P22G	20
19	3325									PW16H6	PW16H6	PW16H6		PU or P20G	PU or P22F	PU or P22G	PU or P22H	22
20	3500									PW16H6	PW16H6			PU or P22F	PU or P22G	PU or P22G	PU or P22H	22
22	3850													PU or P22G	PU or P22H	PU or P22H	PU or P22H	24
24	4200													PU or P22H	PU or P22H	PU or P22H	PU or P27H	24
26	4550													PU or P22H	PU or P22H	PU or P27G	PU or P27H	24
28	4900													PU or P24H	PU or P27G	PU or P27H	PU or P27H	26
30	5250													PU or P27G	PU or P27H	PU or P27H	PU or P27H	26
32	5600													PU or P27G	PU or P27H	PU or P27H	PU or P30J	28
34	5950													PU or P27H	PU or P27H	PU or P30J	PU or P30J	28
36	6300													PU or P27H	PU or P30H	PU or P30J	PU or P30K	28
38	6650													PU or P30H	PU or P30J	PU or P30J	PU or P30K	30
40	7000													PU or P30J	PU or P30J	PU or P30K	PU or P30K	30

Courtesy Acme Engineering & Manufacturing Corp.

* Based on 3½-foot clearance between canopy and appliance.

Note: Refer to the following numbers or letters in the model number of each Centrimaster exhauster to determine the motor horsepower.

76—1/20	A—1/20	C—1/8	E—1/4	G—1/2	J—1
82—1/10	B—1/12	D—1/6	F—1/3	H—3/3	K—1-1/2

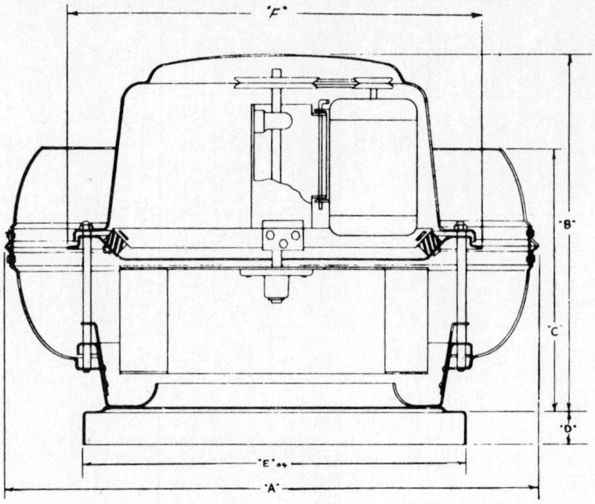

Fig. 15-5. Example of up-blast exhauster.

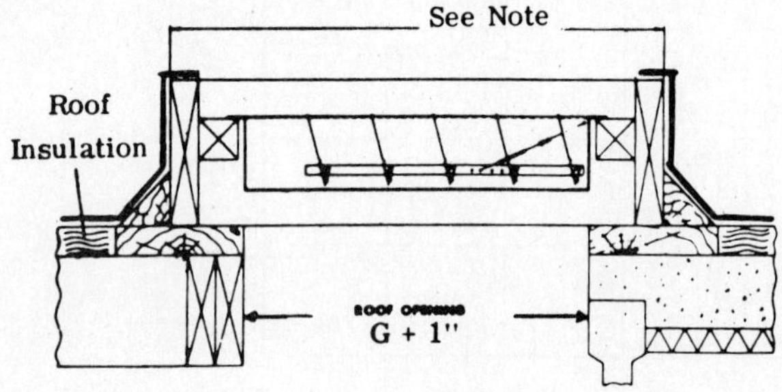

Dimension G recommended type QT damper size.

Note: Outside dimensions of curb base should be 1" to 1 1/2" less than inside curb cap dimension (E) depending on thickness of flashing material used.

Courtesy Acme Engineering & Manufacturing Corp.

selects an up-blast exhauster which will eject the exhausted air high above the roof. Referring to Table 15-2 in the "exposed perimeter open to air flow" column, we find that, for a perimeter of 16 feet, a model PRU14F6 at 2800 CFMs should be used. In the last column, we find that a duct 20 inches in diameter is recommended.

The exhausters in Table 15-2 refer to model numbers of units manufactured by Acme Engineering & Manufacturing Corp. However, the table may be used for similar equipment manufactured by other companies, provided that the same types—up-blast exhauster roof exhauster, wall exhauster, etc.—are used. Figs. 15-5 through 15-8 show these various types and the characteristics of each.

ADDITIONAL RECOMMENDATIONS

On long canopies, two small exhausters usually provide a more uniform air distribution than one large unit. Two exhausters are recommended for island-type canopies over 12 feet long and for wall-type canopies over 18 feet long.

Another advantage of dual exhausters is that they permit the operation of only one unit during low-load periods, which will reduce loss of heat in winter and loss of cooling in summer.

Underside ducts increase static-pressure losses, and the recommended duct sizes in Table 15-2 should be used to hold such losses to a minimum.

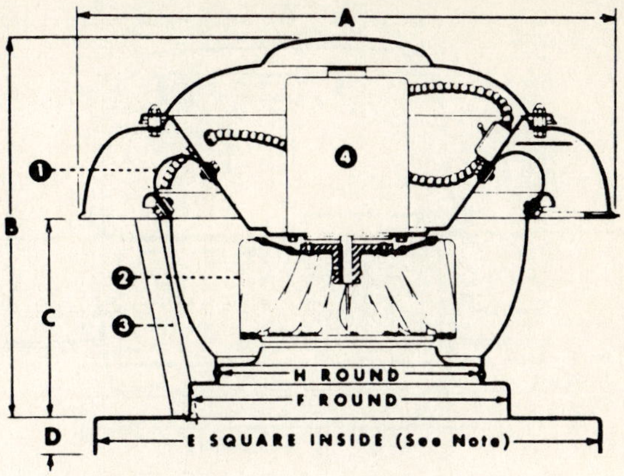

Fig. 15-6. An example of a roof exhauster.

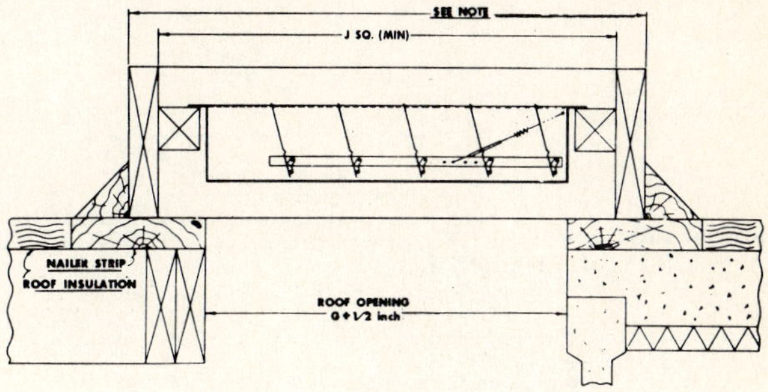

Dimension G recommended type QT damper size.

Note: Outside dimension of curb base should be 1" to 1-1/2" less than the inside curb cap dimension (E) depending on thickness of flashing material used.

Courtesy Acme Engineering & Manufacturing Corp.

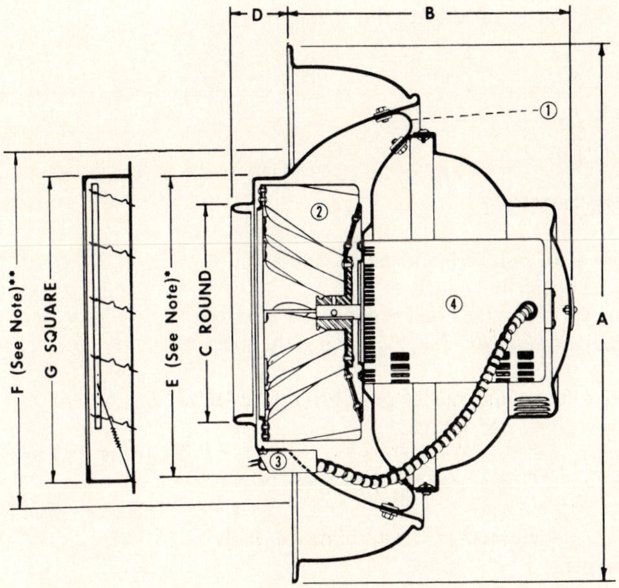

Note:
*For round duct connection to unit without damper, cut a circular hole in the wall to dimension E.

**For square damper in wall installation, cut square hole in wall to dimension F and use 1" thick furring strip to block in the damper.

Courtesy Acme Engineering & Manufacturing Corp.

Fig. 15-7. An example of a wall exhauster.

Canopies should extend from 6 inches to 8 inches beyond the appliance they cover to ensure the best capture of heat and vapors. However, some local ordinances require that the canopy extend beyond the appliance by 25 percent of the width of the appliance. Local authorities should be consulted.

Avoid crossdrafts from open doors and windows or from pedestal-type fans because they will disrupt the "capture velocity" effect and impair the performance of the exhaust system.

MAKEUP AIR

Makeup air is outside air that is introduced into a building to replace air that has been exhausted. If the makeup air is confined to a given area, such as a kitchen exhaust canopy, it is usually not necessary to heat the outside air. However, in other applications, the outside air should be heated to a comfortable temperature before entering a conditioned space.

The floor plan in Fig. 15-11 shows the floor plan of a small drive-in restaurant kitchen. Notice the exhaust hood over the cooking equipment. Because of the long length of this hood, three exhaust fans were used to provide a more uniform air distribution over the entire length of the hood.

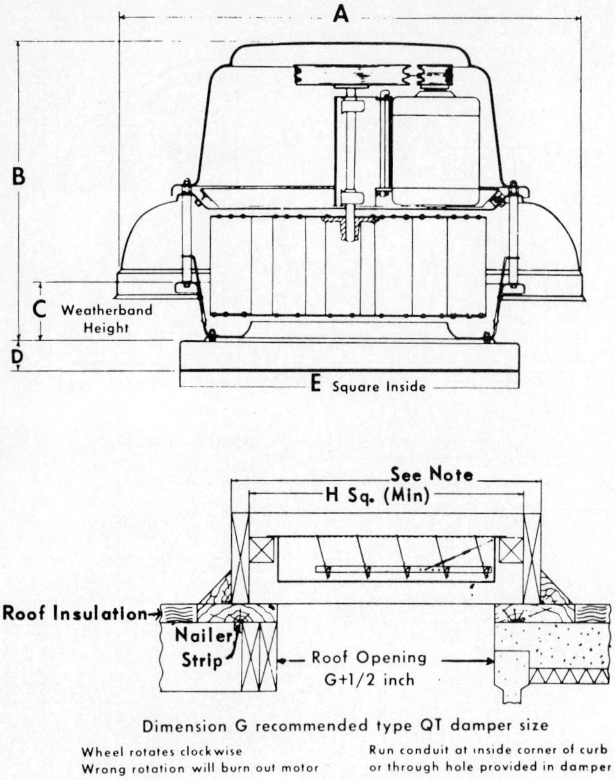

Note: Outside dimensions of curb base should be 1" to $1\frac{1}{2}$" less than inside curb cap dimension (E) depending on thickness of flashing material used.

Courtesy Acme Engineering & Manufacturing Corp.

Fig. 15-8. Example of a belt-driven roof exhauster.

Two makeup air units are mounted on the roof, approximately above the center of the kitchen. These units preheat the makeup air to a comfortable temperature before it enters the kitchen area.

The kitchen hood is also equipped wtih fire extinguishing equipment in case a grease fire should occur. Also, a special relay breaks the fan circuits should a fire occur.

Fig. 15-10 shows detail drawings of the kitchen hood in Fig. 15-9. These drawings include a plan view and an elevation view, complete with dimensions and notes.

The kitchen floor plan shown in Fig. 15-11 is that of a kitchen for a modern vocational school. The exhaust is from the center of the hood and is connected by a 34-by-34-inch duct to a power roof ventilator. Makeup air is provided by a 40-by-40-inch intake head connected to outlets on the hood by a 36-by-36-inch duct. Other details that appear on the working drawings are shown in Figs. 15-12 through 15-21.

SUMMARY

Good ventilation is essential in all commercial kitchens in order to maintain efficient and sanitary operations.

Massive overventilation, however, imposes needless expense in equipment and operating costs.

It is essential to provide sufficient air velocity in the open area surrounding the exposed perimeter of the kitchen hood or canopy so that heat, odors, and vapors do not escape into the kitchen.

UNIT 15 — ASSIGNMENT

1. Prepare a sheet of 11-by-17-inch tracing paper with border lines and title block.
2. Turn to Fig. 15–11. Redraw the floor plan to a scale of ¼ inch = 1 foot.
3. If you examine Fig. 15–11, you will see that the outline of the hood is not drawn to proper scale. Redraw this hood on the plan you completed in No. 2 above, using the proper scale. Use Fig. 15–20 as a guide.
4. Using your knowledge of the methods described in this unit, size the power roof ventilator for this kitchen hood. Also determine the necessary CFM rating.
5. Answer the following True or False questions.

	True	False
a. It is always worth the extra expense to provide as much ventilation as possible.	____	____
b. The first step in the selection of ventilation equipment is to determine the exposed perimeter for the appropriate type canopy.	____	____
c. On long canopies, two small exhausters are better than one large one.	____	____
d. A disadvantage of dual exhausters is that they permit the operation of only one unit during low-load periods.	____	____
e. Underside ducts increase static-pressure losses and these losses must be kept to a minimum.	____	____

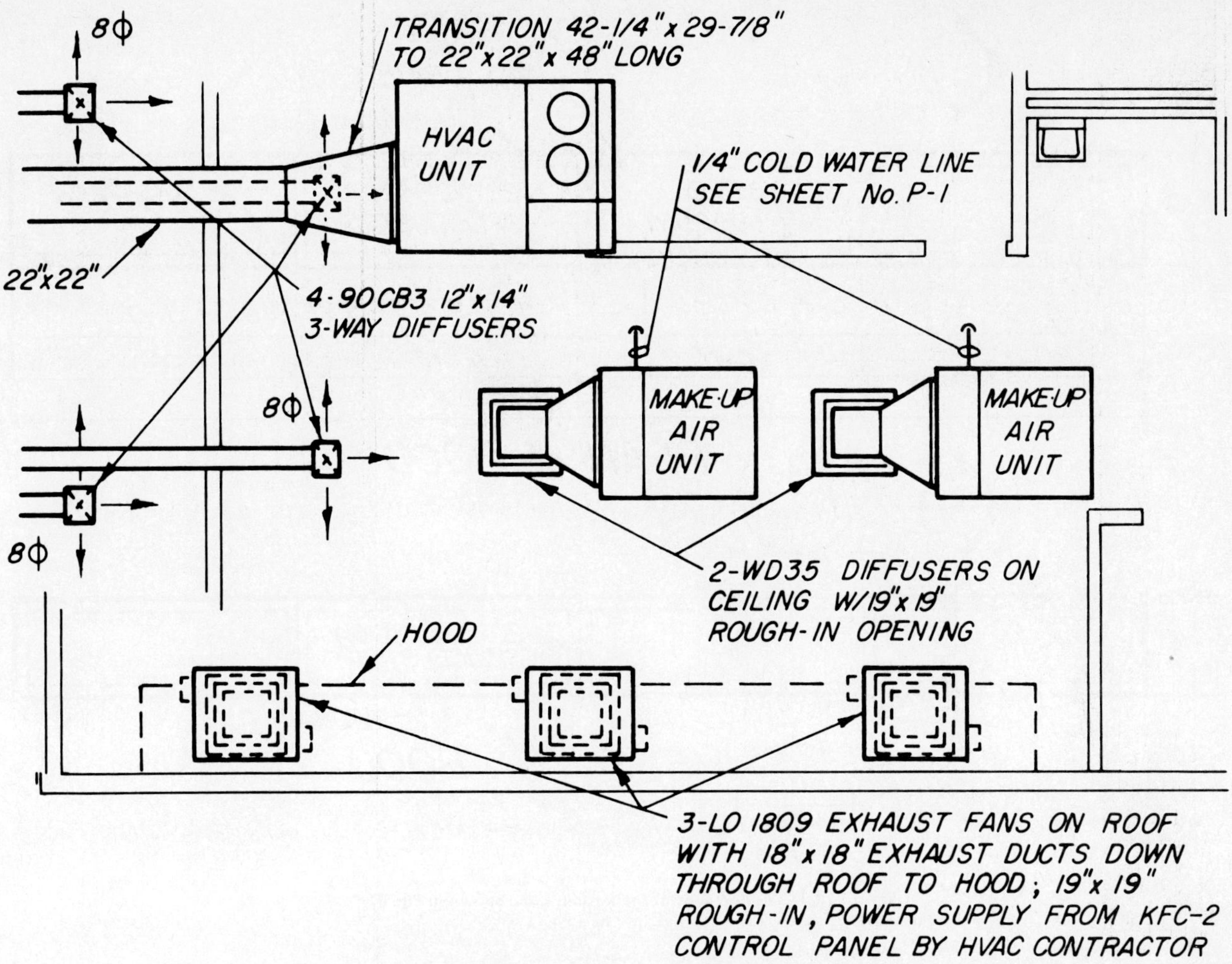

Fig. 15-9. Floor plan of a kitchen for a small drive-in restaurant.

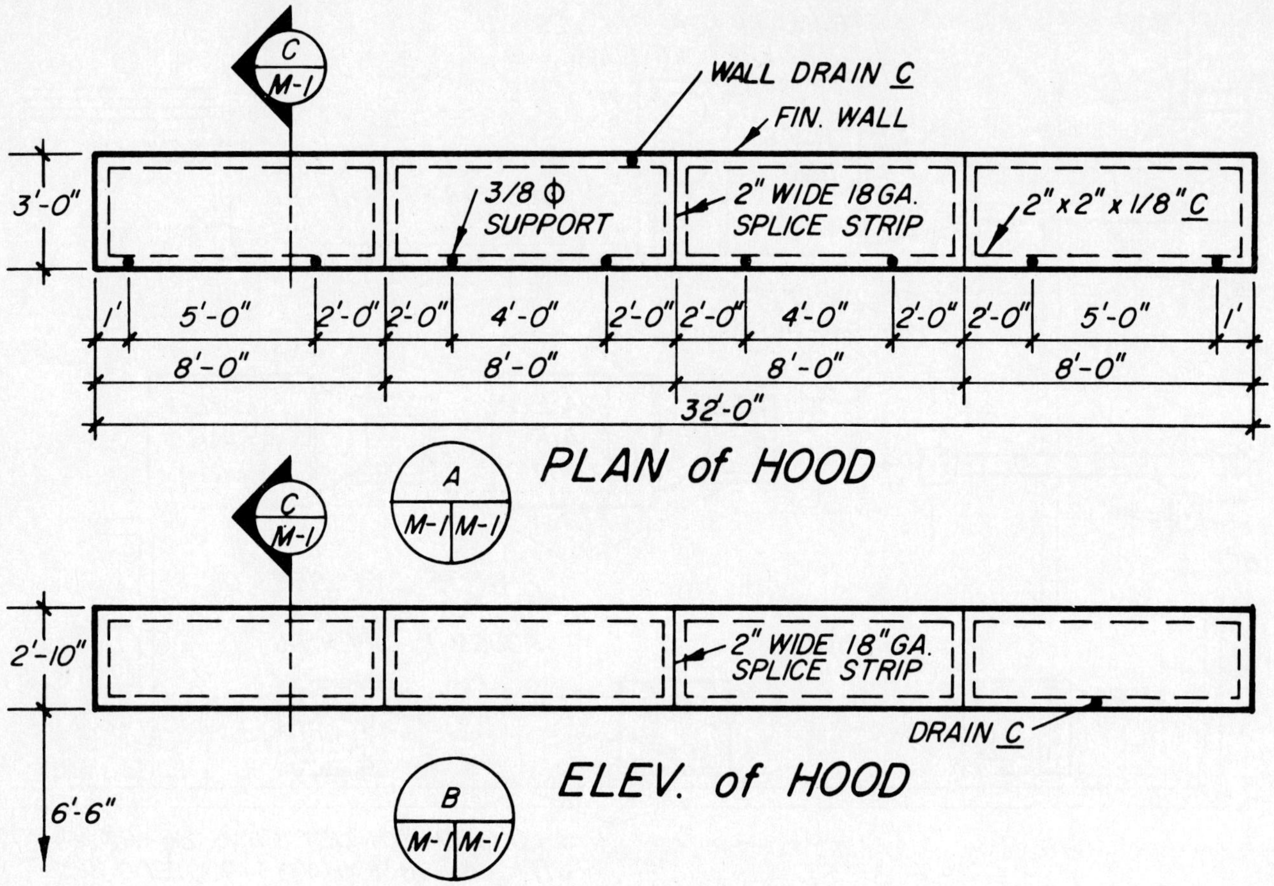

Fig. 15-10. Detail drawings of the kitchen hood shown in Fig. 15-9.

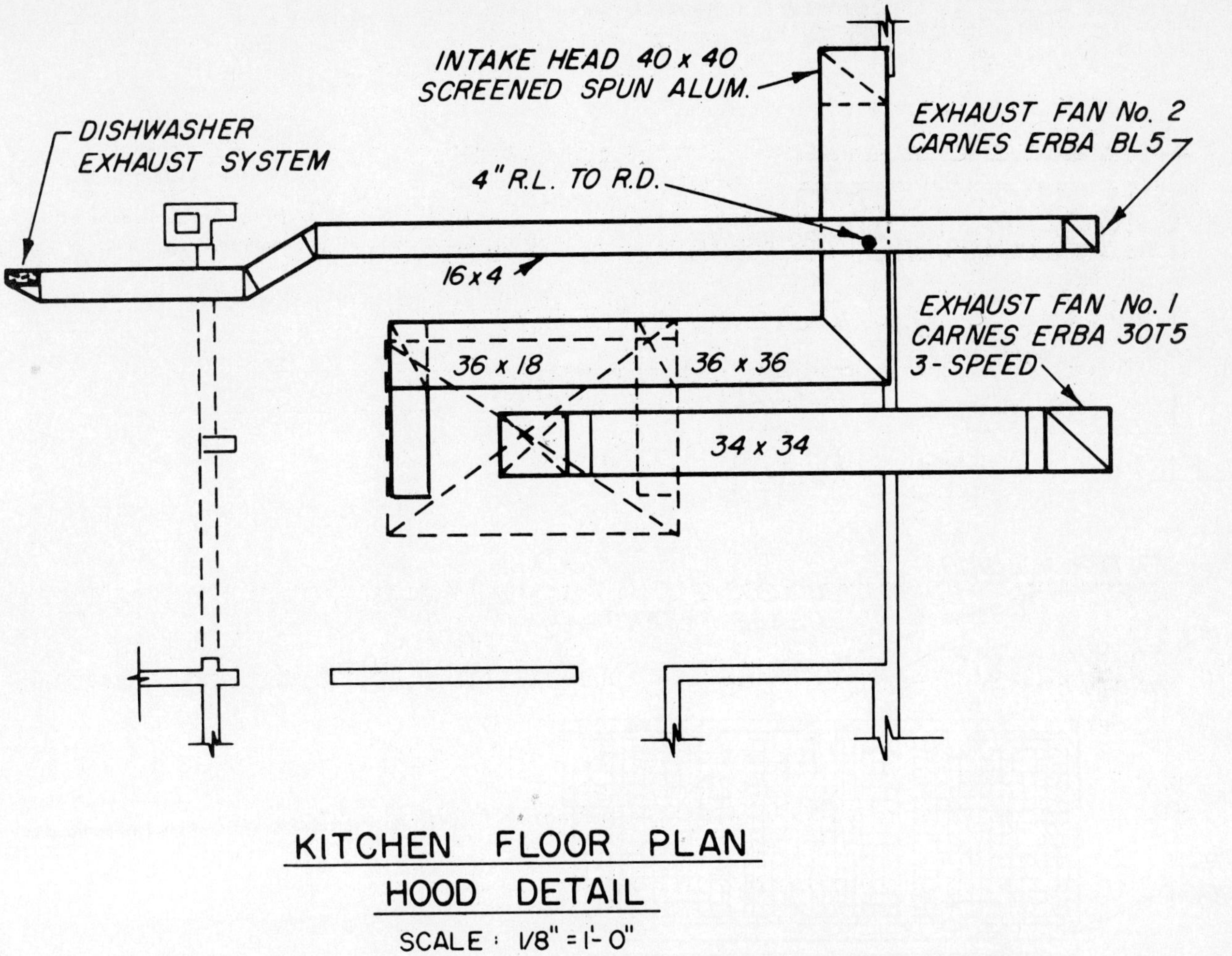

Fig. 15-11. A floor plan of a kitchen for a modern vocational school.

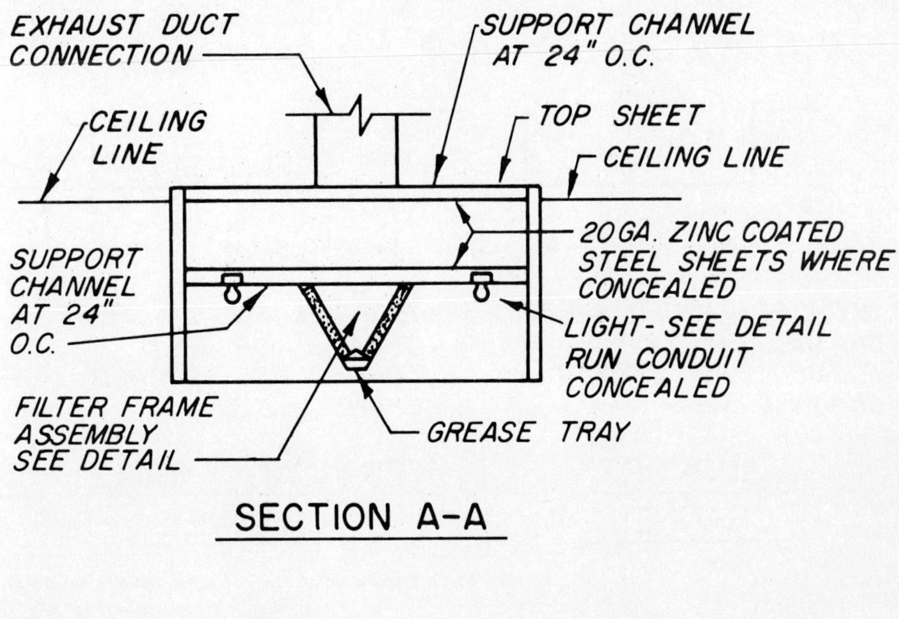

Fig. 15-12. A sectional view of the hood in Fig. 15-11.

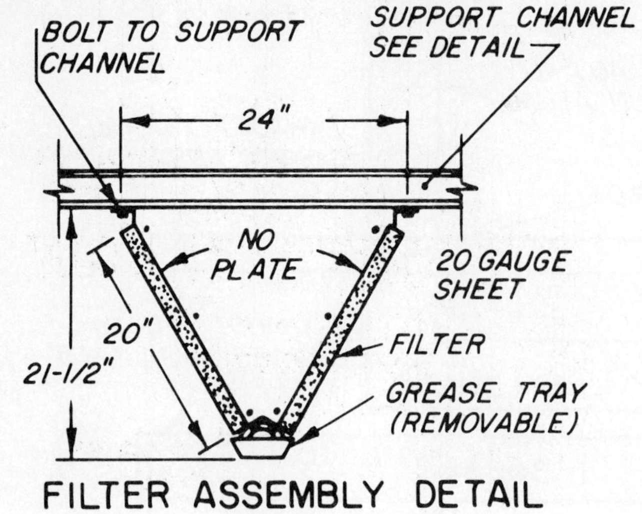

Fig. 15-13. The filter assembly detail of the kitchen hood in Fig. 15-11.

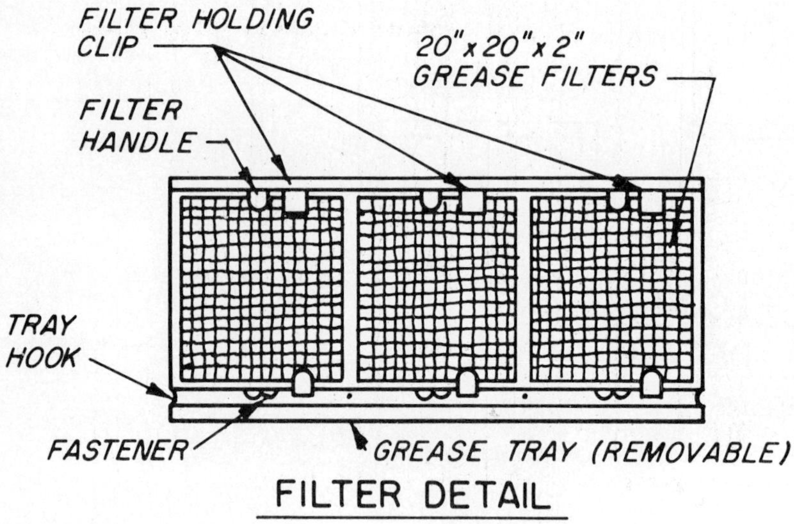

Fig. 15-14. Filter detail of the kitchen hood in Fig. 15-11.

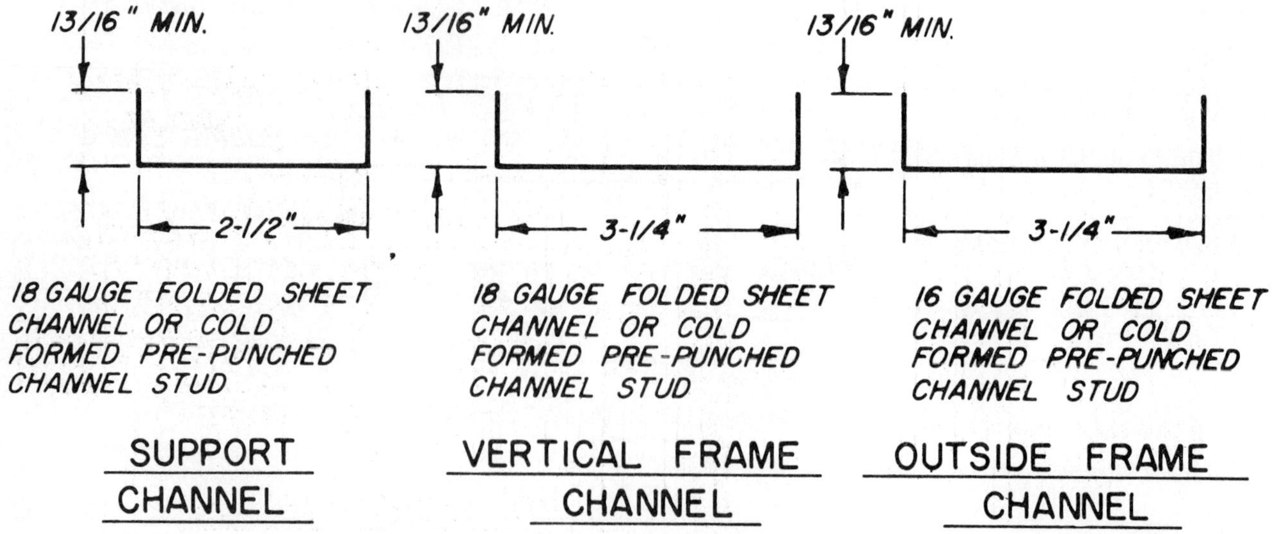

Fig. 15-15. Channel detail for the kitchen hood in Fig. 15-11.

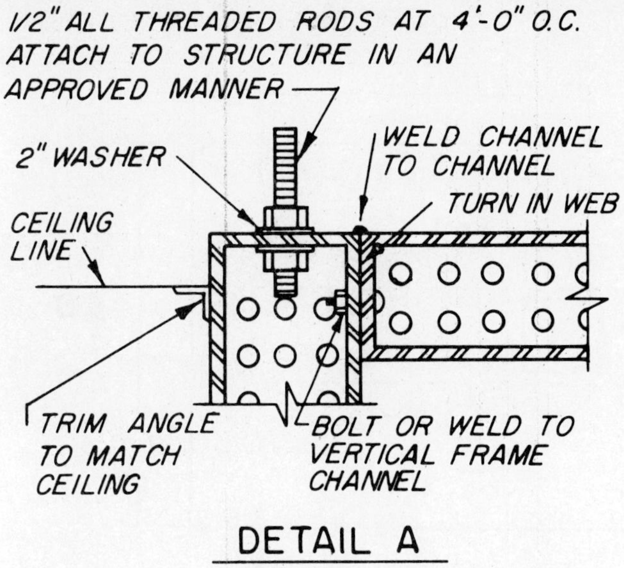

Fig. 15-16. Sectional detail of the kitchen hood in Fig. 15-11.

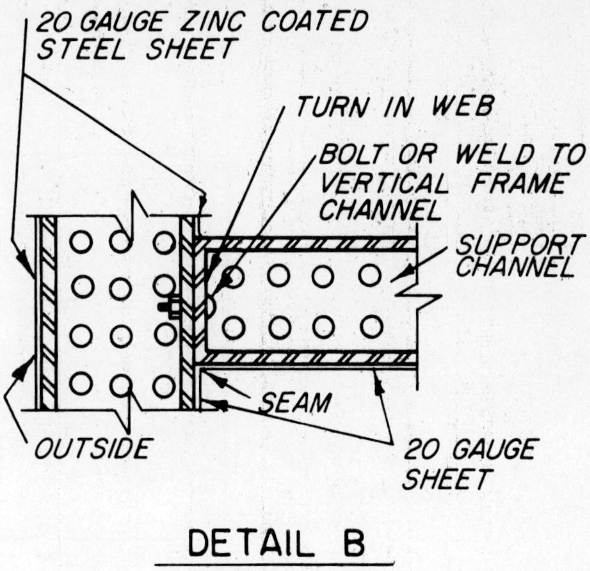

Fig. 15-17. Joint detail of the kitchen hood in Fig. 15-11.

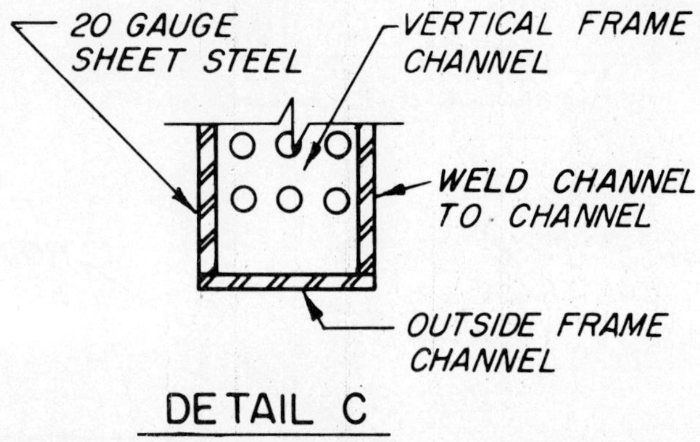

Fig. 15-18. Another sectional detail of the kitchen hood in Fig. 15-11.

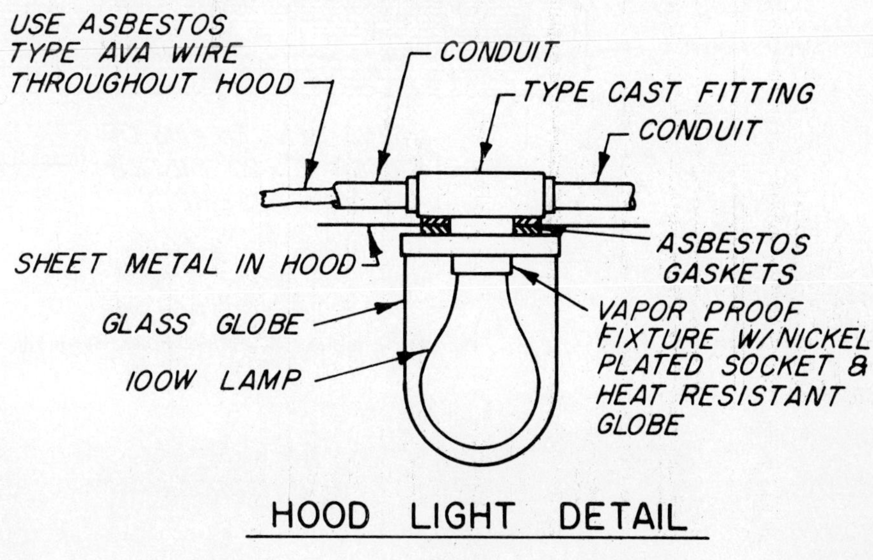

Fig. 15-19. Kitchen-hood-light detail.

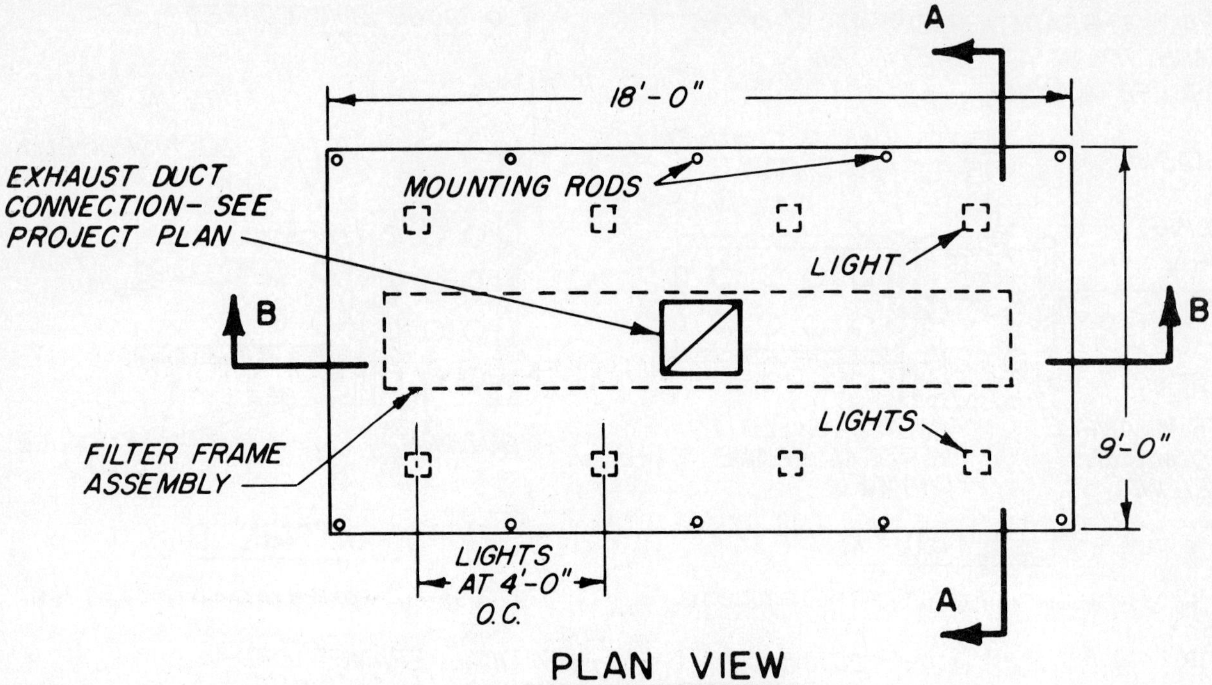

Fig. 15-20. A plan view of the kitchen hood in Fig. 15-11.

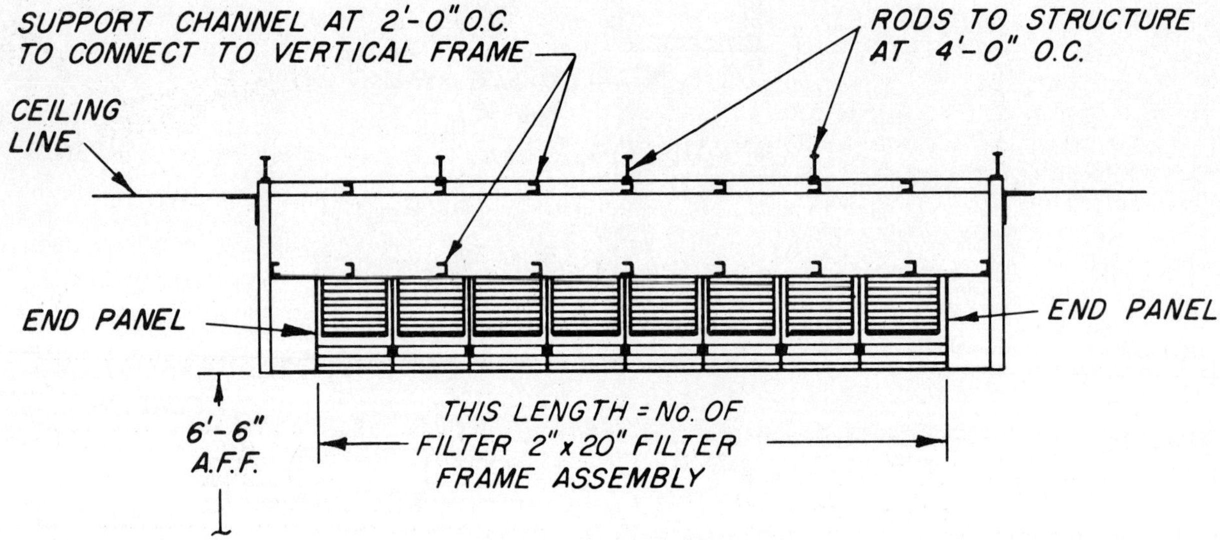

Fig. 15-21. A sectional view of the kitchen hood in Fig. 15-11.

UNIT 16

Ventilation for Garages and Paint Shops

OBJECTIVE: To discuss the problems involved in ventilating commercial garages and paint shops

GENERAL

The action of CO_2 (carbon dioxide) on the human system is harmless in small quantities, but in large amounts, it causes difficulty in breathing. In commercial garages where exhaust fumes from autos are common, CO_2 combined with CO (carbon monoxide) can cause serious illness and even death, if proper ventilation is not provided. Fumes from paints and lacquers used in automobile body or paint shops can also cause illness if the area is improperly ventilated.

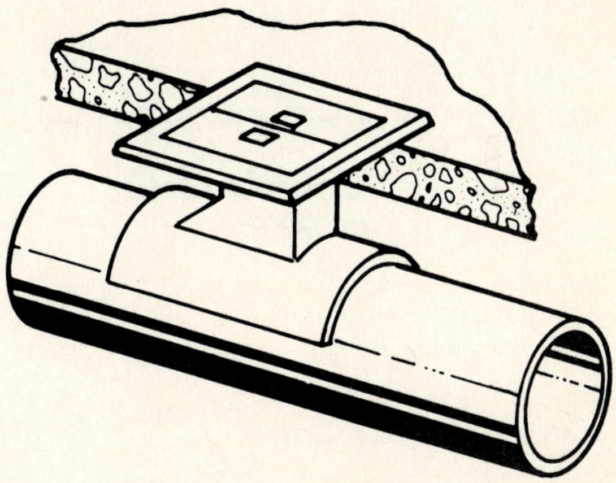

Fig. 16-1. A section of transite duct.

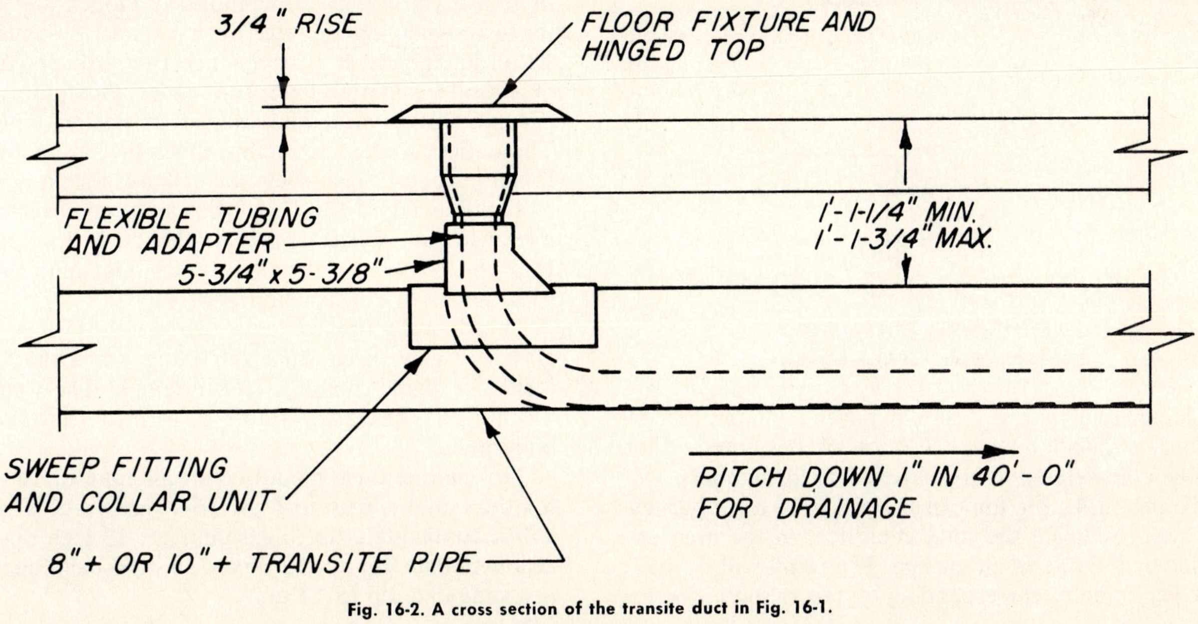

Fig. 16-2. A cross section of the transite duct in Fig. 16-1.

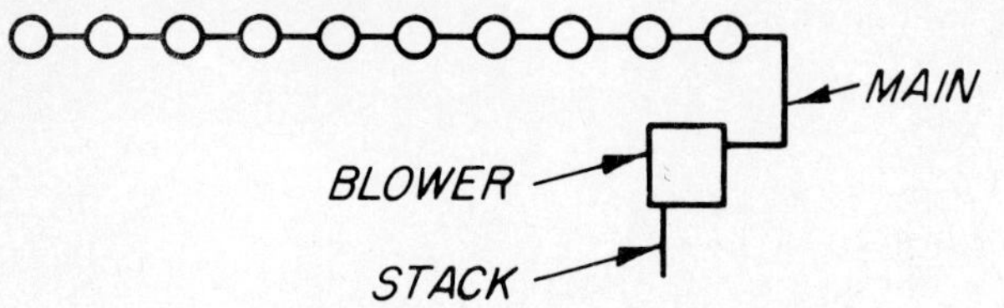

(A) Single-branch-type system.

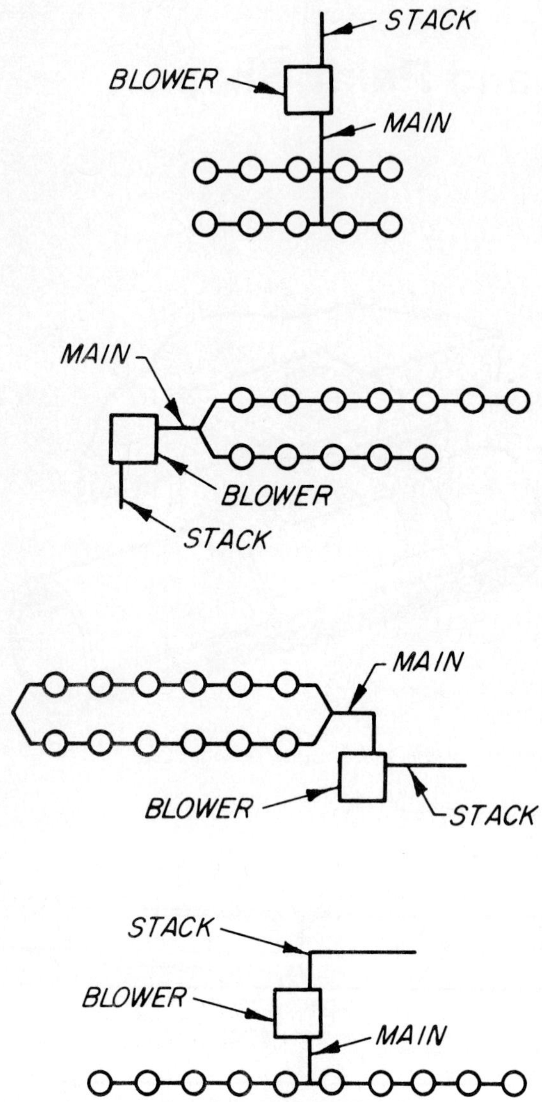

(B) Multiple-branch-type systems.

Fig. 16-3. A variety of branch layouts.

Commercial garages should have a minimum continuous circulation of 1 CFM of air per square foot, or one complete air change every 5 to 10 minutes.

To determine the fan capacity required for a particular area, calculate the total cubic feet in the area and divide by the rate of air change. The result will be cubic feet per minute, corresponding to fan ratings. For example, A garage, 32 feet wide, 120 feet long, and 16 feet high, is to be ventilated. How much air must be moved per minute?

$$32 \times 120 \times 16 = 61{,}440 \text{ cubic feet}$$

$$\frac{61{,}440}{5} = 12{,}288 \text{ CFM}$$

In many commercial garages, firehouses, and similar areas where carbon monoxide fumes are present, ventilation is provided by one or more power roof ventilators or wall fans. This method is desirable when it is impractical to attach tubes directly to the exhaust pipes of the vehicles. Although some odor will be present as the air moves from the vehicle exhaust pipe to the exhaust fan, the fumes do not linger long enough to cause harm to human beings, provided the system is designed properly.

In areas where it is practical, direct-connecting flexible tubes are the safest method to use. In garage areas, maintenance shops, firehouses, etc., these flexible hoses can be connected directly to the vehicle exhaust pipe prior to starting the engine. This method consumes the fumes directly and leaves much less odor than the method previously mentioned. However, in the case of firehouses, the flexible hose connection is designed to break free from the exhaust pipes as the truck leaves the building and, of course, some fumes will be present as this disconnection occurs. Therefore it is recommended that a second means of exhaust, such as a power roof ventilator, be provided for this purpose.

The remaining portion of this unit will deal with the direct-connection type of garage exhaust systems.

UNDERFLOOR EXHAUST SYSTEMS

Underfloor exhaust systems are ideal for garage ventilation because they are completely hidden and out of the way when not in use. Such exhaust systems consist of an exhaust fan from which round ducts (usually of transite material) are run under the floor of the garage. Outlets are then tapped at desired intervals. The outlets contain retracting tubes that are connected to auto exhaust pipes when the exhaust system is in use.

The drawing in Fig. 16-1 shows a section of transite duct with an outlet tapped to the flush floor fixture. Fig. 16-2 shows a cross section of this same drawing with details of construction.

The floor fixture contains two stainless steel, fiberglass, or galvanized dual retracting tubes that are 3 inches in diameter and 6 feet in length. The frame for the hinged self-closing doors is formed from ¼-inch boilerplate.

The transite duct should be a minimum of 10 inches for the outlet shown in Figs. 16-1 and 16-2. This size will accommodate up to 10 units. A 12-inch duct will accommodate up to 12 units and 14-inch duct will accommodate up to 18 units.

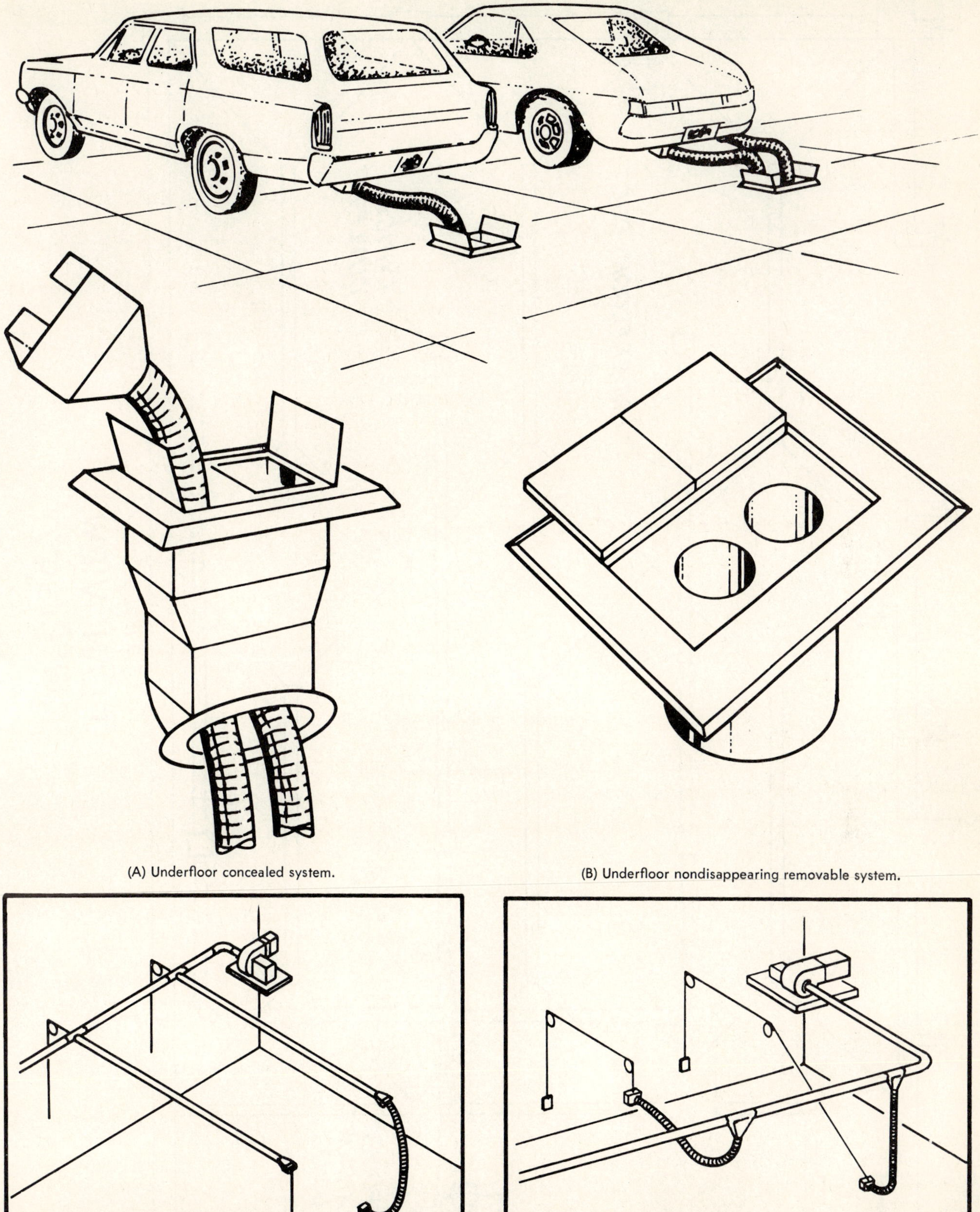

(A) Underfloor concealed system.

(B) Underfloor nondisappearing removable system.

(C) Overhead disappearing system.

(D) Overhead exposed system.

Fig. 16-4. Four types of carbon-monoxide fume exhaust systems.

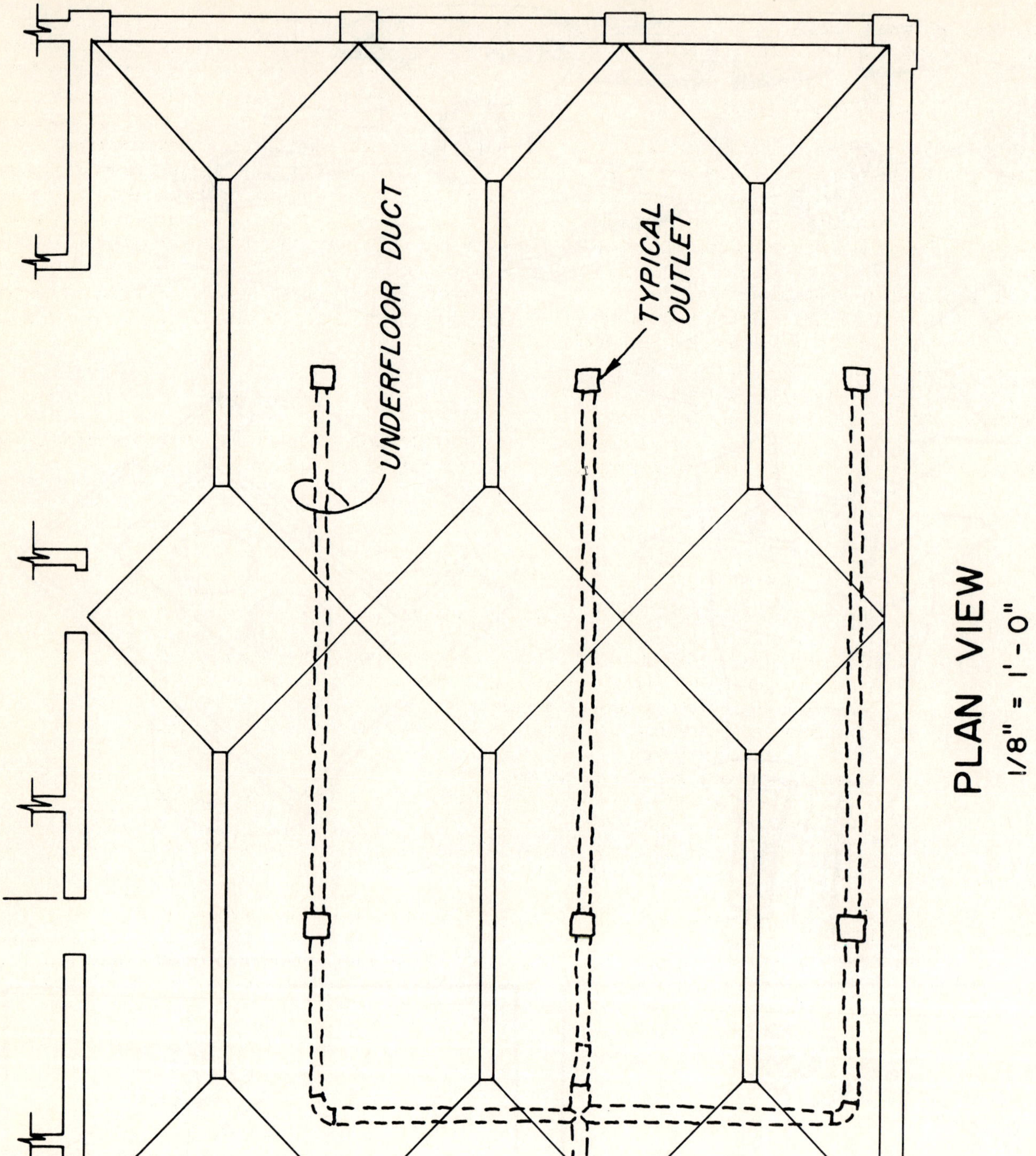

Fig. 16-5. Engineer's sketch of a rescue squad building.

Fig. 16-3 illustrates five different layouts (in schematic form) of underfloor exhaust systems.

OVERHEAD SYSTEMS

Overhead exhaust systems for garages are well suited for existing buildings where installation of an underfloor system would require breaking up the existing concrete floor. The overhead exhaust system can result in a neat installation and has the following features:

1. No tubes hang in the open.
2. The tubes retract within their own chambers by means of automatically closing dampers which prevent loss of heated air from the building.
3. The tubes have steel rollers that enable the tube heads to roll up and down a flat floor.

Overhead exhaust systems are the most practical system to use on vertical stock trucks and tractors. They provide an economical and efficient system in any existing building where exhaust fumes are a problem, such as in fire stations, commercial garages, rescue squad buildings, tractor-trailer maintenance buildings, vocational schools, and gasoline-engine repair shops.

It is recommended that an additional exhaust system be provided in areas using the duct and tube system; the added exhaust system rids the area of fumes from the time the tube is detached from the engine exhaust until the engine is removed from the area. However, the added exhaust system, such as a ceiling exhaust fan, need not be nearly as large as it otherwise would have to be.

SUMMARY

The four illustrations in Fig. 16-4 show four types of carbon-monoxide-fume exhaust systems.

UNIT 16 — ASSIGNMENT

1. Prepare a sheet of 11-by-17-inch tracing paper by drawing border lines and lettering the title block.
2. Turn to the engineer's sketch in Fig. 16–5. This is a design for a rescue squad building. Using this rough sketch as a guide, draw a working drawing. Be as neat as possible. Be sure to include all windows, doors, door swings, etc.
3. Add a single-line duct diagram with floor outlets, the blower and stock, and all notes and dimensions that are necessary.
4. Answer the following True or False questions.

	True	False
a. Proper ventilation is very important because fumes from automobile exhausts and paint can cause serious illness or even death.	____	____
b. Commercial garages should have a minimum of 5 CFM of air per square foot.	____	____
c. In areas where it is practical, direct-connecting flexible tubes are the safest method to use.	____	____
d. Underfloor exhaust systems are the most practical to use on vertical stock trucks and tractors.	____	____
e. An advantage of an overhead exhaust system is that no tubes hang in the open.	____	____

UNIT 17

Ventilation for Toilets

OBJECTIVES: To describe the need for proper ventilation in bathrooms and toilets

To show how toilet ventilation details are shown on working drawings

GENERAL

In the average bathroom, the ordinary act of washing and bathing can add as much as 10 pounds of water a day to the air within the area. This can prove to be highly expensive since excessive moisture will stain

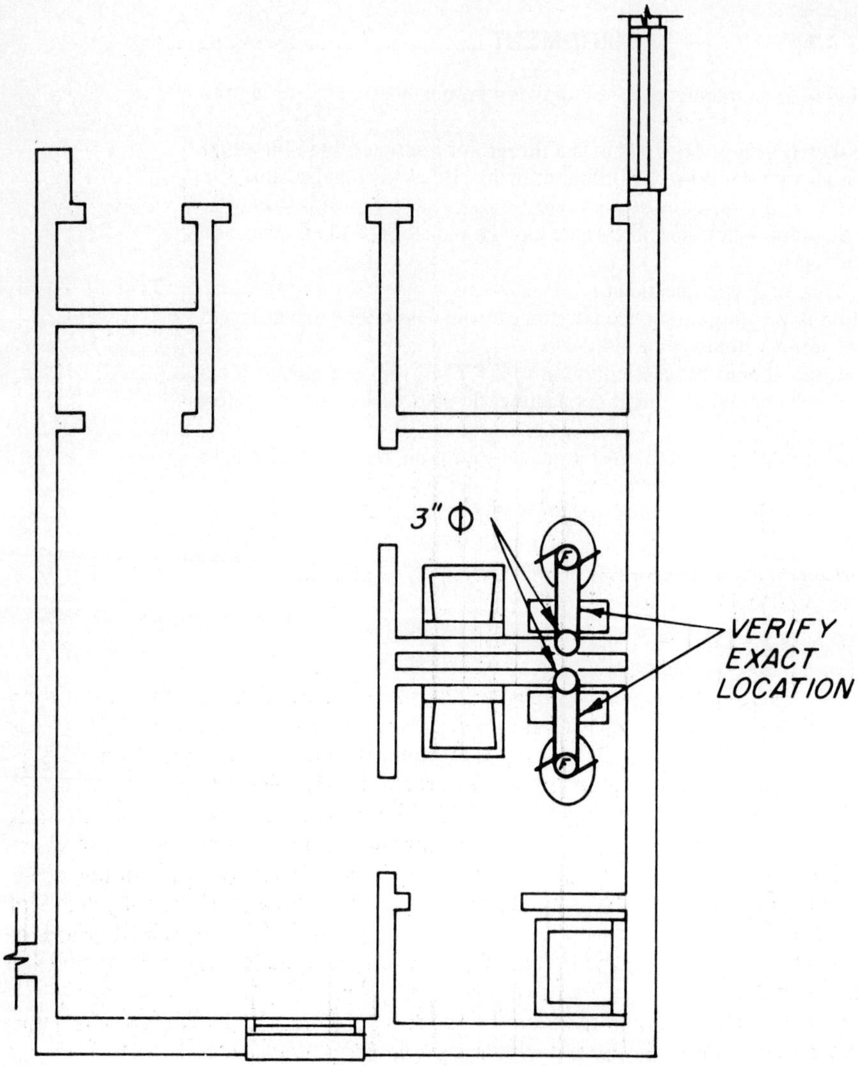

Fig. 17-1. Floor plan of toilet exhaust fans and related ductwork.

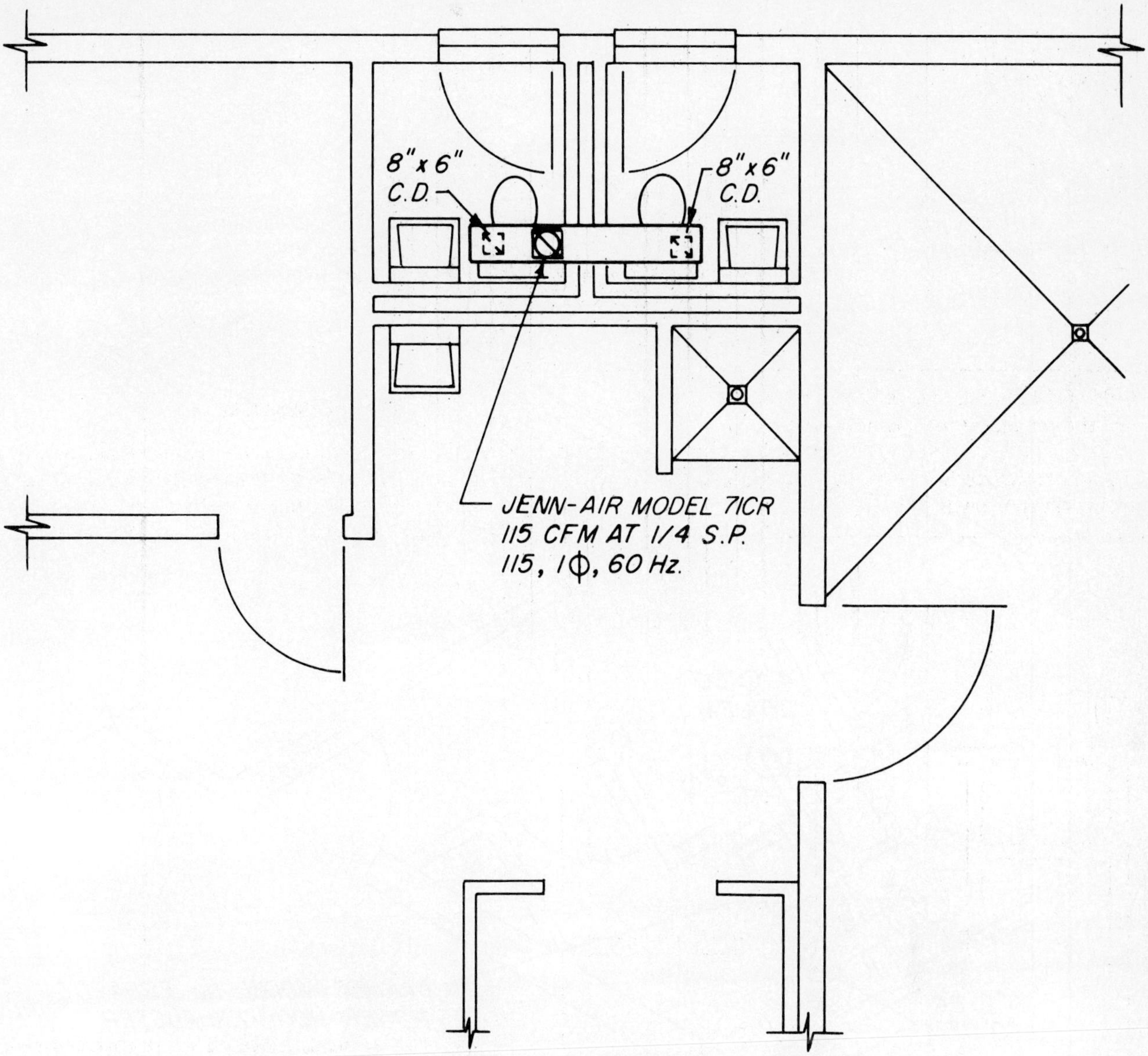

Fig. 17-2. Example of two toilets being ventilated with one exhaust fan and ductwork.

plaster, blister and peel paint, rot wood, rust metal, and cause tile to become unglued. It is obvious that this kind of problem must be solved.

VENTILATING FANS

The use of ventilating fans in bathrooms and toilets is the easiest way to eliminate excessive moisture. The fans are usually installed in either the ceiling or the wall and as close as possible to the source of moisture.

The size of the fan needed depends upon the size of the room to be ventilated, and the room size is found by multiplying the room width times the height times the length to determine the number of cubic feet within the room. The recommended amount of air to be removed per minute is then determined from design guides and the proper fan is selected.

The ASHRAE Handbook recommends that 2 cubic feet of air per minute should be exhausted for every square foot of floor space. Therefore, the amount of air that should be exhausted from each of the 5-by-4.5-foot toilets in Fig. 17-1 is $5 \times 4.5 \times 2 = 45$ CFM. Each of the ceiling fans shown on the floor plan in Fig. 17-1 should exhaust 45 CFM of air.

Again, by referring to Fig. 17-1, you can see that each fan uses a 3-inch-diameter duct to carry the air

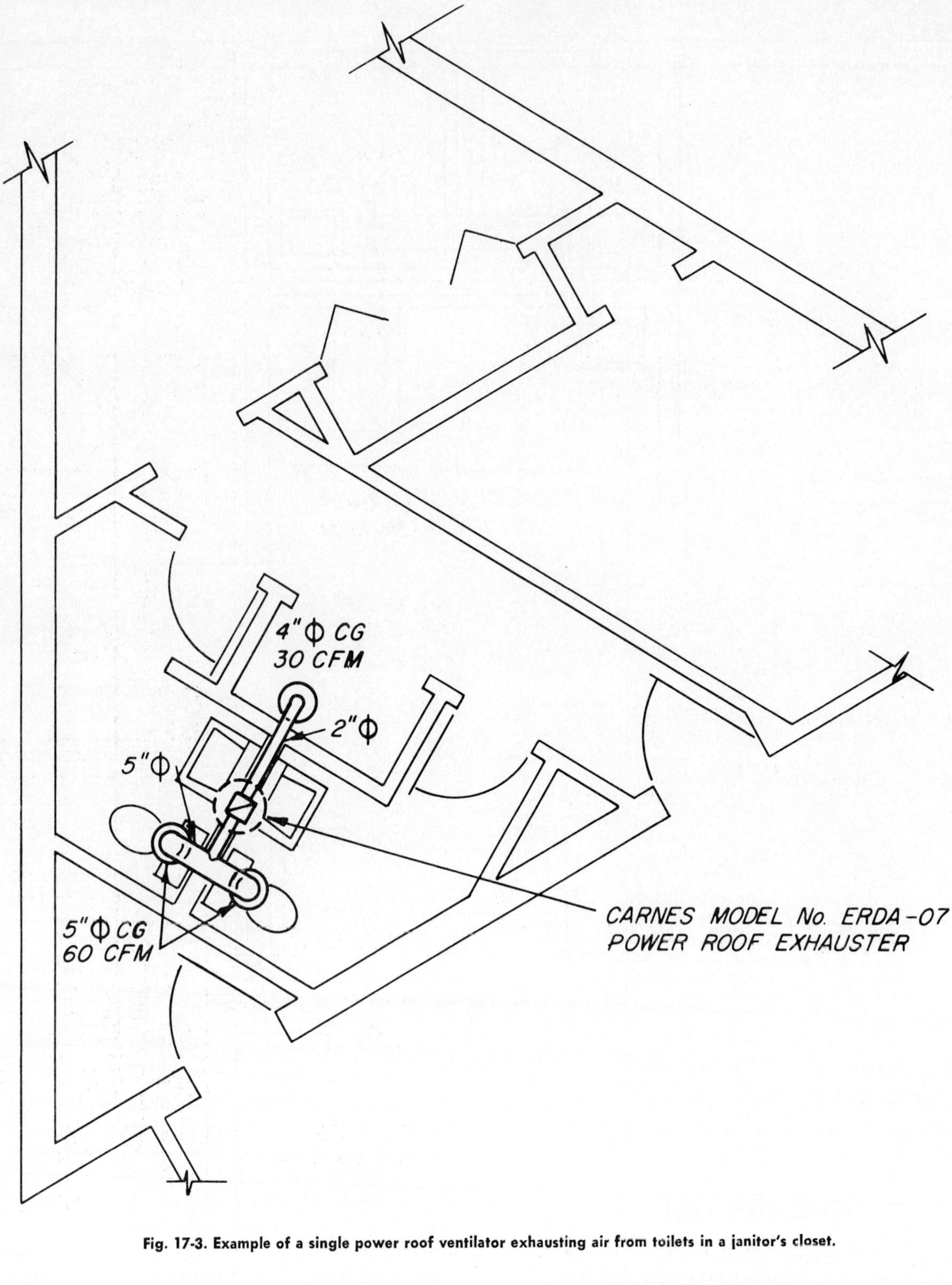

Fig. 17-3. Example of a single power roof ventilator exhausting air from toilets in a janitor's closet.

from the fan; each of the two ducts then turns up in a wall chase and terminates above the roof in a weatherproof roof cap. The latter is not shown on the floor plans but should be shown on the architect's roof plan.

The two toilets shown on the floor plan in Fig. 17-2 are approximately the same as those shown in Fig. 17-1. However, these toilets utilize only one fan to remove the air from the bathrooms. A section of duct is connected to the fan and is run to both areas. Ceiling

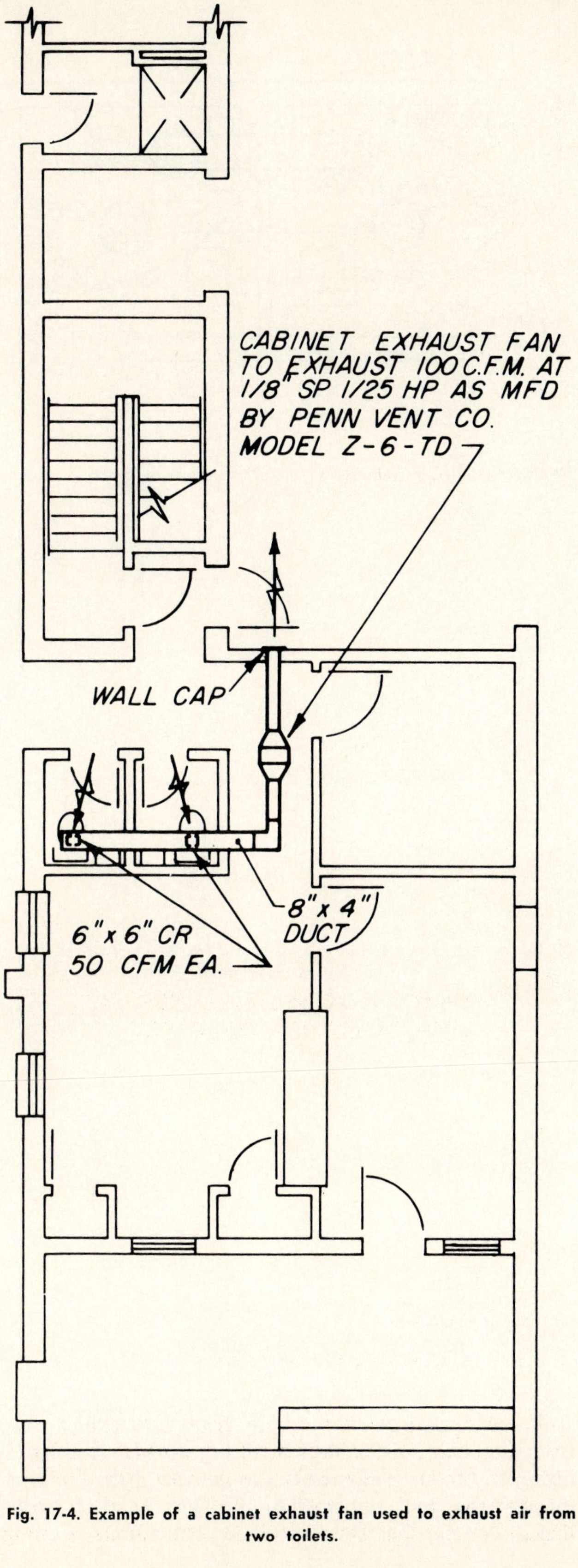

Fig. 17-4. Example of a cabinet exhaust fan used to exhaust air from two toilets.

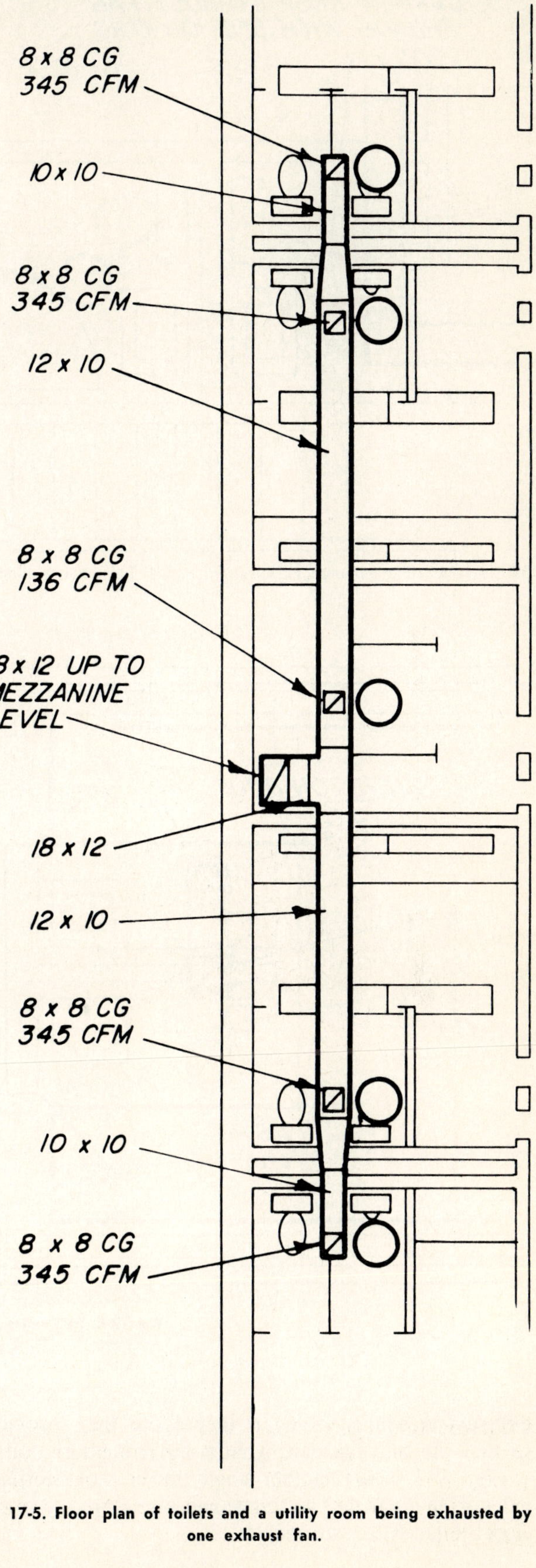

Fig. 17-5. Floor plan of toilets and a utility room being exhausted by one exhaust fan.

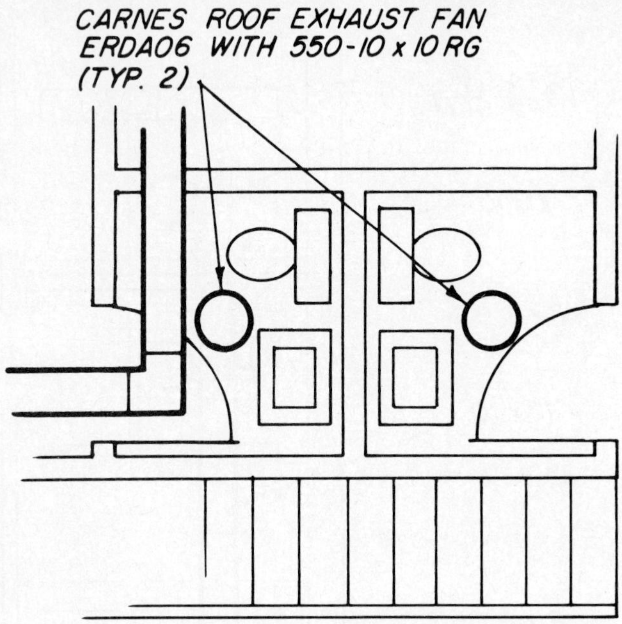

(A) Using a ceiling-mounted exhaust fan in each toilet for ventilation.

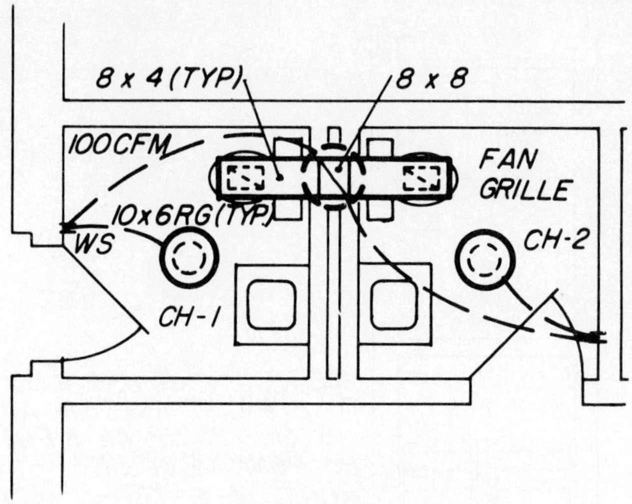

(B) Using a single power roof ventilator, ductwork, and two air outlets for ventilation.

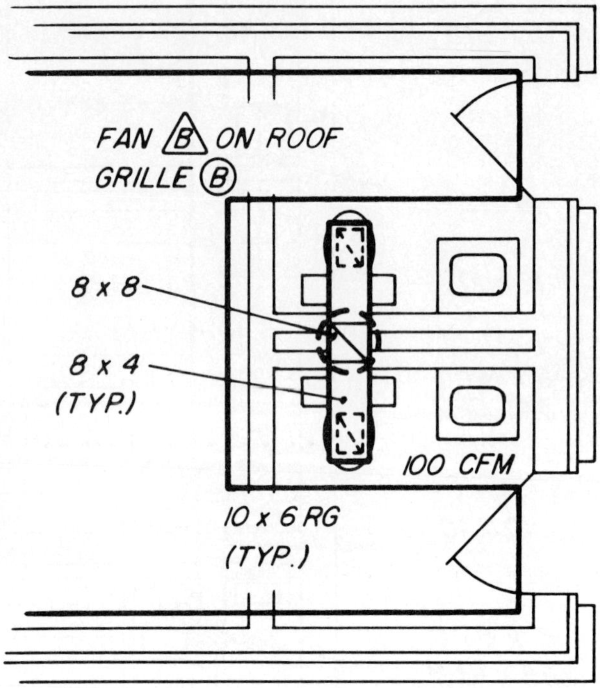

(C) Using a single power roof ventilator, ductwork, and two ceiling air outlets for ventilation.

Fig. 17-6. Two toilets using various types of ventilation.

diffusers measuring 8 by 6 inches are then provided so that the one fan can exhaust air from both toilets.

Fig. 17-3 shows another single power roof ventilator exhausting air from two toilets and also from a janitor's closet adjacent to the toilets.

A single cabinet exhaust fan is used to exhaust air from the two toilets shown in Fig. 17-4. This time, however, the air is carried to the outside of the building through the wall rather than through the roof. Note that a wall cap has been specified at this point.

The floor plan shown in Fig. 17-5 shows several toilets and a utility room exhausted by one exhaust fan. This fan is not shown on the floor plan since it is a power roof ventilator, but the 18-by-12-inch duct, shown on the drawings, turns up through the mezzanine area and then connects to the fan on the roof. Unit 14 shows some power-roof-ventilator details which should be referred to at this time.

Fig. 17-6 shows floor plans of other toilet exhaust fans.

SUMMARY

The use of ventilating fans in bathrooms and toilets is the easiest way to eliminate excessive moisture and maintain clean air within the area.

Most toilet fans are exhausted through the roof or wall.

The size of the fan needed for bath or toilet depends upon the size of the area to be ventilated.

The ASHRAE Handbook recommends that 2 CFM of air be exhausted for every square foot of floor space.

UNIT 17 — ASSIGNMENT

1. Prepare a sheet of 11-by-17-inch tracing paper with borders and title block; turn to Appendix D.
2. Calculate the required fan size (in CFM) needed for the floor plans.
3. On the left side of your paper, draw the bathroom only of the residence to a scale of 1 inch = 1 foot. Also design a ceiling-fan layout for this bath similar to the one in Fig. 17–1 and draw it on the floor plan.
4. On the right side of your paper, draw the two restrooms shown on the floor plan.
5. Design a bath exhaust system and draw it on the floor plan. You may use Fig. 17–2 as a guide.
6. Answer the following True or False questions.

	True	False
a. Ordinary activities such as bathing can add amounts of water to the air that can damage plaster and tile.	____	____
b. The easiest way to eliminate excess moisture in bathrooms is to use ventilating fans.	____	____
c. Most toilet fans are exhausted through the floor.	____	____
d. The size of the fan needed depends on the size of the area to be ventilated.	____	____
e. It is recommended that 1 CFM of air be exhausted for every square foot of floor space.	____	____

UNIT 18

Ventilation for Household Appliances

OBJECTIVES: To describe some requirements for the ventilation of household appliances

To show how the ventilation of these appliances is shown on HVAC working drawings

To teach the student how to draw working drawings for household-appliance ventilation systems

INTRODUCTION

There are several residential appliances that require ventilation for proper and safe operation. The most widely used of these appliances are clothes dryers, electric ranges and barbecue units and central cleaning systems.

CLOTHES DRYER

The pictorial drawing in Fig. 18-1 shows a clothes dryer installed in the corner of a room inside a house. A 4-inch flexible duct is connected to the discharge outlet on the dryer and run to a wall cap which discharges the air and lint to the outside of the building. Note that the wall cap has a shield to protect the duct from the outside weather.

The drawing in Fig. 18-2 shows how a dryer vent and wall cap are shown on the floor plan of an actual

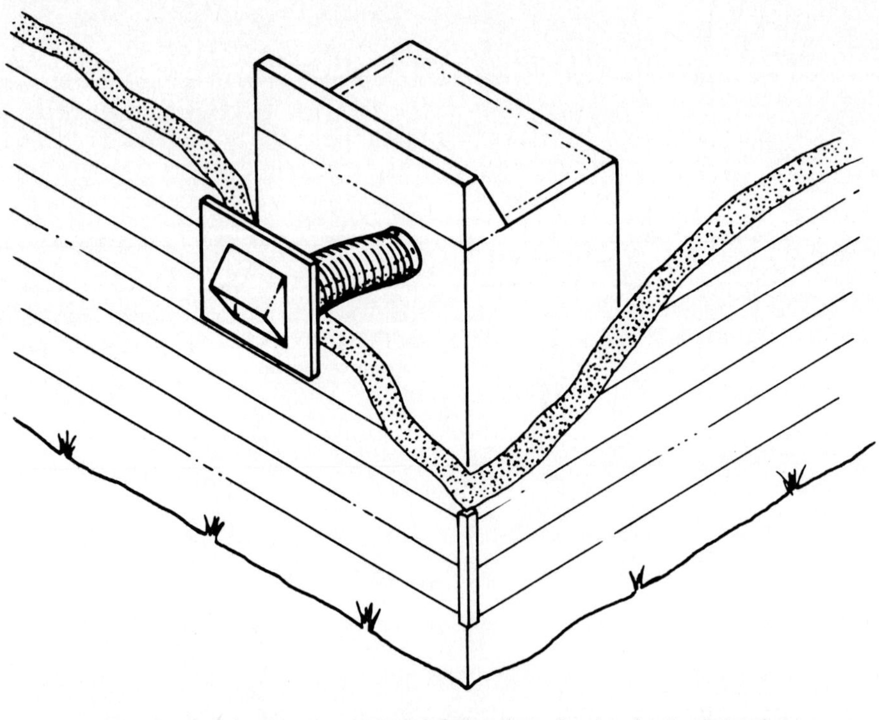

Fig. 18-1. Pictorial drawing of a residential clothes dryer showing details of air exhaust.

working drawing. This application is similar to the one shown in Fig. 18-1, except that the exhaust outlet is on the side of the dryer and not the back.

Fig. 18-3 shows another floor plan with the clothes dryer vented. This time the dryer is exhausted to the crawl space area under the house and not to the outside. Such an installation should be avoided whenever possible, but it is acceptable if no other means of venting the dryer is possible.

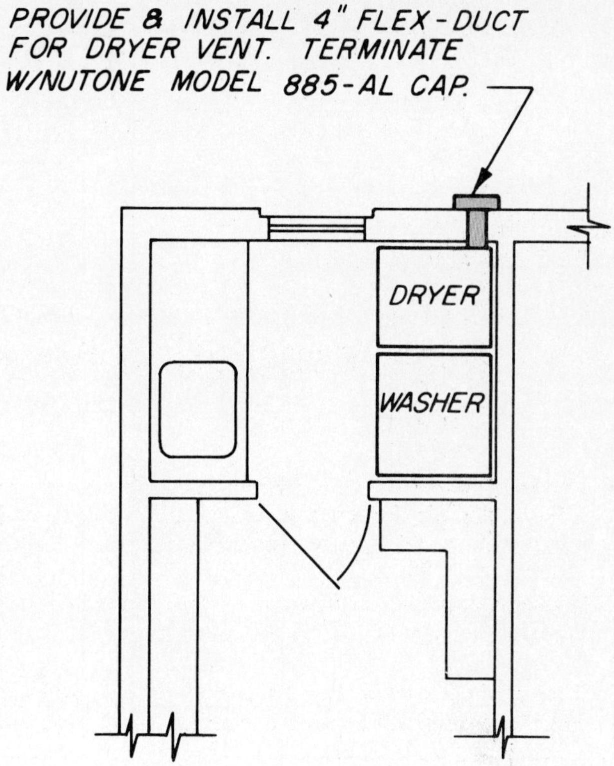

Fig. 18-2. Floor plan of a laundry room showing dryer exhaust.

ELECTRIC RANGE AND BARBECUE UNIT

A wall elevation of a residential kitchen can be seen in Fig. 18-4. A stone fireplace is shown approximately in the center of the base cabinet. This will house a factory-built electronic barbecue unit. A counter is shown on the left of the stone fireplace; the space to the right will contain an electric range.

A custom-built copper hood is shown extending the entire width of the wall. This hood will provide ventilation for the barbecue and the electric range by means of a factory-built blower and a 10-inch-diameter exhaust duct, as shown by the broken lines.

Fig. 18-5 shows this same kitchen wall in a floor plan view. The elevation of this wall, as shown in Fig. 18-4, would normally appear in the architectural section of the drawings, while the floor plan (Fig. 18-5) would usually appear on the HVAC drawings with a note referring the users of the plans to the elevation in the architectural section.

CENTRAL CLEANING SYSTEM

While not usually thought of as a ventilating system the central cleaning system installed in many modern residences today uses air to make household chores much easier.

A central power unit, shown in Fig. 18-6, is installed in a central location within the house, usually in the basement, the garage, or a closet. These units contain a one- or two-horsepower motor, and most utilize a 6-gallon disposable soil bag which is easily replaced. A complete loss of vacuum tells the user that the bag needs replacing.

Two-inch-diameter semirigid tubing is run from the central unit to automatic inlets throughout the house. When the flexible cleaning hose is attached to the inlets, the central unit is automatically energized and cleaning can take place anywhere and in any direction within 25 feet of the inlet. Fig. 18-7 shows a pictorial view of such a system in operation.

Exhaust to the outside is recommended (as a guard against possible odors from cigarette butts and/or food particles collected in the dust bag) but is not necessary for efficient operation.

SUMMARY

Many household appliances need to be ventilated in order to remove odors, harmful fumes, smoke, etc., and to allow the appliances to operate properly.

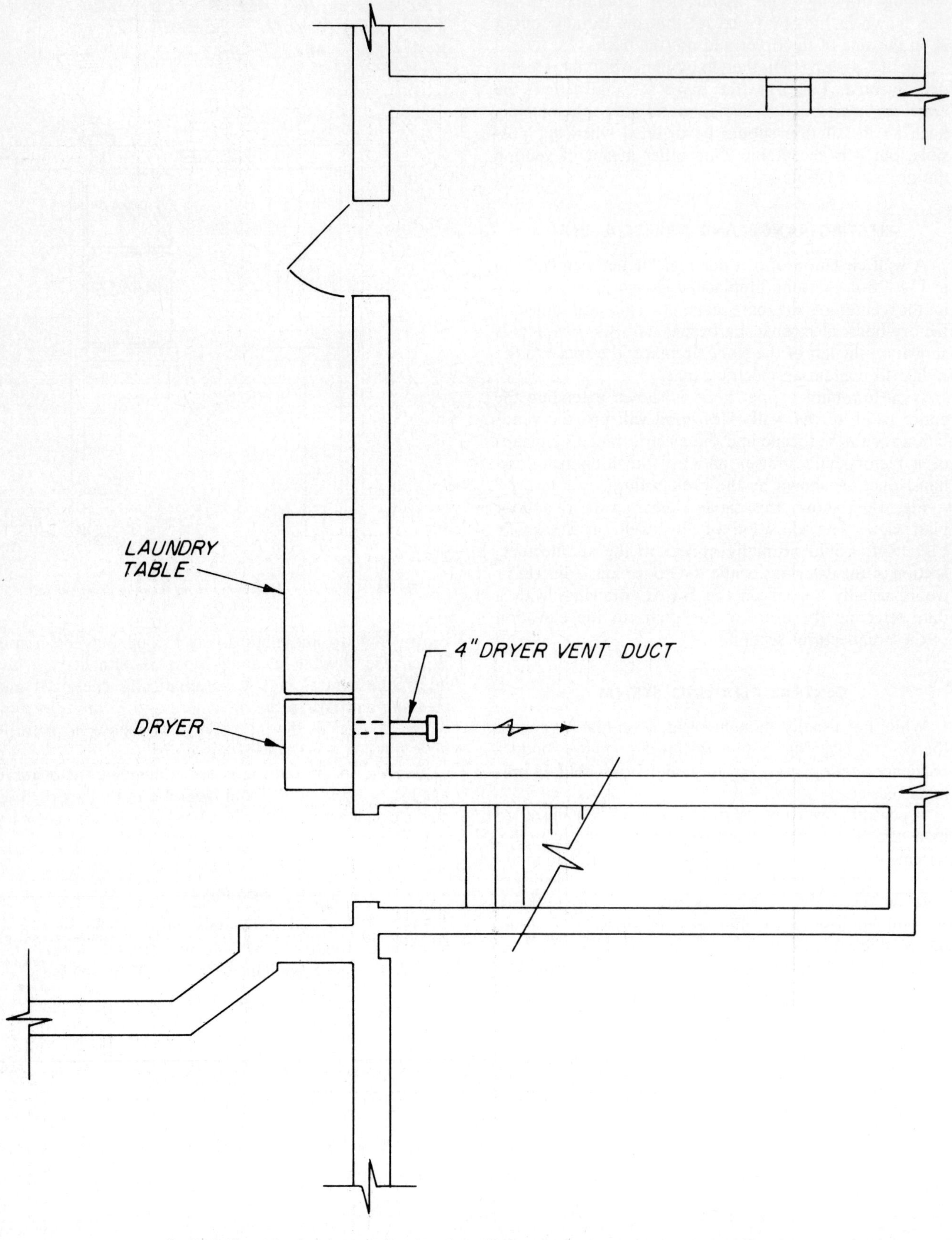

Fig. 18-3. Floor plan showing a clothes dryer being exhausted into the crawl space under the house.

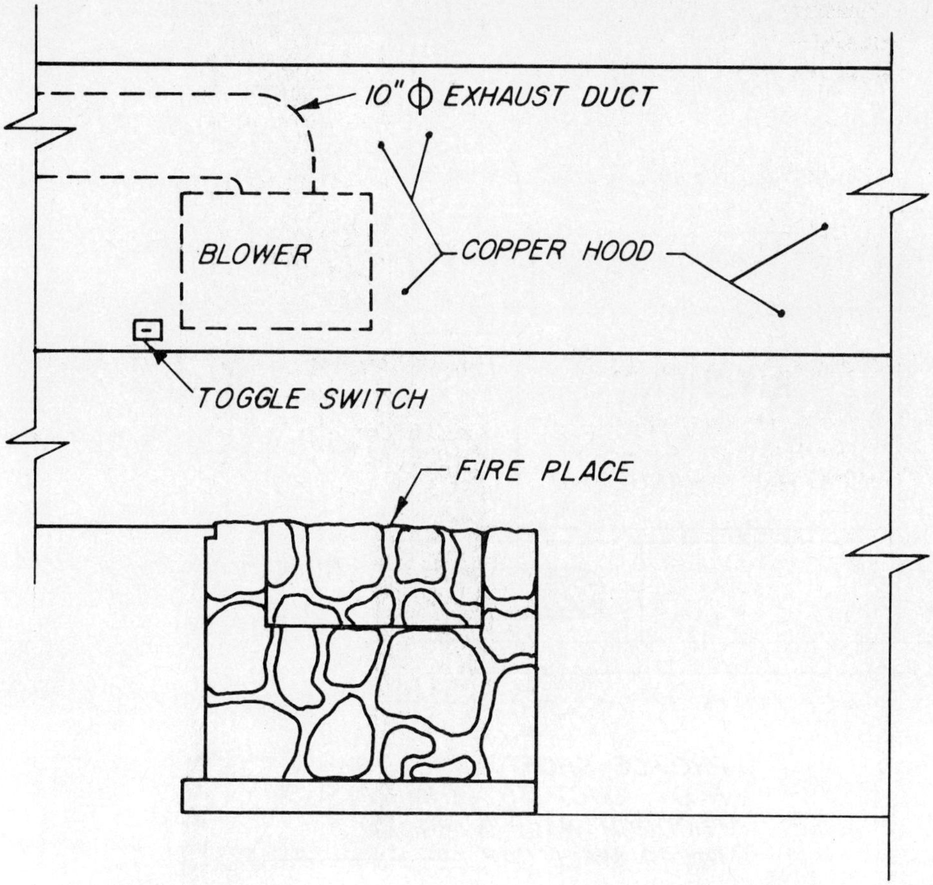

Fig. 18-4. Sectional drawing showing an exhaust system for an indoor barbecue.

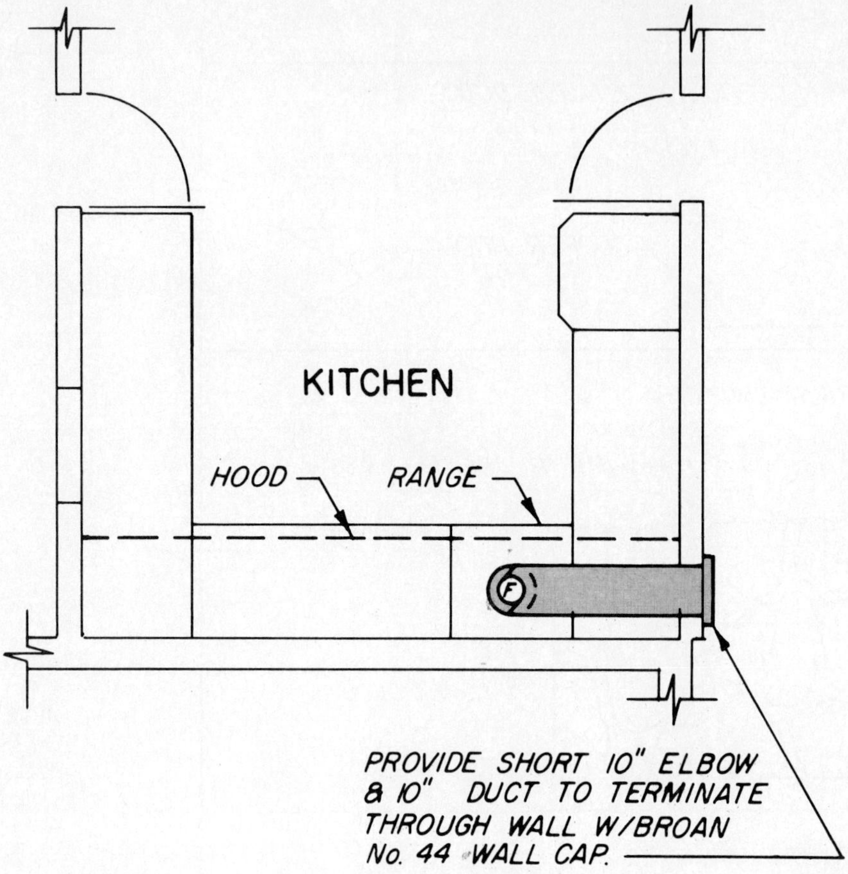

Fig. 18-5. Plan view of Fig. 18-4.

Fig. 18-7. Pictorial view of a central exhaust system in use.

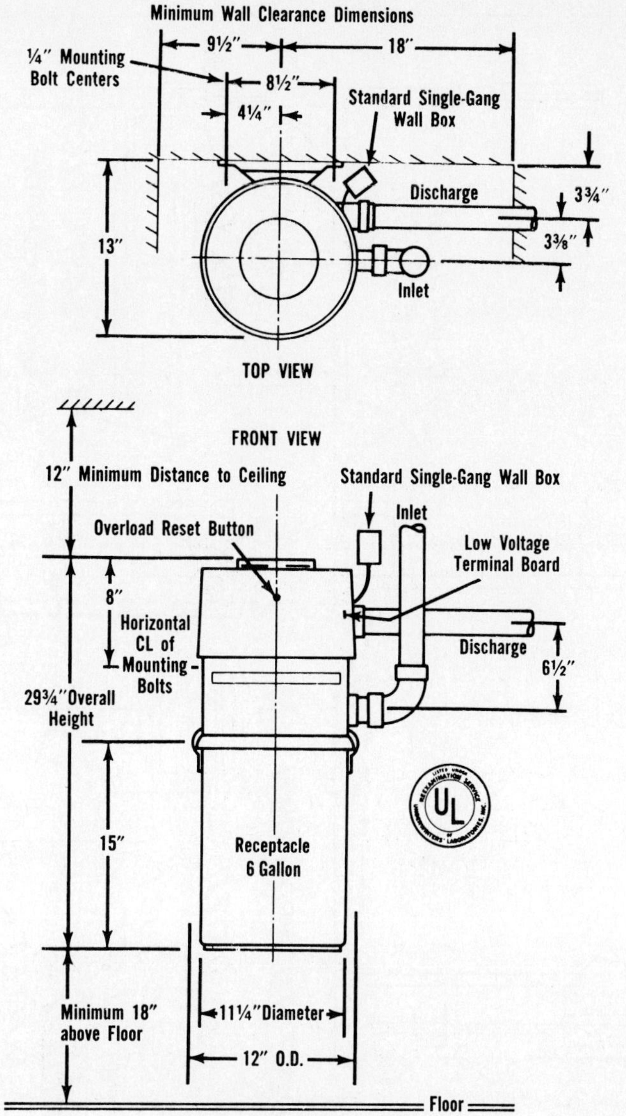

Courtesy Nutone, Division Scoville Manufacturing Co.

Fig. 18-6. Plan and sectional views of a central cleaning system.

UNIT 18 — ASSIGNMENT

1. Prepare one sheet of 11-by-17-inch tracing paper with border lines and title block; turn to Appendix D.
2. Trace the background of the residence that is shown.
3. Add the central cleaning system, clothes dryer vent, and range hood vent to the background you just completed. Use the sketch in Fig. 18–8 as a reference.
4. Answer the following True or False questions.

	True	False
a. Exhausting a clothes dryer to the crawl space under the house is always best if possible.	_____	_____
b. It is recommended that a central cleaning system be exhausted to the outside but it isn't necessary for efficient operation.	_____	_____

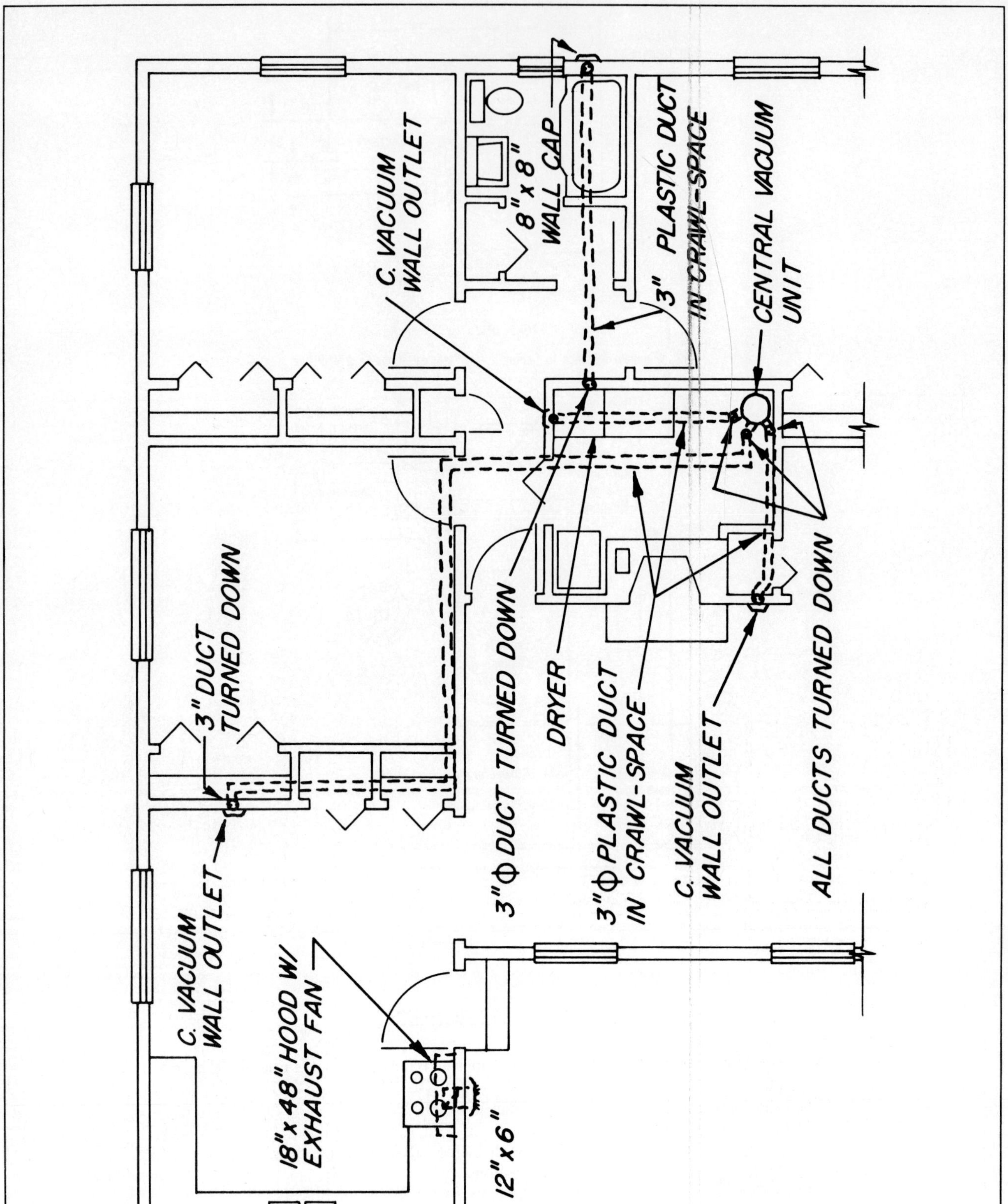

Fig. 18-8. Engineer's sketch for the assignment in this unit.

UNIT 19

Air-Handling Units

OBJECTIVES: To discuss the types of air-handling units used to distribute air in HVAC systems

GENERAL

Fans are used in air-handling units for the circulation of air in HVAC systems. They are manufactured in the four general types illustrated in Fig. 19-1.

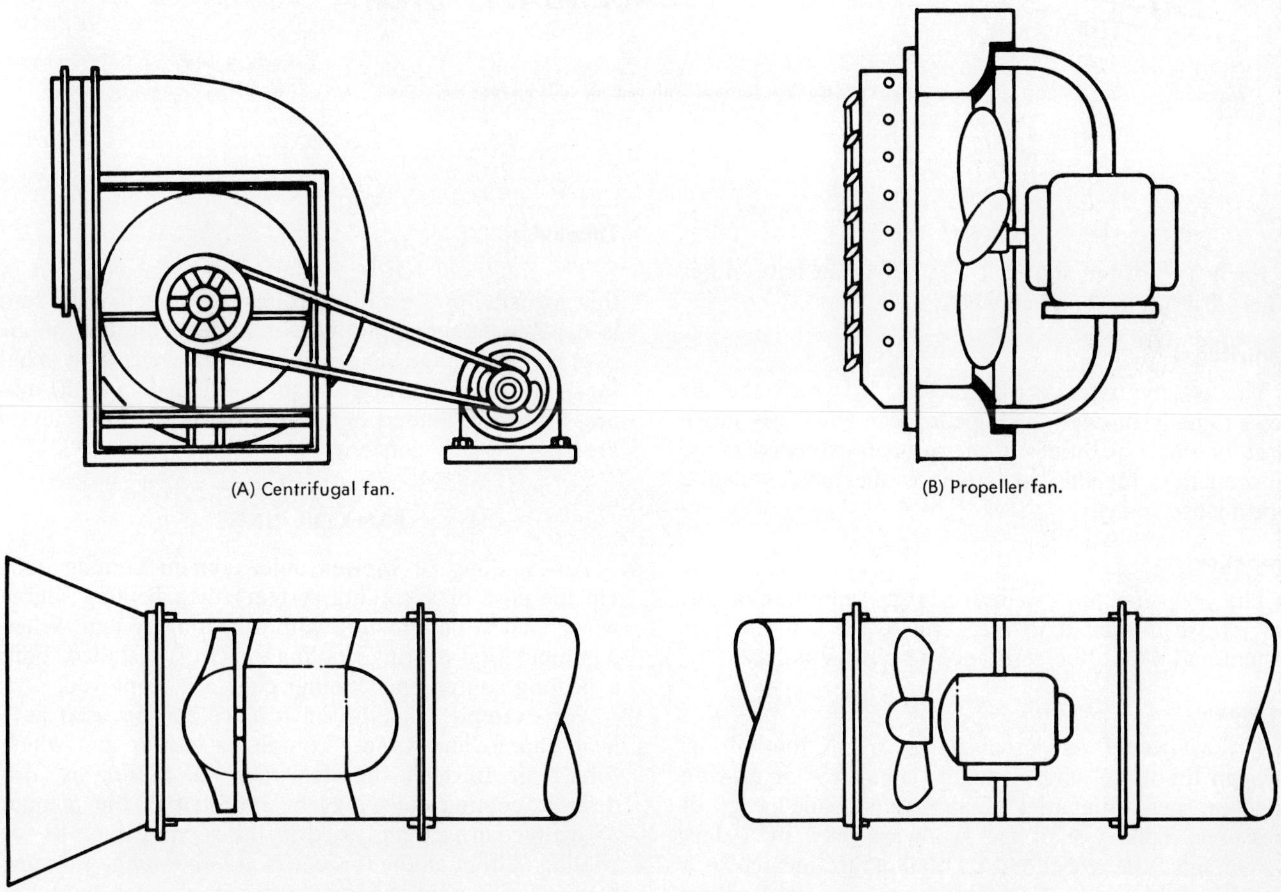

(A) Centrifugal fan.

(B) Propeller fan.

(C) Vaneaxial fan.

(D) Tubeaxial fan.

Fig. 19-1. Four general types of fans used in HVAC systems.

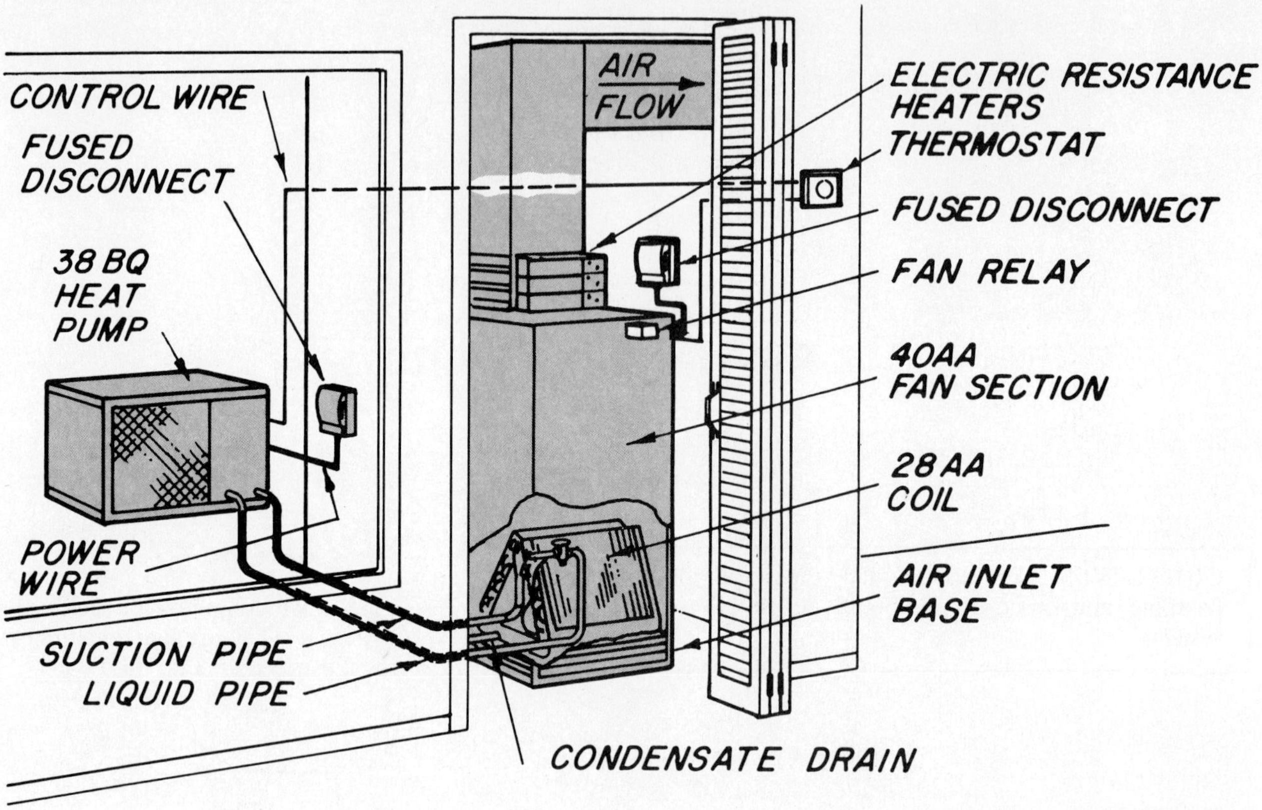

Courtesy Carrier Air Conditioning Co.

Fig. 19-2. Up-flow furnace with cooling coils on return-air side.

FANS

Each type of fan shown in Fig. 19-1 can have either a belt drive or a direct connection.

Centrifugal

The widely used *centrifugal fan,* in which the air flows radially through the impeller, can efficiently move large or small quantities of air over an extended range of pressures; for this reason, it is the most versatile type in use today.

Propeller

The *propeller fan* can move large quantities of air, but its use is limited to areas where there is no duct system and where the resistance to airflow is low.

Vaneaxial

The *vaneaxial fan* produces an axial flow of air through the wheel and blades. It is capable of moving large or small quantities of air over a wide range of pressures. This type of fan is always used in a duct system since the wheel and its blading are located in a cylindrical housing. Air-guide vanes are used either before or after the wheel.

Tubeaxial

The *tubeaxial fan* is similar to the vaneaxial fan in that the air flows axially through the impeller. It, too, is capable of moving small or large quantities of air over a wide range of pressures. However, most axial fans produce more noise than centrifugal fans do and are, therefore, limited to use in areas where noise levels are of secondary concern.

FAN-COIL UNITS

Air-handling or blower units contain cooling coils (in the case of a cooling system) or a heating source when used in conjunction with a heating system. When a combination heating/cooling system is installed, both a heating source and cooling coils are employed.

For example, an oil-fired furnace can be used as a heat source. The furnace contains a blower unit which forces air through the ductwork. If cooling is also desired, cooling coils may be installed in the plenum above the furnace and used in the summer months for cooling. Although the furnace is not operating, the same blower fan is used to force "cool" air through the ductwork.

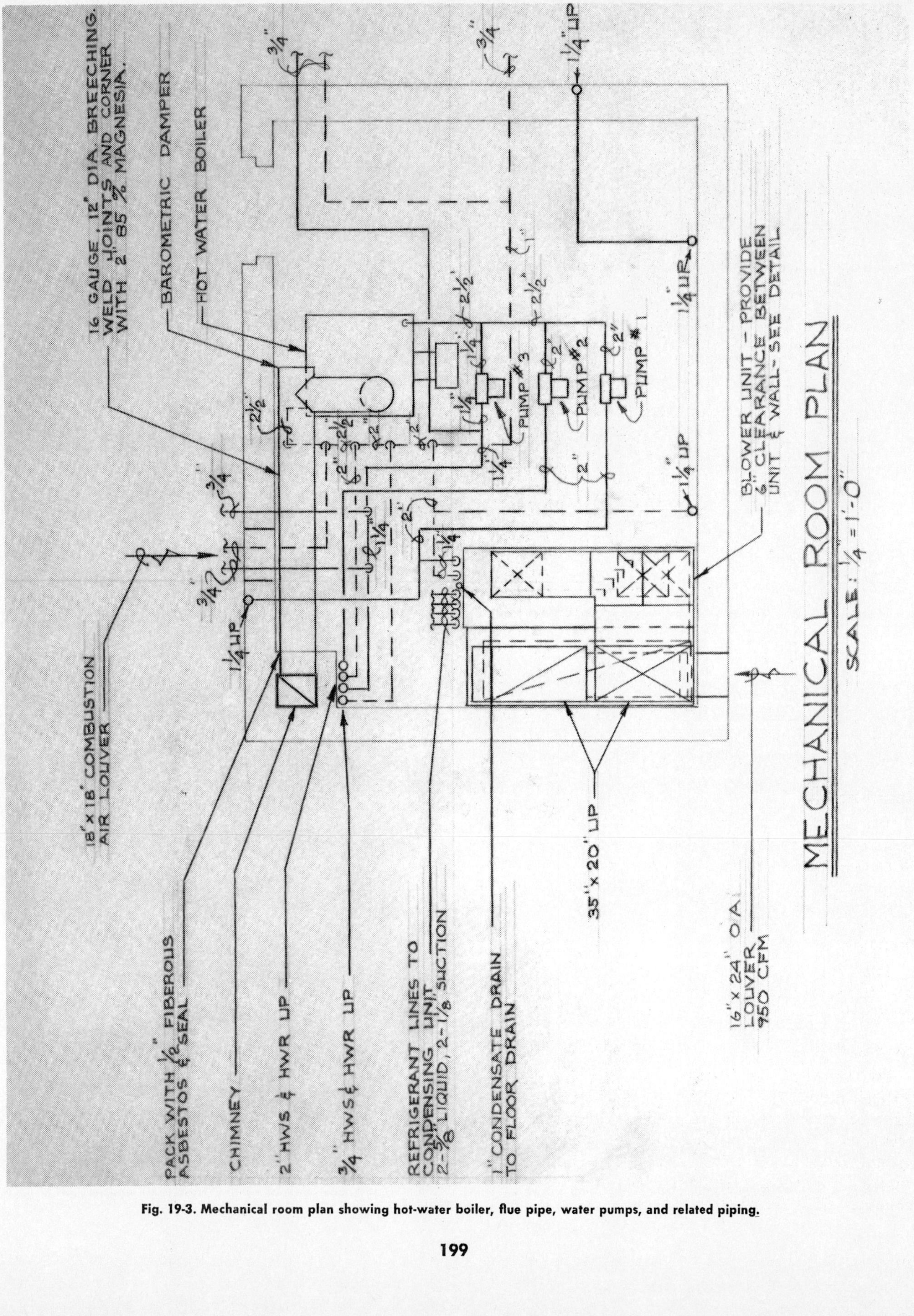

Fig. 19-3. Mechanical room plan showing hot-water boiler, flue pipe, water pumps, and related piping.

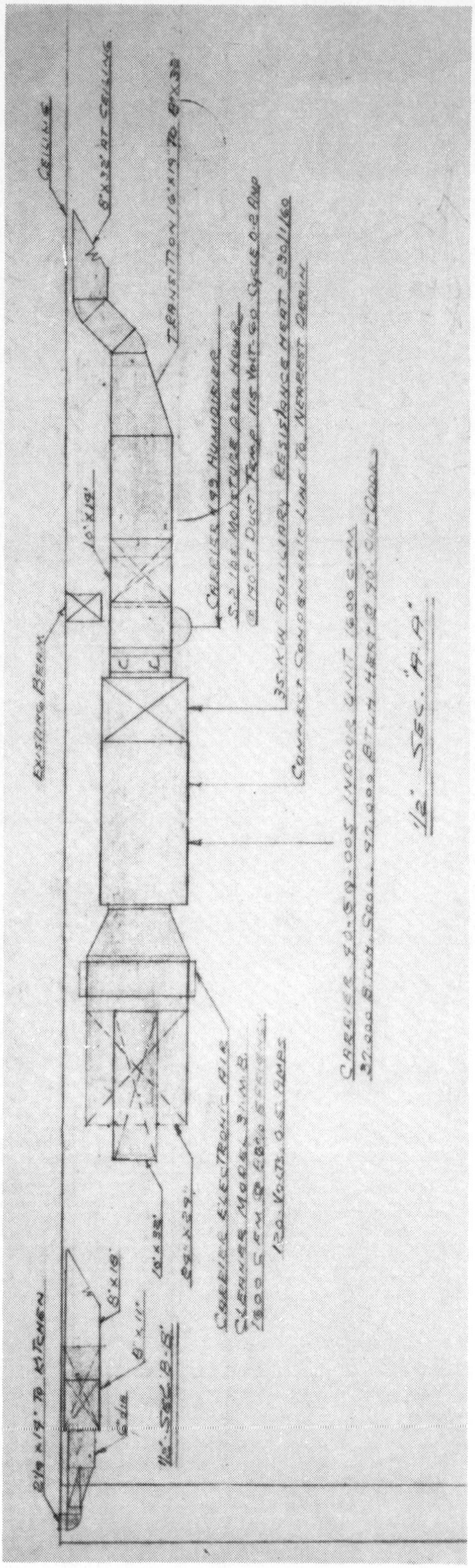

Fig. 19-4. Sectional view of ceiling-hung air-handling unit with related ductwork.

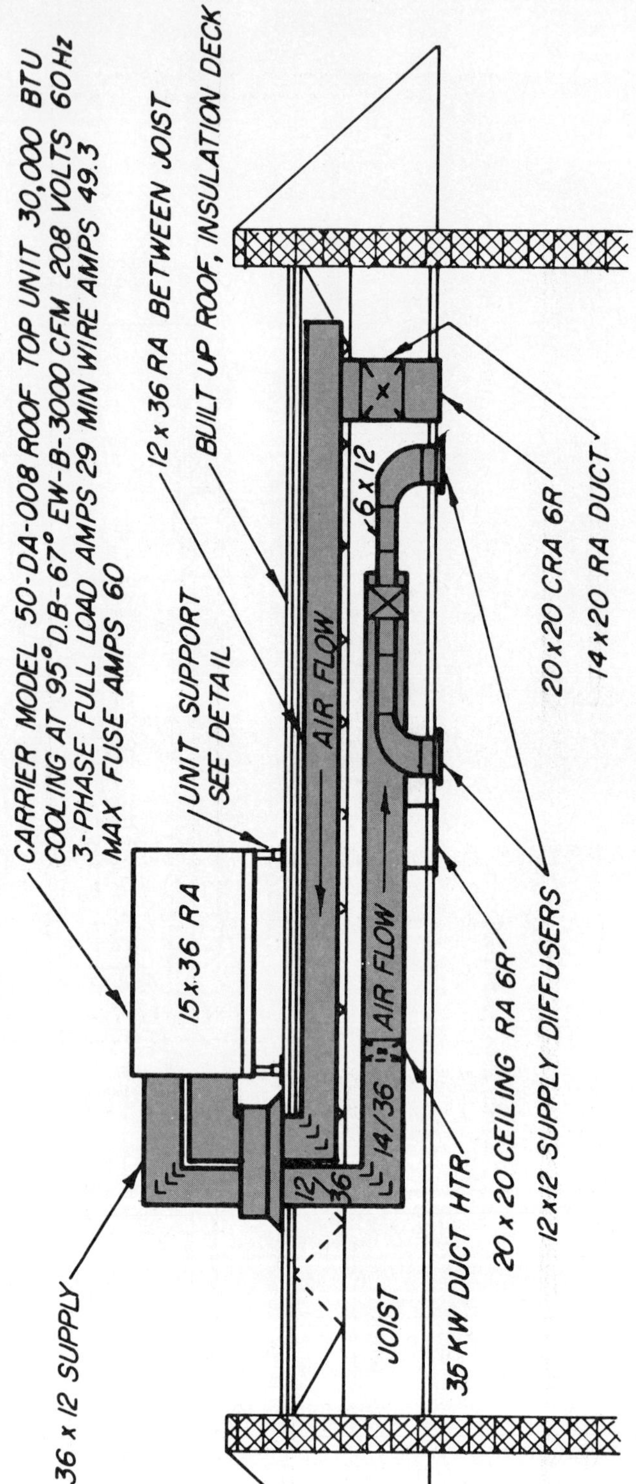

Fig. 19-7. Sectional view of rooftop HVAC unit with related ductwork, diffusers, grilles, etc.

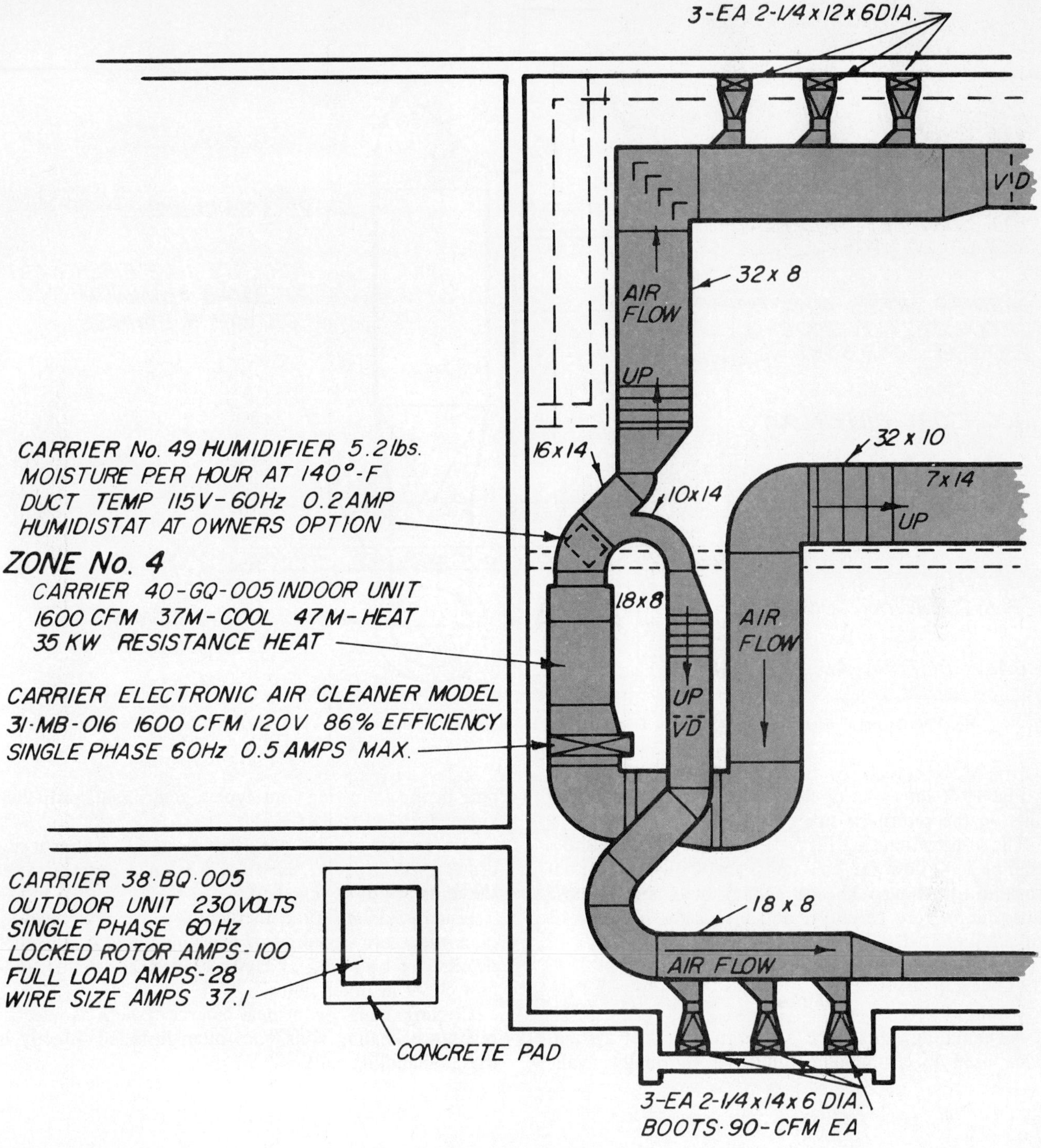

Fig. 19-5. Plan view of Fig. 19-4.

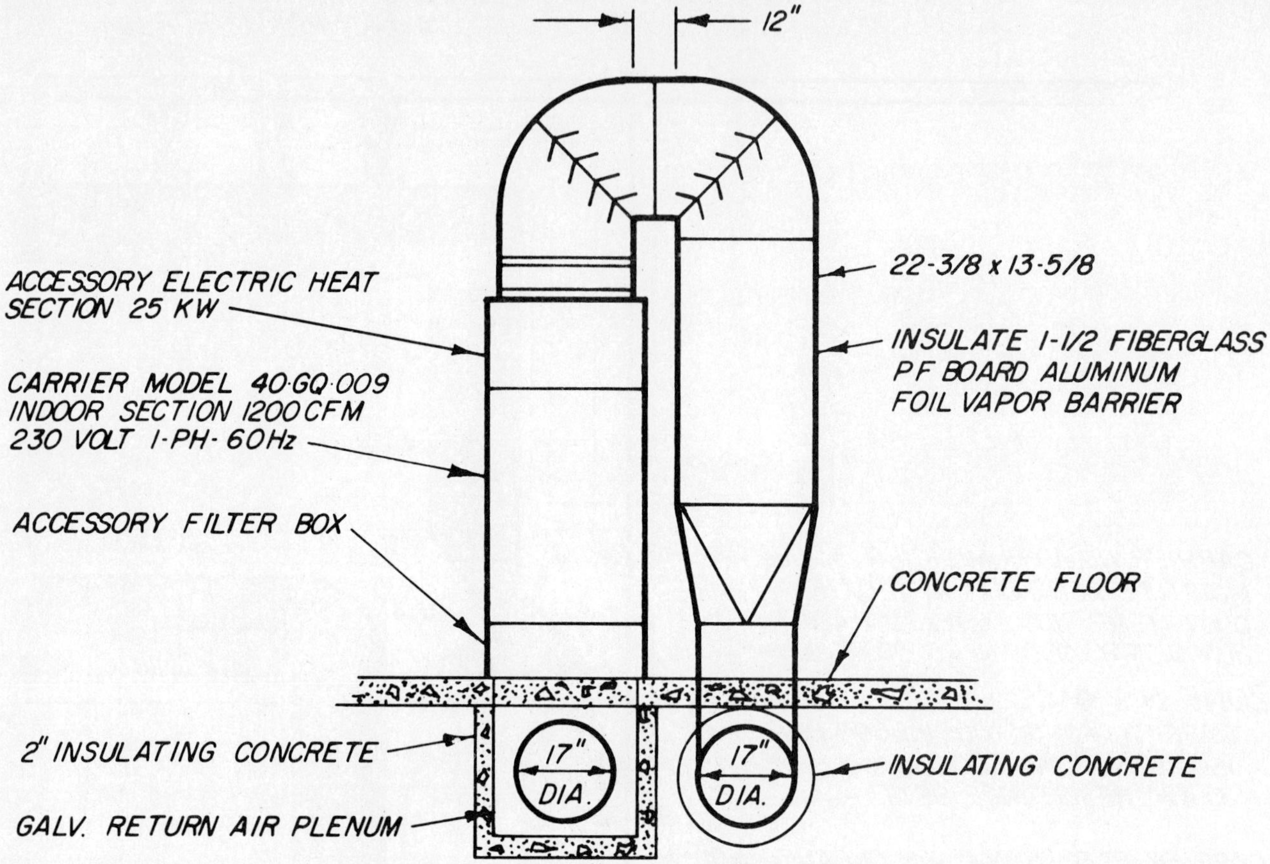

Fig. 19-6. Sectional view of an up-flow air-handling unit with related ductwork turning down and running under concrete floor.

Fig. 19-2 shows an up-flow furnace with the cooling coils on the return-air side of the furnace.

The illustrations in Figs. 19-3 through 19-7 show examples of how fan-coil units are shown on actual working drawings. The student should study these illustrations very carefully, and also refer to Unit 13 for additional details of fan-coil units.

SUMMARY

Air-handling units force air through a duct system. Fans used in air-handling units are normally available in the following four types: centrifugal, propeller, vaneaxial, and tubeaxial.

Air-handling units are shown on HVAC working drawings in order to show the workmen exactly where the units are to be installed.

Usually, details of construction, other than a floor plan view are required for air-handling units; such details can be found in Units 13 and 14 of this book as well as in other units.

Cooling coils or a heat source (such as electric resistance heating coils) are often installed directly in the air-handling unit.

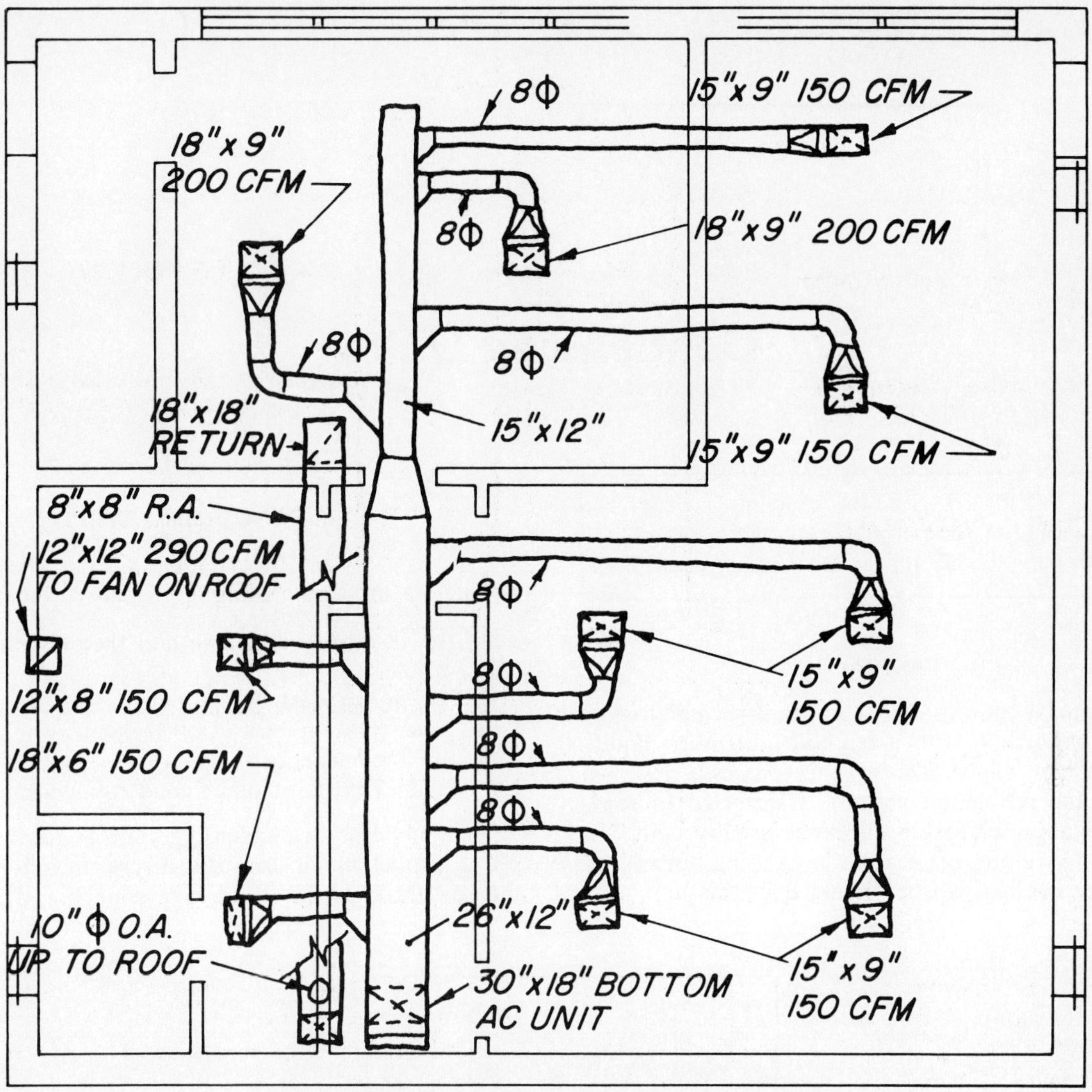

Fig. 19-8. Engineer's sketch for use in the assignment in this unit.

UNIT 19 — ASSIGNMENT

1. Prepare a sheet of 11-by-17-inch tracing paper with border lines and title block; turn to Fig. 19–8.
2. The engineer's sketch you see in Fig. 19–8 is typical of those made by designers in consulting engineers' offices. It is from this type of sketch that HVAC draftsmen make neat working drawings. Use the sketch as a guide, and produce a neat drawing of the floor plan, air-handling unit, and all related ductwork. The scale should be ¼ inch = 1 foot.
3. Answer the following fill-in-the-blank questions.
   - **a.** __________ are used in air-handling units for the circulation of air in HVAC systems.
   - **b.** Air moves radially through the impeller in a(n) __________ fan.
   - **c.** The __________ fan and the __________ fan are similar in that the air flows axially through the impeller.
   - **d.** The __________ fan can move large quantities of air, but its use is limited.
   - **e.** Each type of fan can have either a belt drive or a(n) __________.

UNIT 20

Condensing Units

OBJECTIVES: To give the student a general understanding of the purpose for condensing units and how they should appear on working drawings

GENERAL

The primary purpose of a condensing unit, when used in conjunction with a fan-coil unit, is to liquify the refrigerant vapor. During this process, the heat added to the refrigerant must be transferred to some other substance, the other substance usually being air or water. Therefore, condensing units using both water and air for cooling will be discussed in this unit.

AIR-CONDENSING UNITS

The use of air for condensing purposes is very popular in modern cooling applications since:

1. Air is readily available and there is no disposal problem.
2. The initial cost of equipment and installation is lower than the cost of other condensing means.
3. Maintenance costs are usually less than maintenance costs of other condensing means.

The modern air-condensing unit is available with either propeller- or centrifugal-type fans for moving the air through the condenser coil. Fig. 20-1 illus-

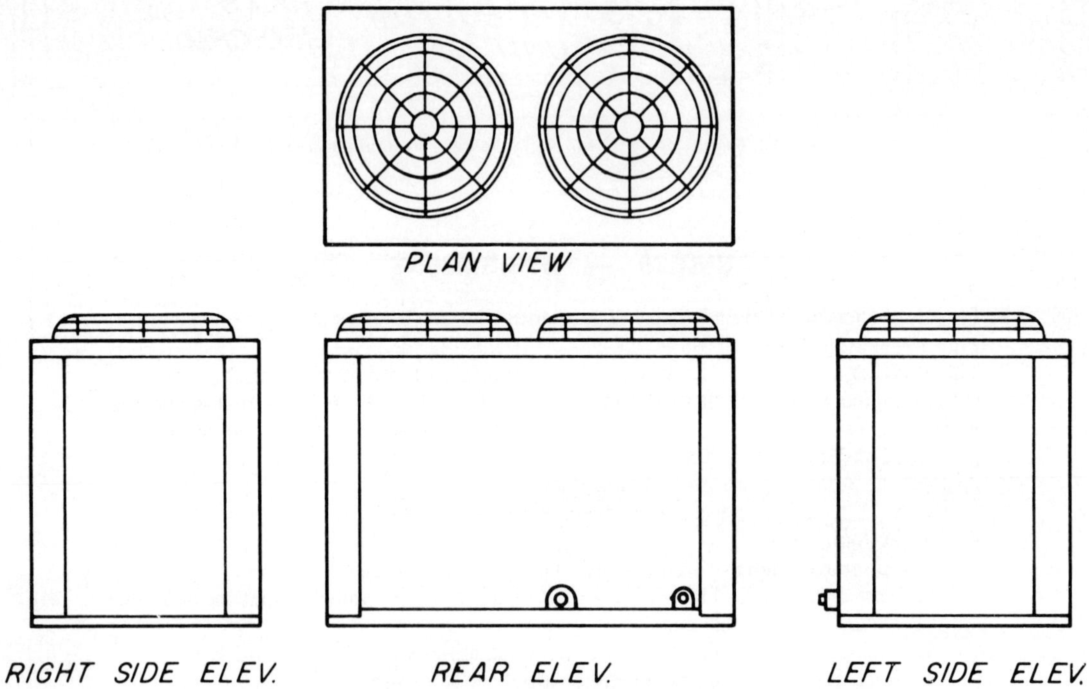

Fig. 20-1. Propeller-type condensing unit.

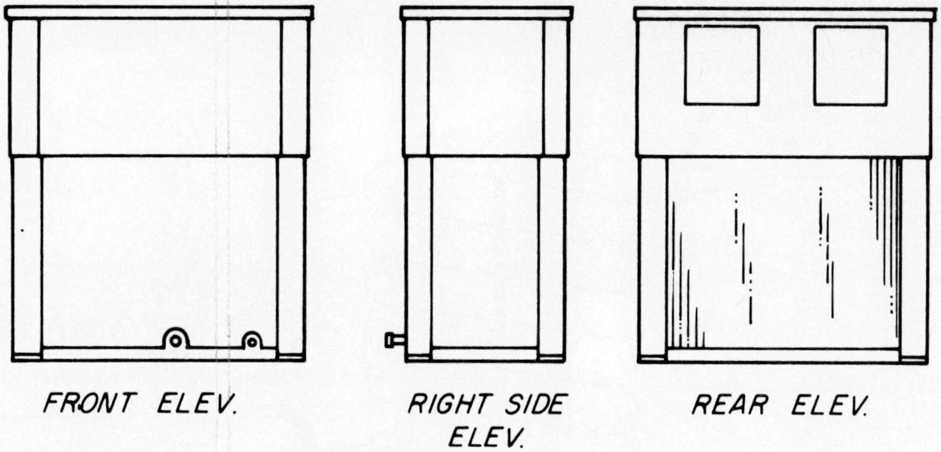

Fig. 20-2. Centrifugal-fan-type condensing unit.

trates a propeller-type unit, while Fig. 20-2 shows a centrifugal-type one. The propeller-type fans are normally used when the condensing units are installed outdoors and the centrifugal-type fans are generally used when the condensing unit is installed indoors with ductwork to carry the outside air to and from the unit.

WATER-COOLED CONDENSING UNITS

The four common types of water-cooled condensers are double-pipe, double-tube, shell-and-coil, and shell-and-tube.

Double-Pipe Condensers

This early type of condenser consists of 1¼-inch pipes inside 2-inch pipes. The water passes through the 1¼-inch pipe while the refrigerant vapor condenses within the annular space between the 2-inch and 1¼-inch pipes. See Fig. 20-3.

Double-Tube Condenser

The principle of the double-tube condenser is very similar to that of the double-pipe. The double-tube condenser is commonly used for condensing units having a capacity of about 7½ tons. Water flows through an inside tube of ¾- to ⅞-inch diameter, while the refrigerant flows through the annular space between this tube and a 1- or 1¼-inch outside tube. This double-tube arrangement is usually formed into a coil as shown in Fig. 20-4.

Shell-and-Coil Condenser

A shell-and-coil condenser consists of a continuous copper coil (in which water flows) mounted inside a steel shell. Refrigerant vapor from the compressor is discharged inside the shell to condense on the outside of the cold water-filled tubes.

Shell-and-Tube Condenser

The shell-and-tube condenser (Fig. 20-5) consists of a large number of ¾- or ⅝-inch tubes installed inside a steel shell. Water flows inside the tubes while the refrigerant vapor flows outside, around the nest of tubes. The vapor condenses on the outside surface of the tubes and drips to the bottom of the condenser, as shown in Fig. 20-5.

WATER REQUIRED

The heat from the hot refrigerant vapor is, of course, transferred to the water, and the water cannot be reused until it has cooled. Therefore, a continuous supply of cold water is necessary for a water-cooled condenser, or else the water must be recirculated through a cooling tower. Water is cooled through a cooling tower by moving the air either by means of natural draft, an induced draft, or a forced draft. A cooling tower using a natural draft is shown in Fig. 20-6.

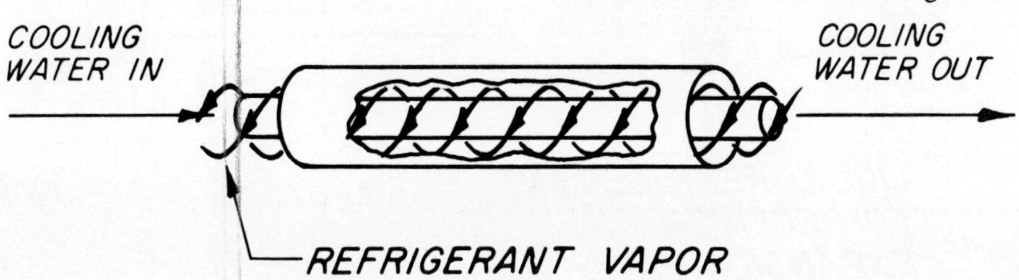

Fig. 20-3. Operating principle of the double-pipe condenser.

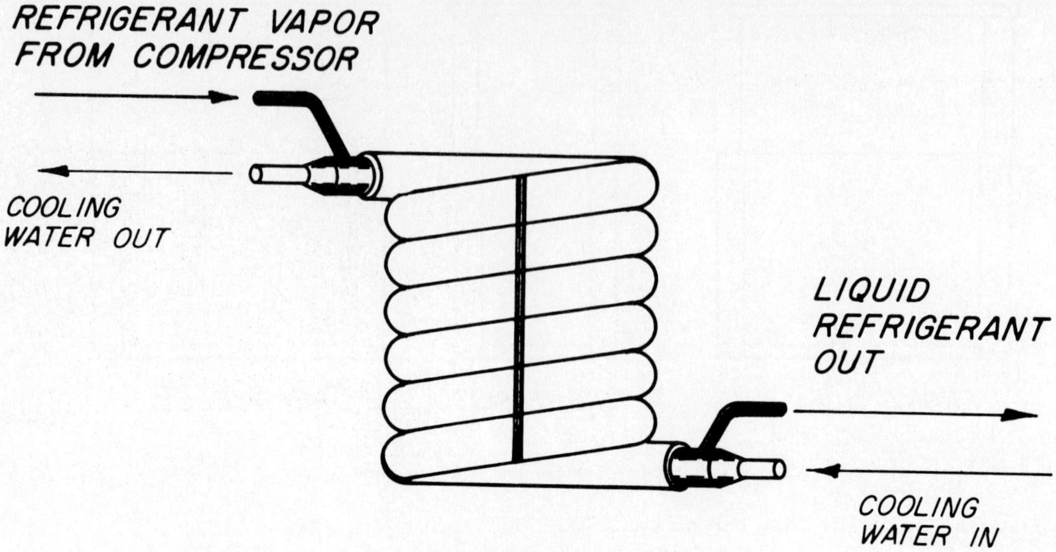

Fig. 20-4. Double-tube condenser.

WORKING DRAWINGS

The HVAC draftsman must show the locations of all condensing units on any given project so that the HVAC contractor and his workmen will know exactly where the engineer or designer intended the units to be installed. While the units need not be drawn in a pictorial view, the units should be drawn on plan views as close to the actual physical shape and size as possible. One example of a condensing unit used for a residence is shown in Fig. 20-7. This unit is drawn to a scale of ¼ inch = 1 foot. The actual unit—as determined from the manufacturer's dimensions—was round in shape (as viewed in plan view) and was, therefore, drawn to this same shape on the plans.

The illustrations in Figs. 20-8 through 20-15 show how condensing units appear on actual working drawings. Some units need elevation drawings to show various details of construction. The reader should study these drawings very carefully before attempting the assignment at the end of this unit.

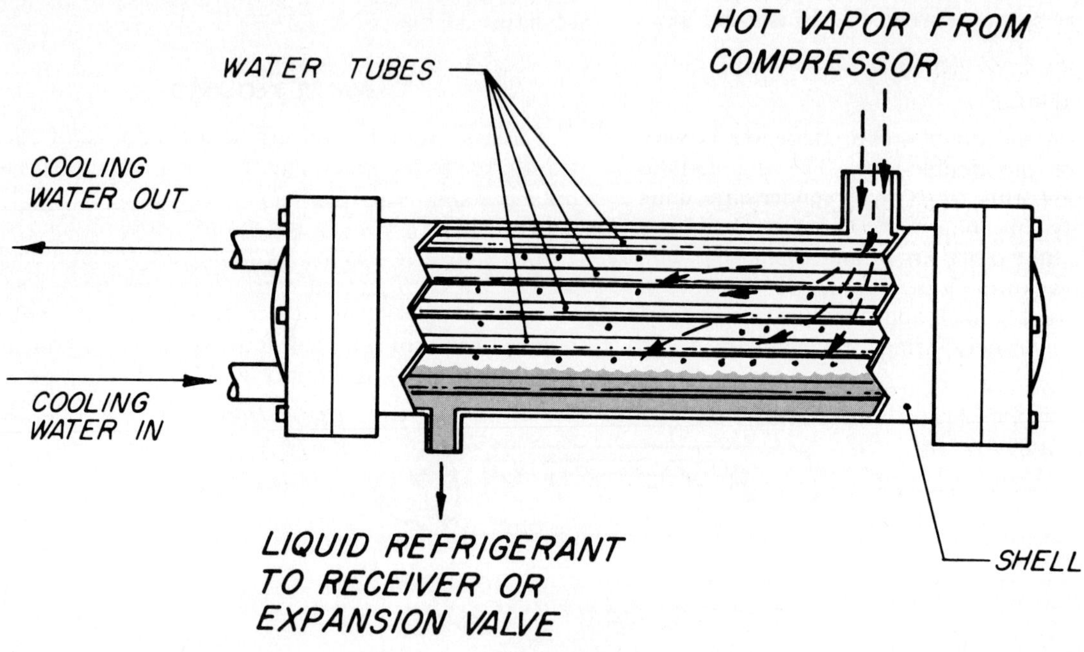

Fig. 20-5. Shell-and-tube condenser.

SUMMARY

Condensing units are used in refrigerant cooling systems to liquify or "cool" the refrigerant vapor.

Most condensing units used either air or water for the transferal of heat.

Modern air-condensing units are available with either propeller or centrifugal fans for moving air through the condenser coil.

Water-cooled condensing units are available in four common types; double-pipe, double-tube, shell-and-tube, and shell-and-coil.

A sufficient amount of cold water must be provided for water-cooled condensing units.

Cooling towers are sometimes used in order to recirculate (and thus conserve) water used in water-cooled condensing units.

Cooling towers utilize air to cool the water. The three most commonly used types are natural draft, induced draft, and forced draft.

Fig. 20-6. Photograph of a "natural-draft cooling tower."

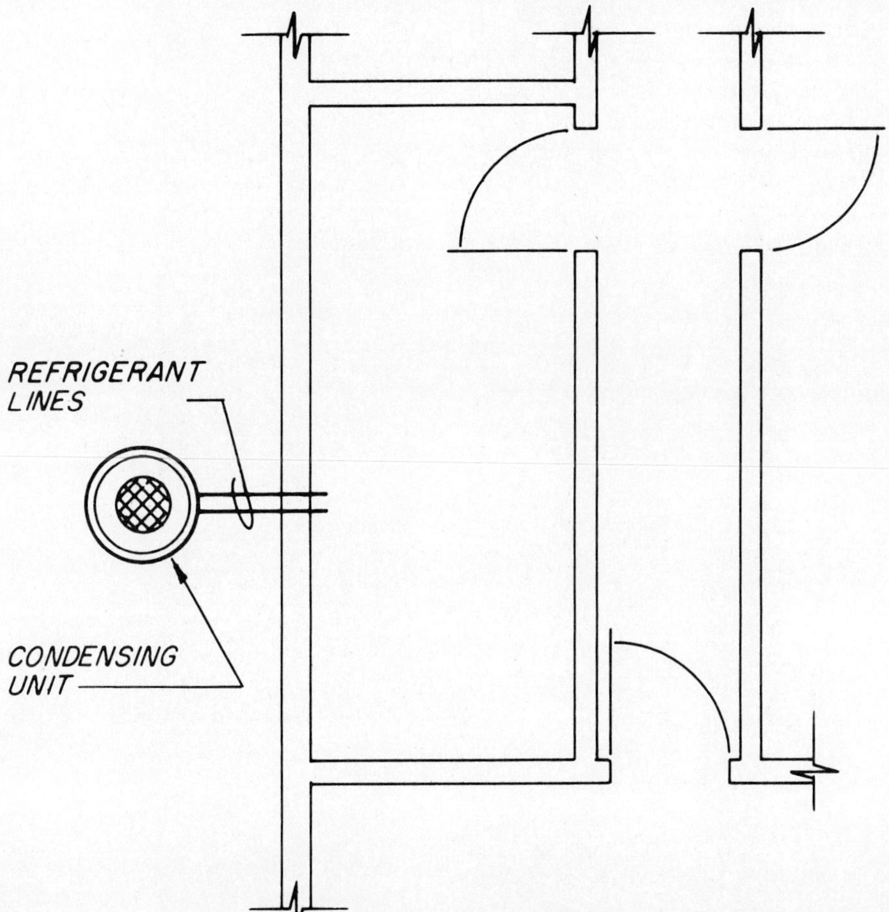

Fig. 20-7. Plan view of a residential condensing unit.

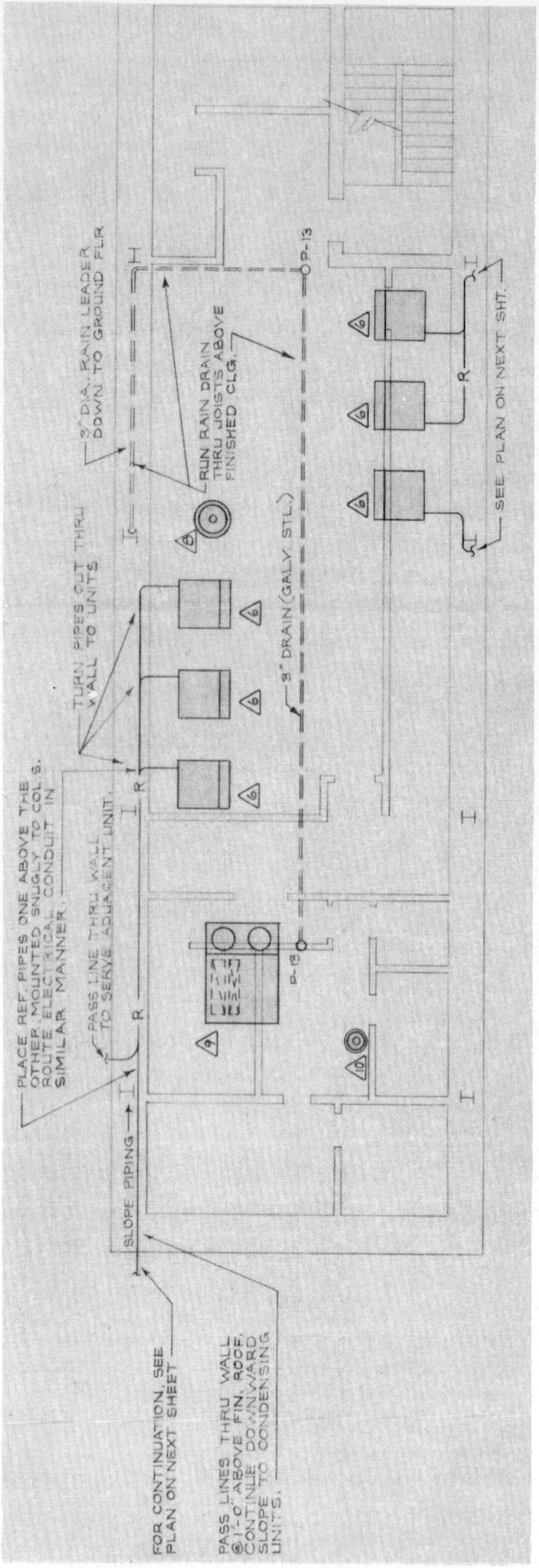

Fig. 20-8. Plan view of a building with several condensing units mounted on the roof.

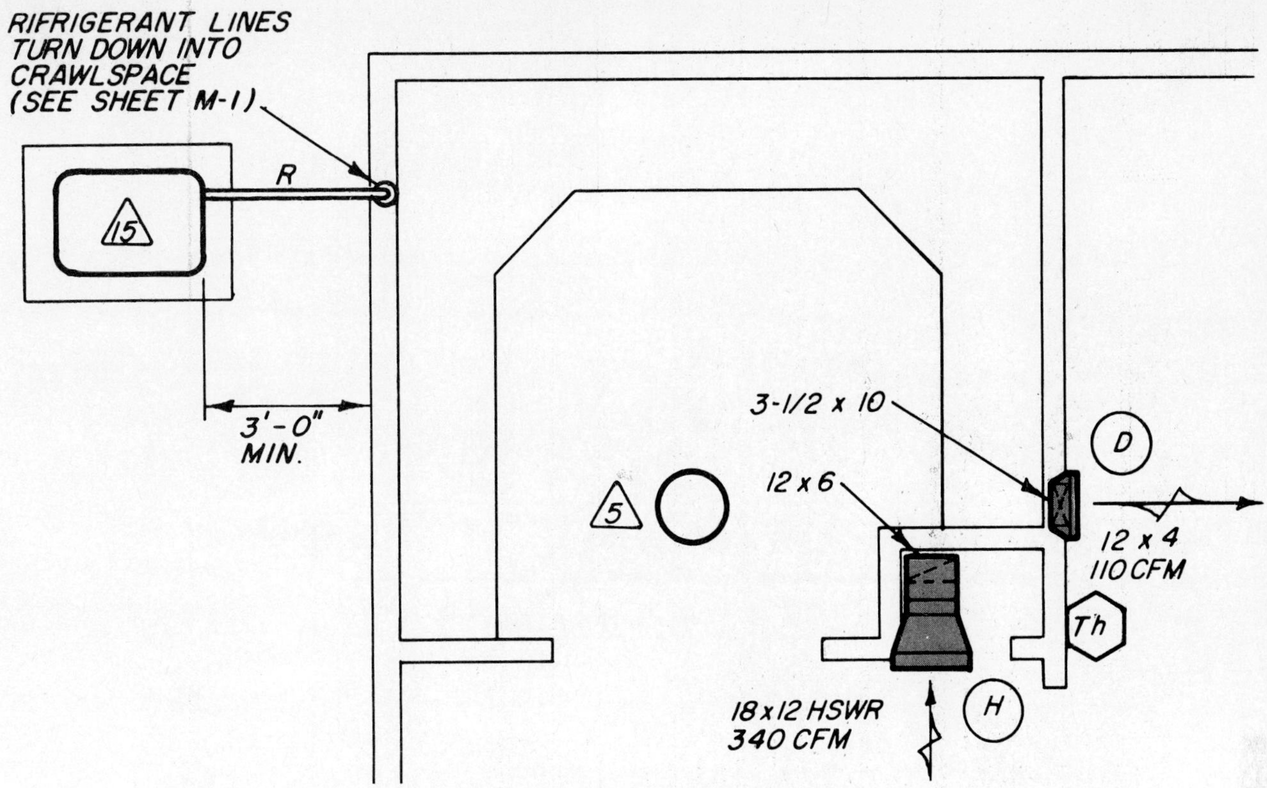

Fig. 20-9. Plan view of a bank building with a pad-mounted condensing unit outside the building.

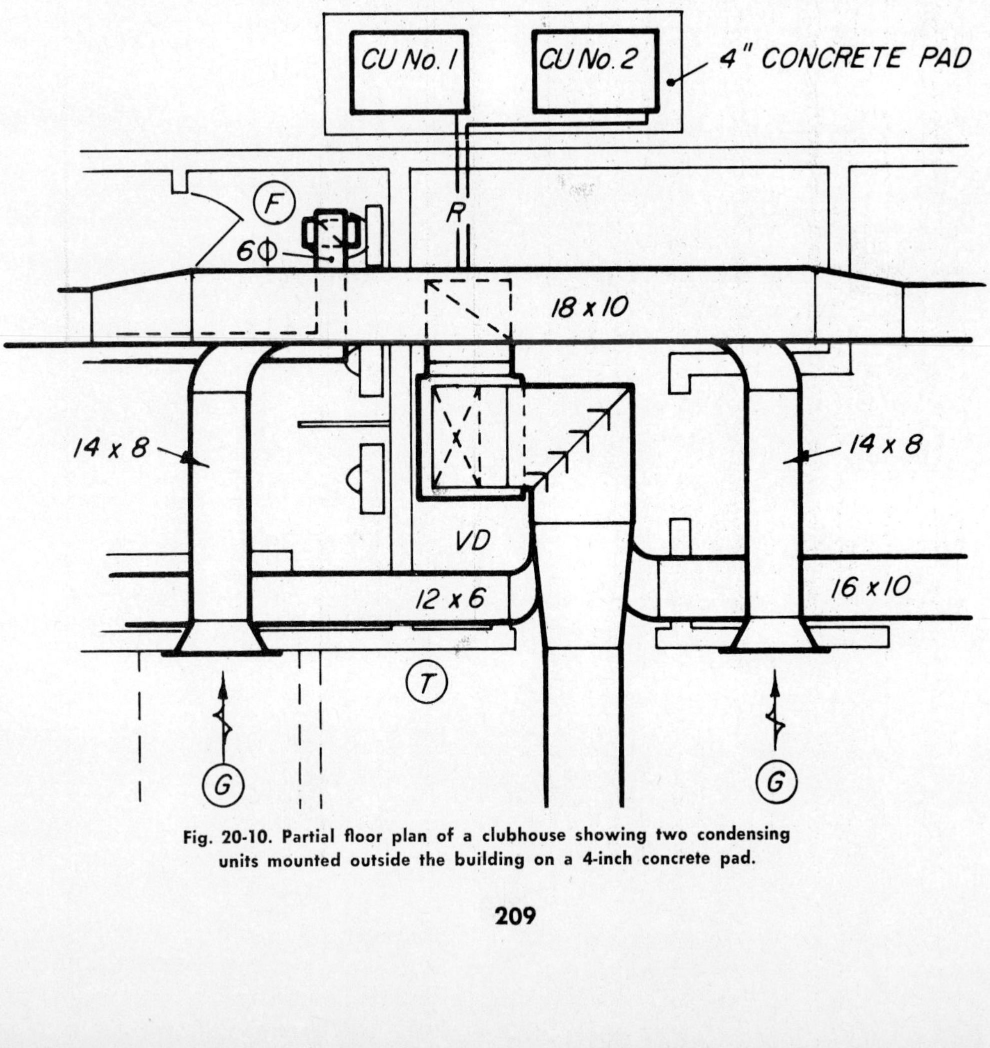

Fig. 20-10. Partial floor plan of a clubhouse showing two condensing units mounted outside the building on a 4-inch concrete pad.

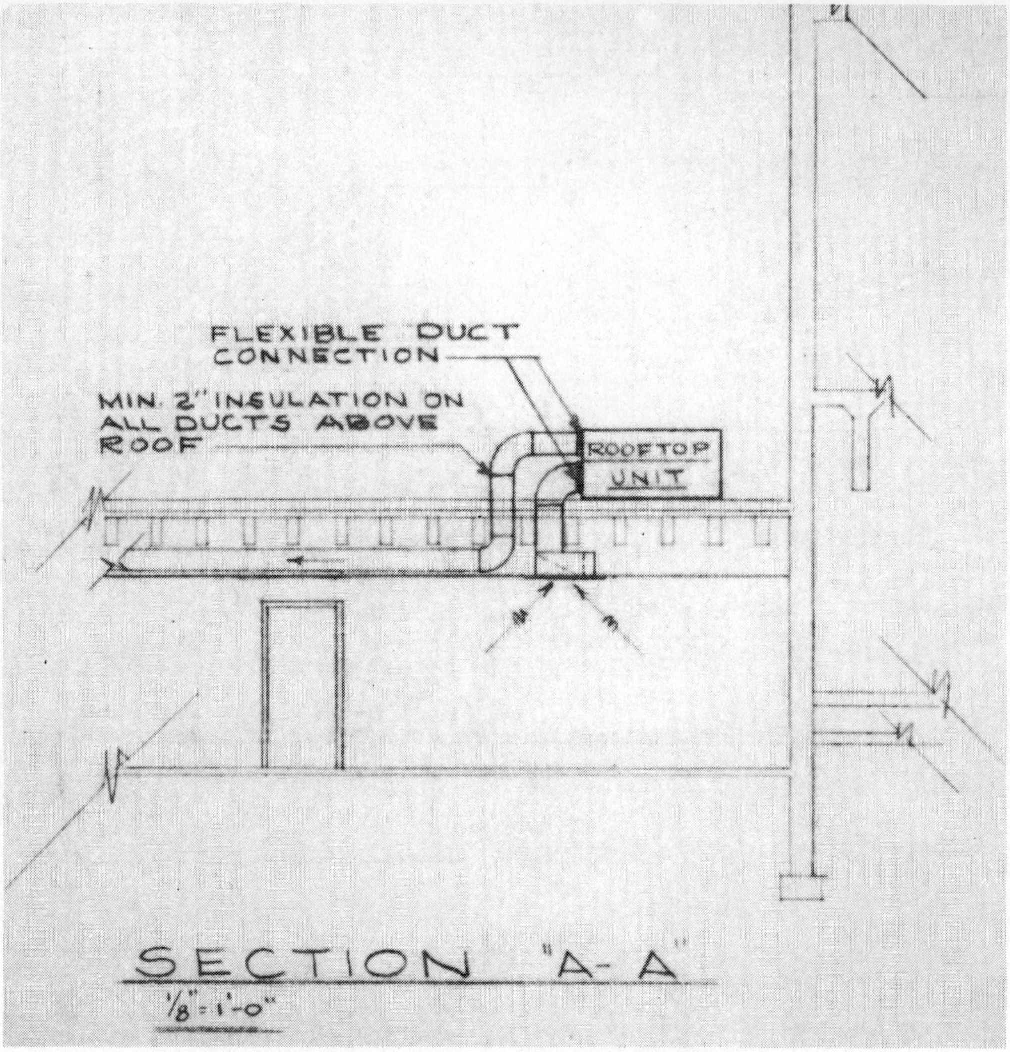

Fig. 20-11. Sectional drawing of a self-contained HVAC rooftop unit.

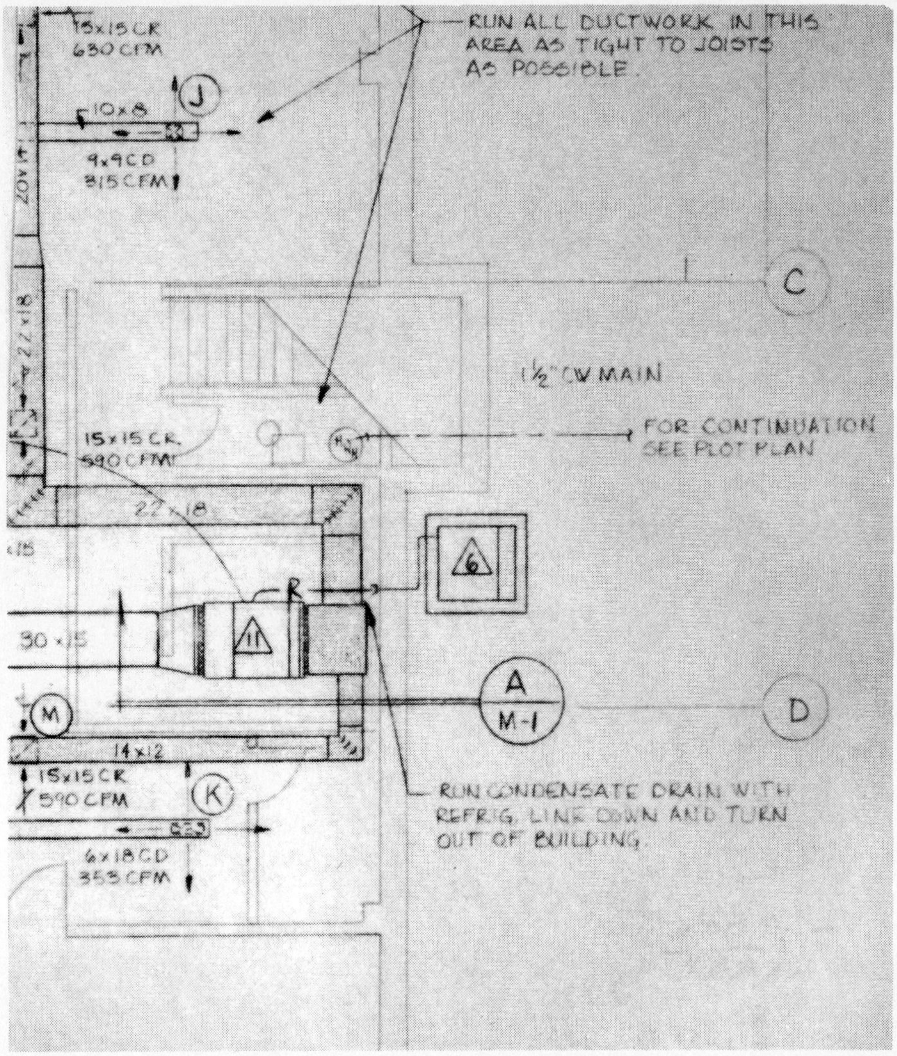

Fig. 20-12. Partial floor plan of a department store showing pad-mounted condensing unit (6) with refrigerant lines running to air-handling unit (11).

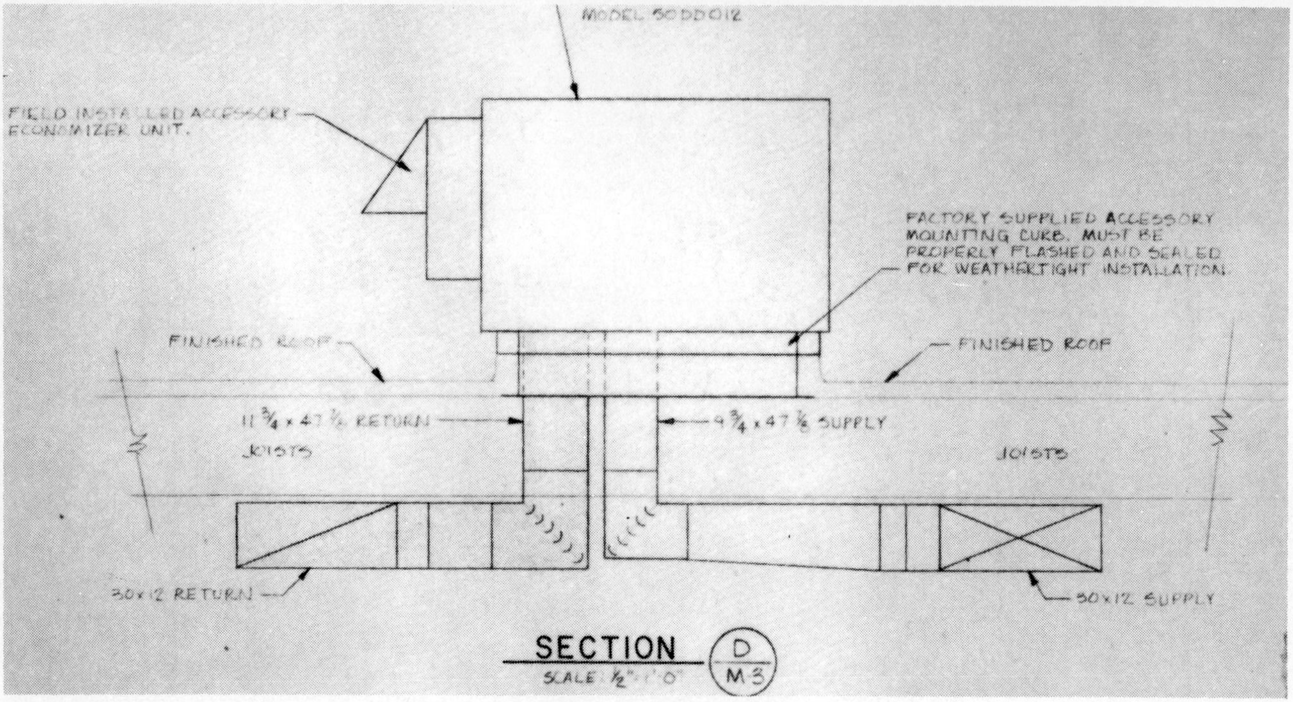

Fig. 20-13. Section of a self-contained HVAC roof-top unit.

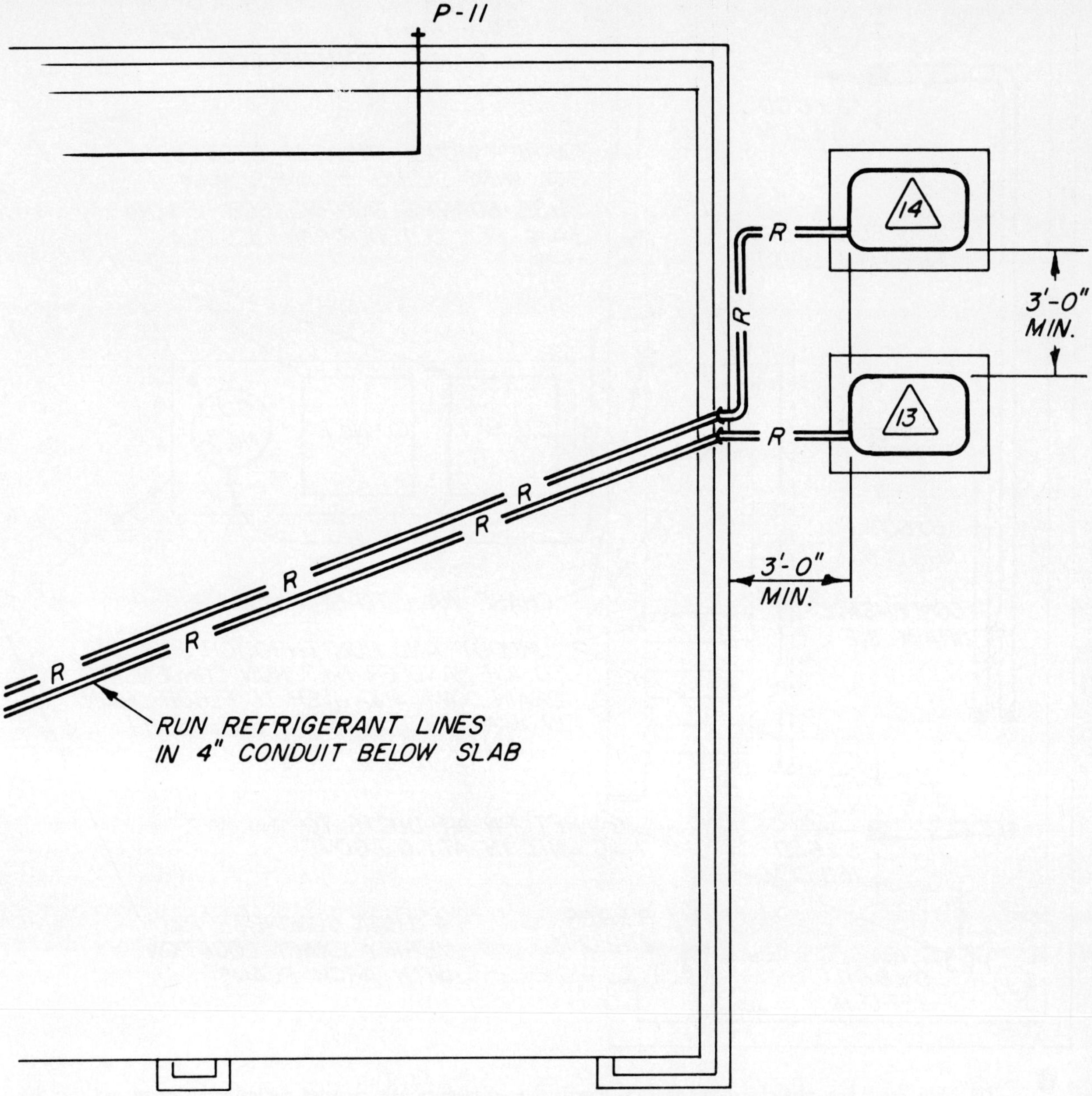

Fig. 20-14. Partial floor plan of a residence showing two pad-mounted condensing units with related refrigerant lines running to cooling coils located in the air-handling units. (The air-handling units are not shown on this plan.)

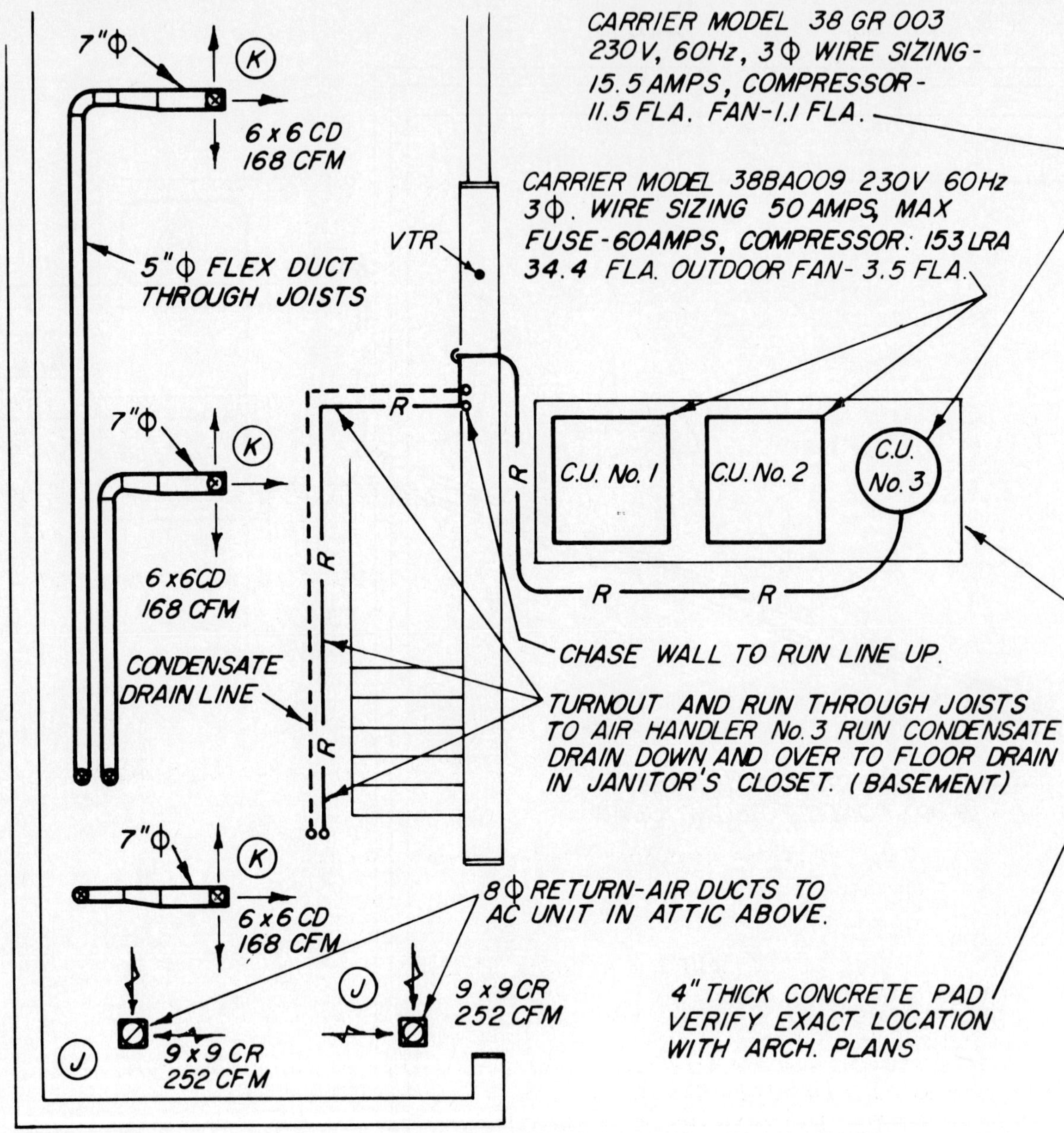

Fig. 20-15. Partial floor plan of a church building showing three condensing units mounted outside on a concrete pad.

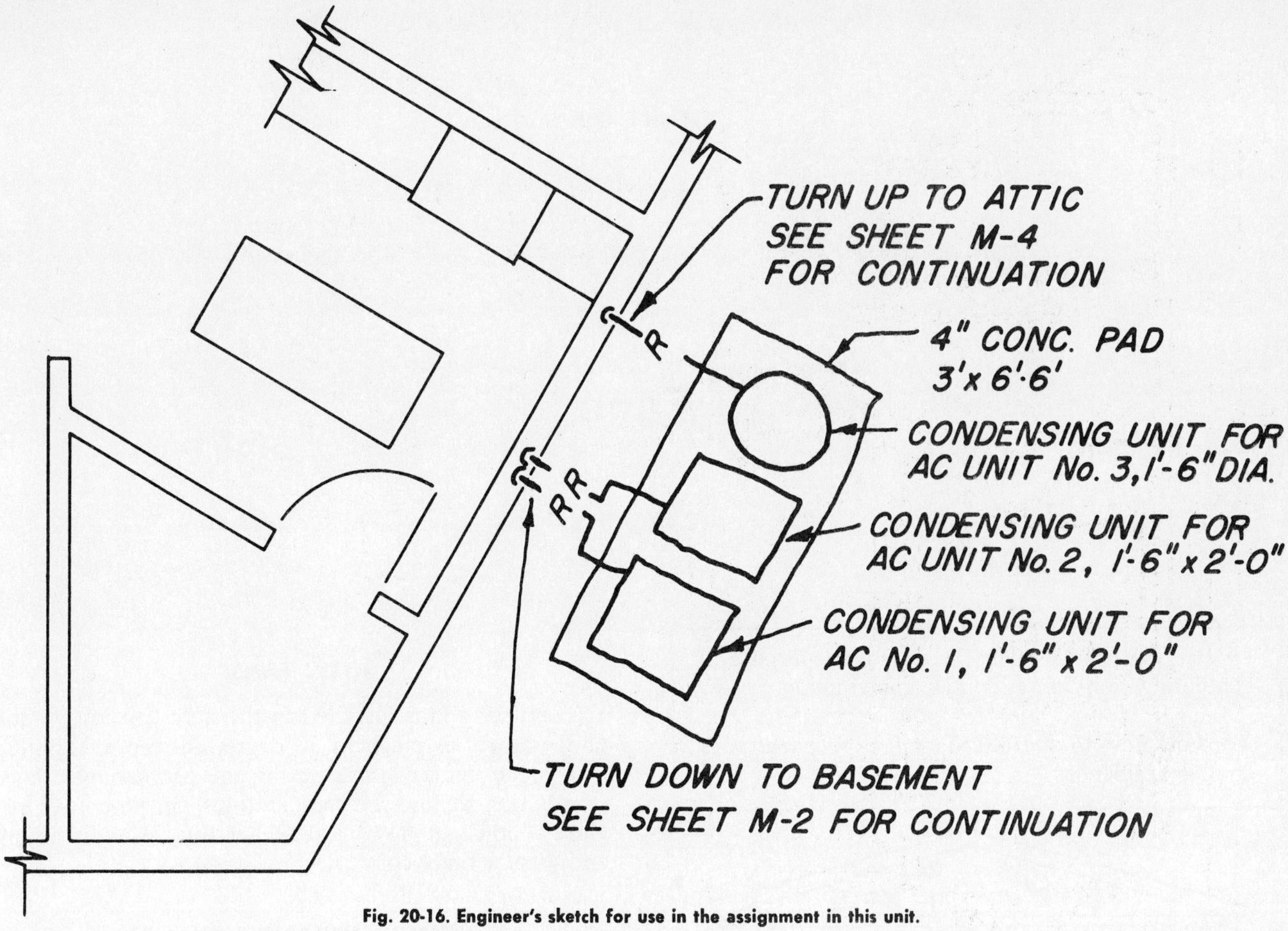

Fig. 20-16. Engineer's sketch for use in the assignment in this unit.

UNIT 20 — ASSIGNMENT

1. Prepare a sheet of 11-by-17-inch tracing paper with border lines and title block; turn to Fig. 20–16.
2. Using the examples in this unit as a reference, draw the three condensing units for the residence in Fig. 20–16. The drawing should be centered and the scale should be ¼ inch = 1 foot. Use 4-H lead for drawing the building portions with a light solid line. The condensing units and related refrigerant piping should be drawn with 2-H lead, and all lettering should be done with H lead.
3. Answer the following fill-in-the-blank questions.
    a. The primary purpose of a condensing unit, when used in conjunction with a fan-coil, is to __________ the refrigerant vapor.
    b. The __________ condenser was the first type.
    c. A(n) __________ condenser consists of a continuous copper coil mounted inside a steel shell.
    d. Cooling towers utilize __________ to cool the water.
    e. __________ fans are normally used when the condensing units are installed outdoors.

UNIT 21

Fans

OBJECTIVES: To teach the reader the purpose of ventilating fans in residential construction

To show how to size attic fans for the purpose of cooling residences

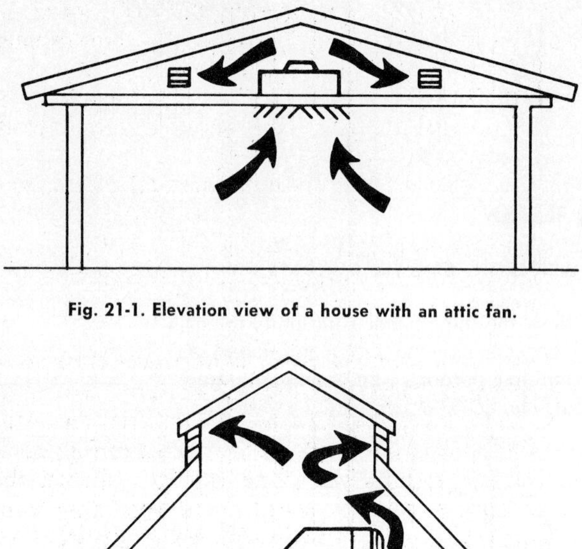

Fig. 21-1. Elevation view of a house with an attic fan.

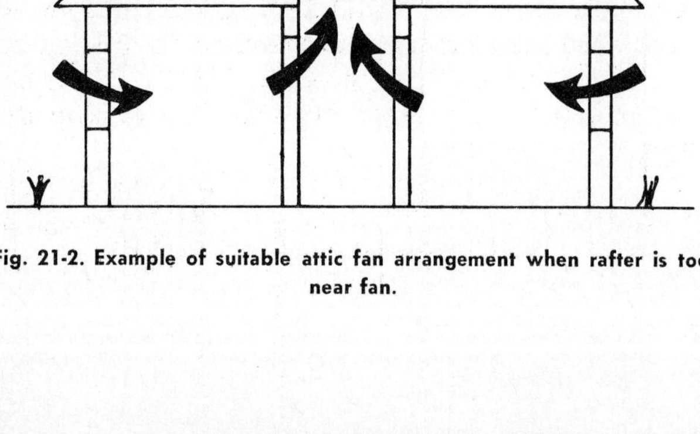

Fig. 21-2. Example of suitable attic fan arrangement when rafter is too near fan.

ATTIC FANS

Ventilating fans installed in the attic can lower attic temperatures as much as 30 degrees or more, resulting in a much cooler house during the hot summer days. In addition to lowered temperatures, an attic fan provides a constant flow of fresh air for easier breathing and greater body comfort.

HOME VENTILATION BY FAN

The size of fan necessary to properly ventilate a home depends on the size of the house and also the number of complete air changes desired each minute. Usually one air change per minute will suffice for most areas throughout the United States, except in the northern states where one air change every 1½ minutes will suffice.

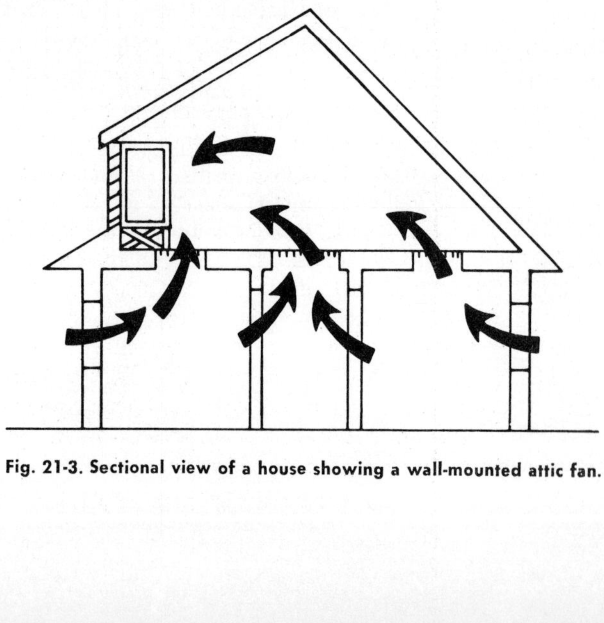

Fig. 21-3. Sectional view of a house showing a wall-mounted attic fan.

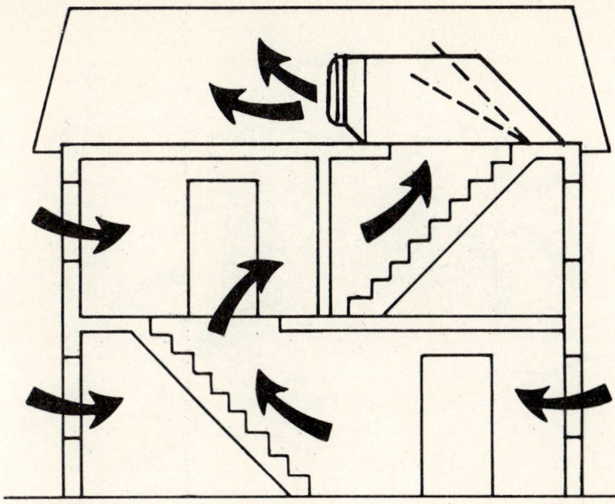

Fig. 21-4. Section of a house showing attic fan installed over stairway.

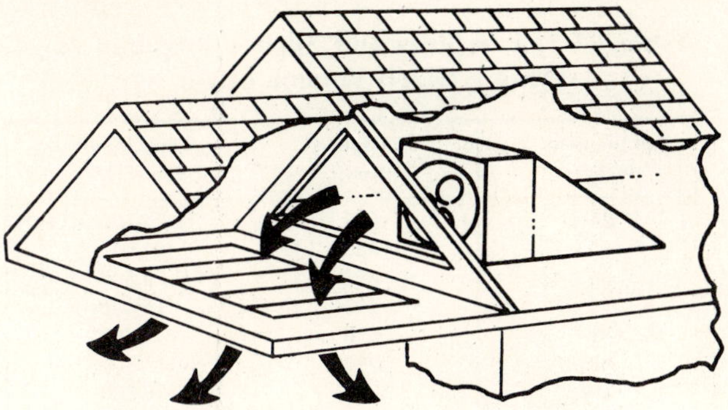

Fig. 21-5. Example of an exhaust grille in a porch roof.

For example, if a one-story home measures 24 by 40 feet and has a ceiling height of 8 feet, the volume of the house will be $24 \times 40 \times 8 = 7680$ cubic feet. Thus, for one air change per minute, the fan must move 7680 CFM of air.

The fan should be located in the center of the house and above the attic ceiling in order to provide the ideal flow of air. Fig. 21-1 shows an elevation view of a house with the attic fan located in the hall. The fan is mounted directly above automatic shutters allowing the air to be pulled through the fan. Exhaust vents are located in the gables.

An exhaust fan located in the attic should be placed below the attic rafters and at a distance at least equal to its own diameter. Therefore, a fan 30 inches in diameter should be placed at least 30 inches below the rafters. This distance will permit the air to flow freely through the fan, whereas a closer distance may cause the fan to overwork and be unable to discharge the amount of air for which it is rated.

Fig. 21-2 shows an attic fan arrangement suitable for either a one- or a two-story house, especially one in which rafters may be too near the fan. This arrangement permits the fan to direct its discharge toward the unobstructed attic area for ready exhaust.

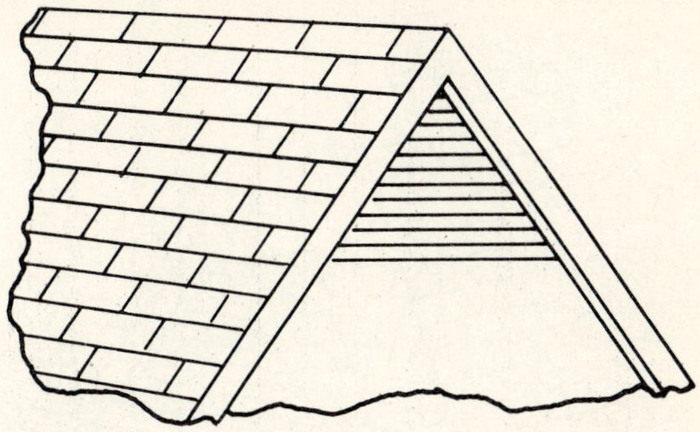

Fig. 21-6. Example of an exhaust grille in gable ends of attic.

If the fan is mounted on a wall, as shown in Fig. 21-3, the entire attic should be airtight, since all of it is used as a suction chamber. Insulation should be installed in the attic rafters as well as in the joists. With this arrangement, the exhaust louvers should be of the adjustable type so that they can be closed in cold weather.

In houses having a stairway to the attic, the suction fan can be installed over the stairway opening, and an airtight door within the chamber will provide access to the attic from below. Fig. 21-4 shows such an arrangement.

Adequate exhaust for the fan discharge must be provided. The actual area necessary for an exhaust opening is governed by several factors:

1. Whether or not mesh screening or hardware cloth is used
2. Whether the louvers are constructed of wood or metal
3. The thickness of the louvers
4. Whether the opening is to be completely unobstructed

A porch roof (see Fig. 21-5) is an excellent location for an exhaust grille since the grille is protected from the weather. If a porch roof is not available, an opening at the gable ends of the attic, as shown in Fig. 21-6, is also good. If only one gable end is available, the exhaust grille should be placed at the end away from the prevailing winds. Louvers are built into dormers of various types as shown in Figs. 21-7 and 21-8.

Flat or shed roof structures present a problem in housing the fan as well as discharging the air. A wall exhaust fan may be the best solution in this case. However, a penthouse, as shown in Fig. 21-9, can be built to house the exhaust fan and also to provide louvers for exhaust air.

If the gable opening of a house is not sufficient for exhausted air, screened openings in the overhanging eaves may be provided. Such an arrangement is shown in Fig. 21-10.

Fig. 21-7. Exhaust grille built into a dormer.

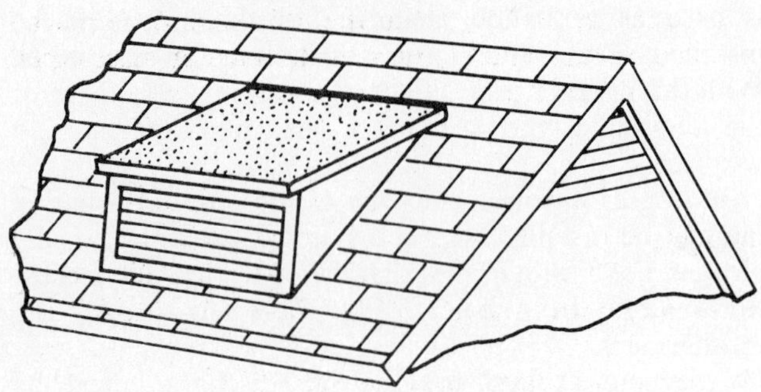

Fig. 21-8. Another example of a grille built into a dormer.

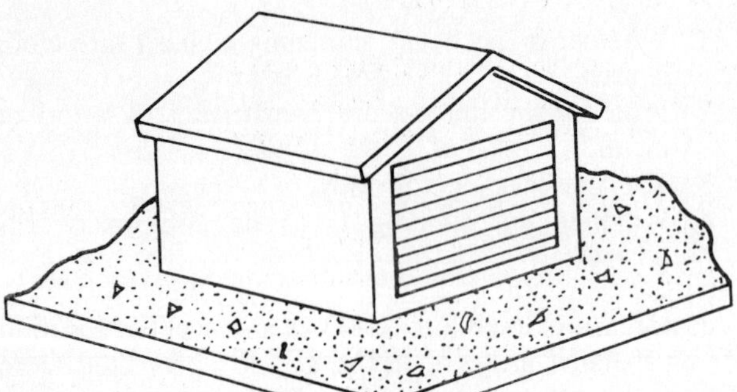

Fig. 21-9. A penthouse, or cupola, housing an exhaust fan; louvers are provided for exhaust air.

SIZING THE FAN

In order to determine the size of fan needed for a given house, scale the floor area of each room on the drawings and multiply this by the ceiling height to determine the cubic feet in each room. Total all room figures to arrive at the total cubic feet of air in the house. Then select a fan to handle this air volume. The fan should provide one air change per minute. Table 21-1 can be used to select the size of fan required to move this volume of air.

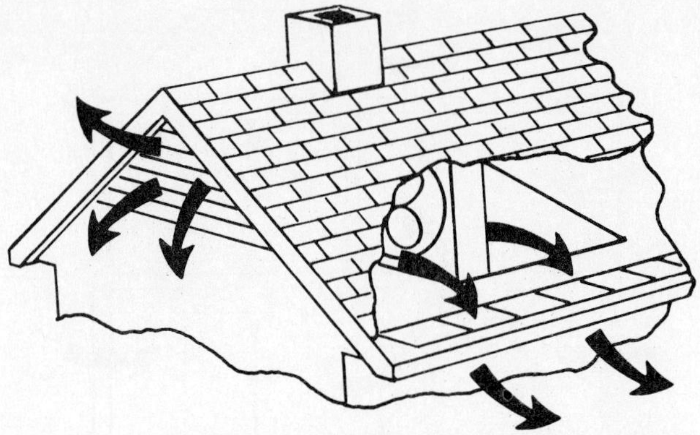

Fig. 21-10. Example of both gable opening and openings in the eaves for exhausting air.

Table 21-1. Fan Sizes and Speeds Required to Move a Given Volume of Air

Fan Diameter (inches)	Fan Speed (rpm)	Capacity of Fan (CFM)
24	600	5,100
30	460	7,400
36	390	9,500
42	290	12,000
42	370	14,700
48	295	17,500
48	335	18,500
54	250	19,400
60	220	27,900

Table 21-2. Approximate Exhaust Opening Sizes

Fan Size (inches)	Type and Size of Opening (sq ft)		
	Full Opening	Wood Louvers	Metal Louvers
22	11	16	14
24	12	18	16
30	18	27	24
36	26	39	33
42	36	54	47
48	46	69	60

OUTSIDE OPENING NEEDED FOR EXHAUST

Table 21-2 lists the approximate size of exhaust opening needed for various types of openings and sizes of exhaust fans.

SUMMARY

A residential attic fan provides a constant flow of fresh air during warm weather and can lower the attic temperature as much as 30 degrees.

Usually, one air change per minute will suffice for most areas throughout the United States.

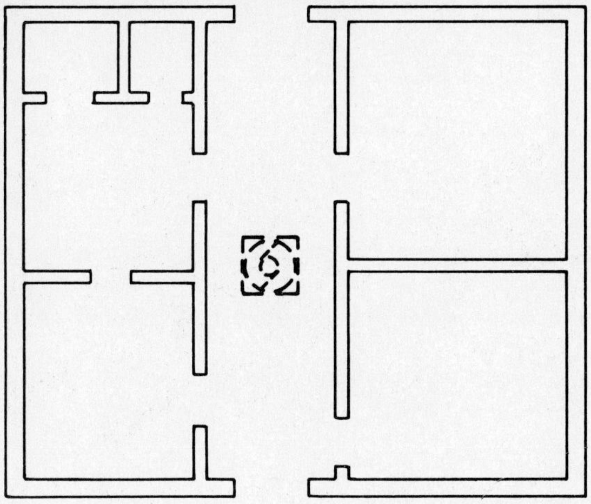

Fig. 21-11. Floor plan of house with attic fan opening in hallway.

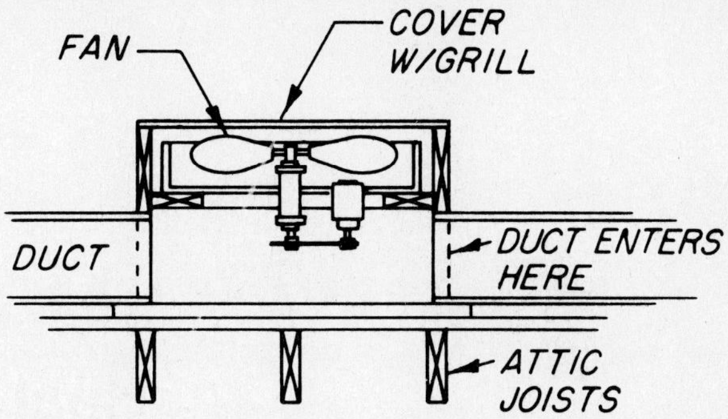

Fig. 21-12. Sectional view of exhaust fan for the assignment in this unit.

The ideal spot to locate an attic exhaust fan is in the center of the house such as in a central hallway. Fig. 21-11 shows a floor plan of a one-story house with an attic fan shown in the hallway.

An exhaust fan located in the attic should be placed below the attic rafters and at a distance at least equal to its own diameter.

UNIT 21 — ASSIGNMENT

1. Prepare a sheet of 11-by-17-inch tracing paper with border lines and title block; turn to Appendix D.
2. Measure the room sizes of the residence floor plan. Use the architect's scale, ¼ inch = 1 foot. Assuming an 8-foot ceiling height throughout the house, calculate the total cubic feet of space within the house.
3. Determine the fan diameter needed to provide one air change per minute.
4. Using your prepared sheet, draw the floor plan of the residence.
5. Using the knowledge you have gained from this unit, locate the best spot for the exhaust fan. With a heavy, solid line, draw the fan and housing onto the floor plan you just completed. Assume the fan housing is square, as illustrated in Fig. 21–12. Also assume that the housing extends four inches beyond the blades. Use Fig. 21–11 as a guide.
6. Answer the following True or False questions.

	True	False
a. Ventilating fans installed in the attic can lower the temperature as much as 30°.	____	____
b. An attic fan 30 inches in diameter should be placed no more than 15 inches below the rafters.	____	____
c. Usually one air change per minute will be enough for most areas in the United States.	____	____
d. The thickness of the louvers has no effect on the size of the needed exhaust opening.	____	____
e. If an attic fan is mounted on the wall, the entire attic should be airtight.	____	____

UNIT 22

Heat Recovery

> OBJECTIVES: To become familiar with the various methods of recovering heat
>
> To practice drawing these systems on HVAC drawings

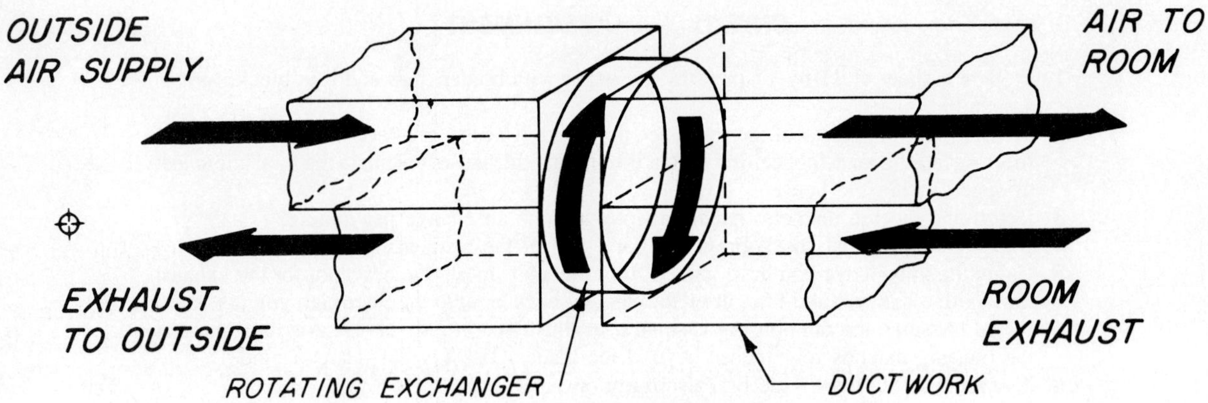

Fig. 22-1. Example showing operating principle of an enthalpy exchanger.

GENERAL

We are all aware of the worldwide energy shortage. The long range solution to this problem is the use of new technology and energy sources. However, for the immediate future, the solution involves using less energy without sacrificing economic well-being. Various heat recovery devices have proven to be some of the most effective ways of gaining economical operation in process heating and cooling, since such devices recover useful energy otherwise lost to the atmosphere.

As much as 50 to 80 percent of wasted energy can be economically salvaged from heated or cooled exhaust air and then reused. This can be accomplished by a number of methods, which will be discussed in this unit.

Another important source of heat capture is the collection of energy from outside air which is then used to heat and cool the inside of a structure. This procedure will be covered in this unit also.

HEAT WHEELS

There are two types of heat wheels in use today. One is the *enthalpy exchanger* which recovers both

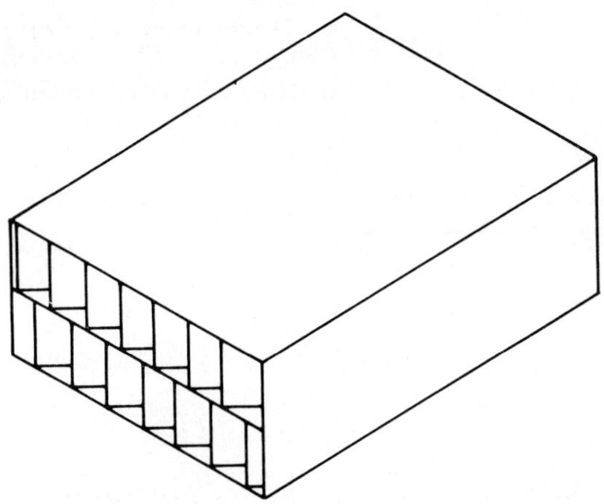

Fig. 22-2. Air-to-air heat exchanger.

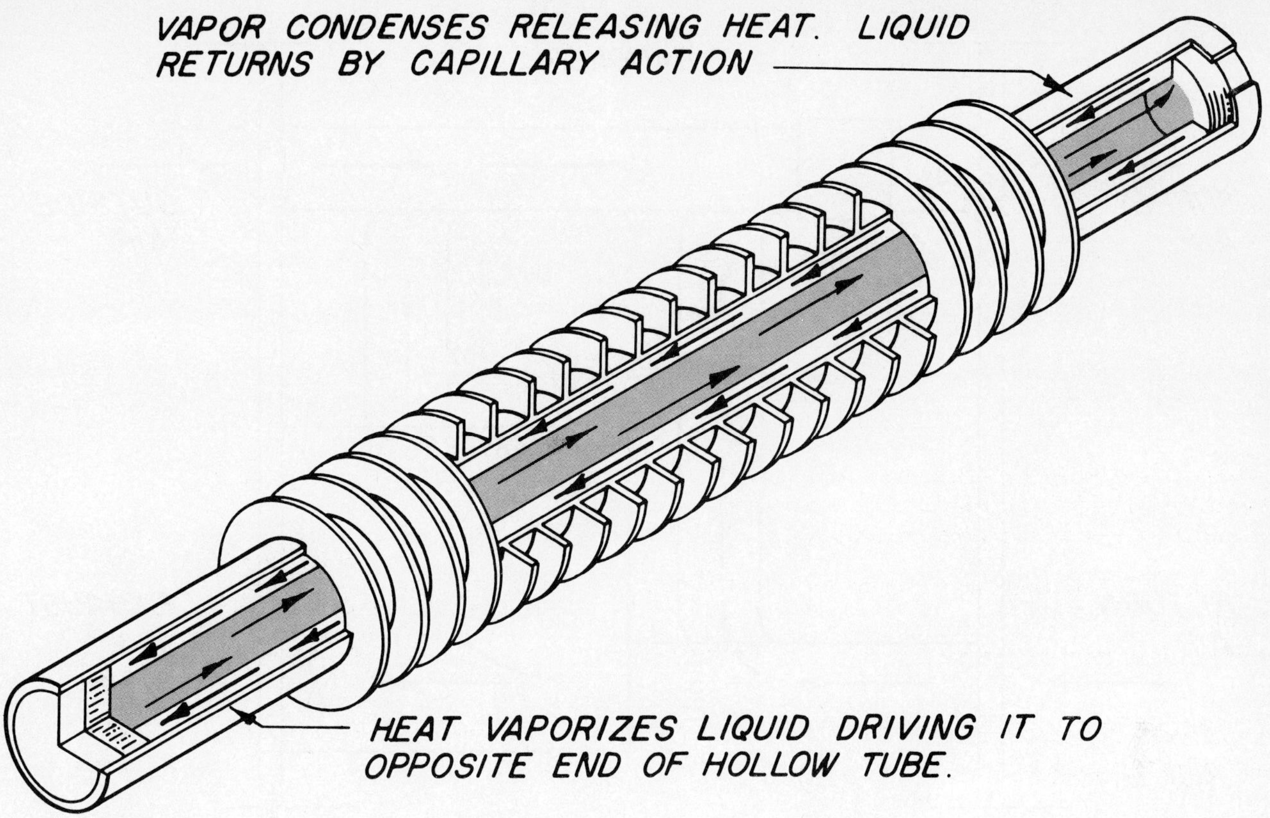

Fig. 22-3. Working principle of a heat pipe.

sensible and latent heat; the other is the *sensible exchanger* which recovers sensible heat only. Both types require that the exhaust and air supply ducts be adjacent.

The enthalpy exchanger is installed in the ductwork, as shown in Fig. 22-1. It slowly rotates between the exhaust and supply air streams. In winter "heat and moisture" recovered from room exhaust air preheat the incoming outdoor air supply. In summer the reverse occurs as cool and dry conditioned air recovered from the roof exhaust precools the incoming outdoor air supply.

The sensible exchanger operates similarly to the enthalpy exchanger except that it now recovers sensible heat only and is particularly useful for processing applications, high temperature exhaust, and drying systems.

STATIC HEAT EXCHANGERS

The air-to-air heat exchanger illustrated in Fig. 22-2 represents a static means for transporting heat between the exhaust and outside airstreams which pass through it in a counterflow fashion. It has no moving parts and is maintenance free except in cases where the exhaust air will be cooled to a temperature below dew point; in that case arrangements for adequate condensate drainage are required.

Like the heat wheel, the static heat exchanger requires that the exhaust and supply air ducts be adjacent to one another (Fig. 22-1) to permit the transfer of heat. The advantages of this unit are:

1. Freedom from maintenance (there are no moving parts)
2. Ruggedness and durability
3. Low installation cost
4. Reduction of heating and cooling loads
5. Efficiency in excess of 70 percent

HEAT PIPES

The heat pipe is an air-to-air heat transfer device. It is radically different from most heat exchanges in that it takes advantage of the enormous quantities of heat that are absorbed by a vaporizing liquid or released by a condensing vapor.

A heat pipe consists of a hollow tube which is sealed at both ends. The inside of the tube is a wicking structure consisting of a fine wire mesh and a charge of refrigerant. When one end of the tube is heated, the liquid refrigerant vaporizes, absorbing the heat.

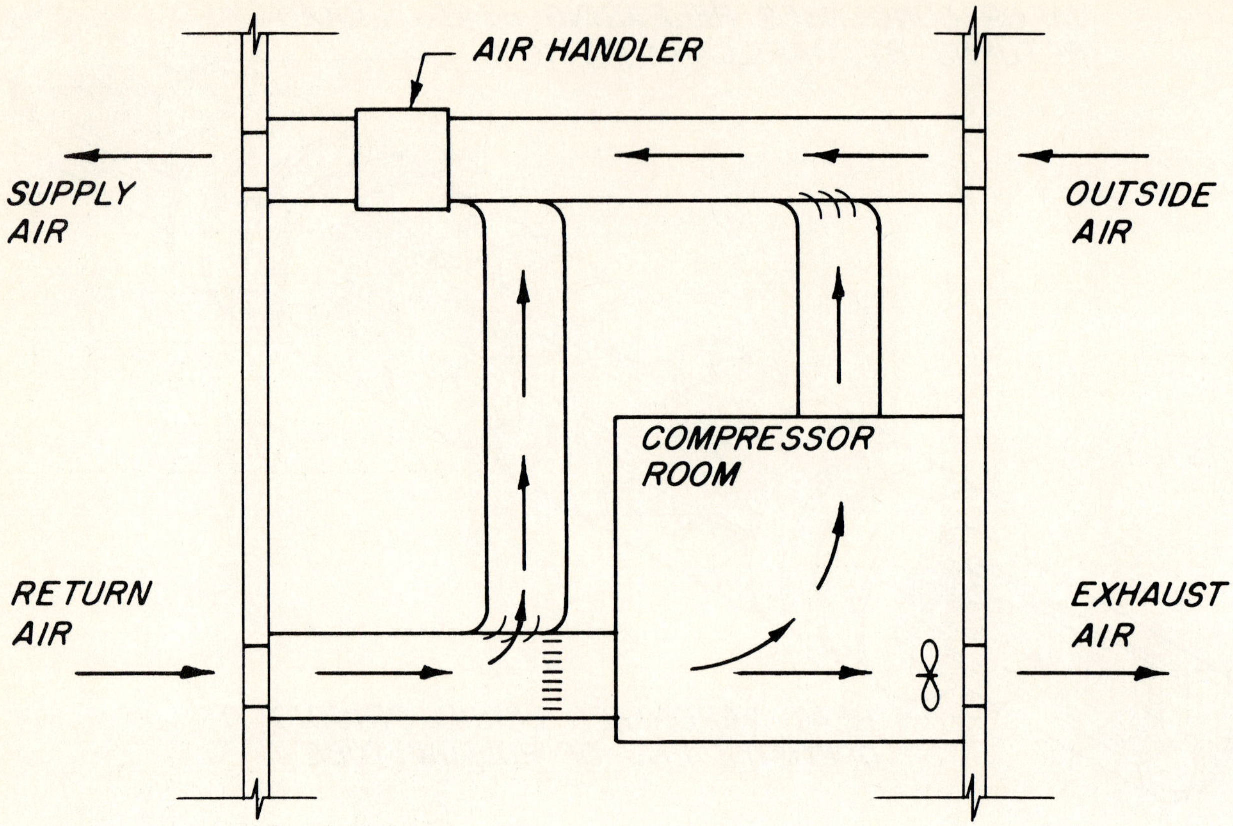

Fig. 22-4. Operating principles of an all-air heat recovery system.

The hot vapor then flows through the hollow center of the wick to the other end of the pipe, releasing the absorbed heat.

Fig. 22-3 illustrates the working principles of a heat pipe.

When a stack of heat pipes are installed in a duct system so that the exhaust air flows through one side of the unit and the supply air counterflows through the opposite side, heat is transferred from one airstream to another.

In winter, heat is transferred from the building exhaust air to the cold makeup air which, in turn, preheats the fresh air. In summer, the process is reversed so the heat will flow in the opposite direction; that is, from the hot makeup air to the cooled building exhaust air, thus precooling the incoming fresh air.

HEAT RECOVERY FROM FOOD SERVICE REFRIGERATION

Frozen food cases are familiar sights in the modern supermarket. The excess heat given off by the compressors and condensers used in the operation of these cases offers a source of heat which can be captured and returned to the sales area to provide comfort heating and humidity control. This method is probably one of the greatest opportunities for heat recovery and energy savings.

The three basic methods of recovering heat from food service refrigeration are all-air, refrigerant-to-water-to-air, and all-refrigerant.

The all-air system is the least complex of the three, and a typical system appears in Fig. 22-4. In this illustration, the return air from the store's sales area is passed trough the compressor room, absorbing the rejected heat which is used to preheat incoming outside ventilation air flowing to the air handler. When the outside air temperature is high enough not to require preheating, the return air is ducted directly to the air handler, bypassing the compressor room.

HEAT PUMPS

The term heat pump, as applied to a year-round air conditioning system, usually denotes a system in which refrigeration equipment is used (1) to extract heat from a heat source and transfer it to a conditioned space when heating is desired and (2) to remove heat from the space and discharge it to a heat sink when cooling and dehumidification are desired.

Heat pumps operate on several systems; the most common are air-to-air, air-to-water, and water-to-air.

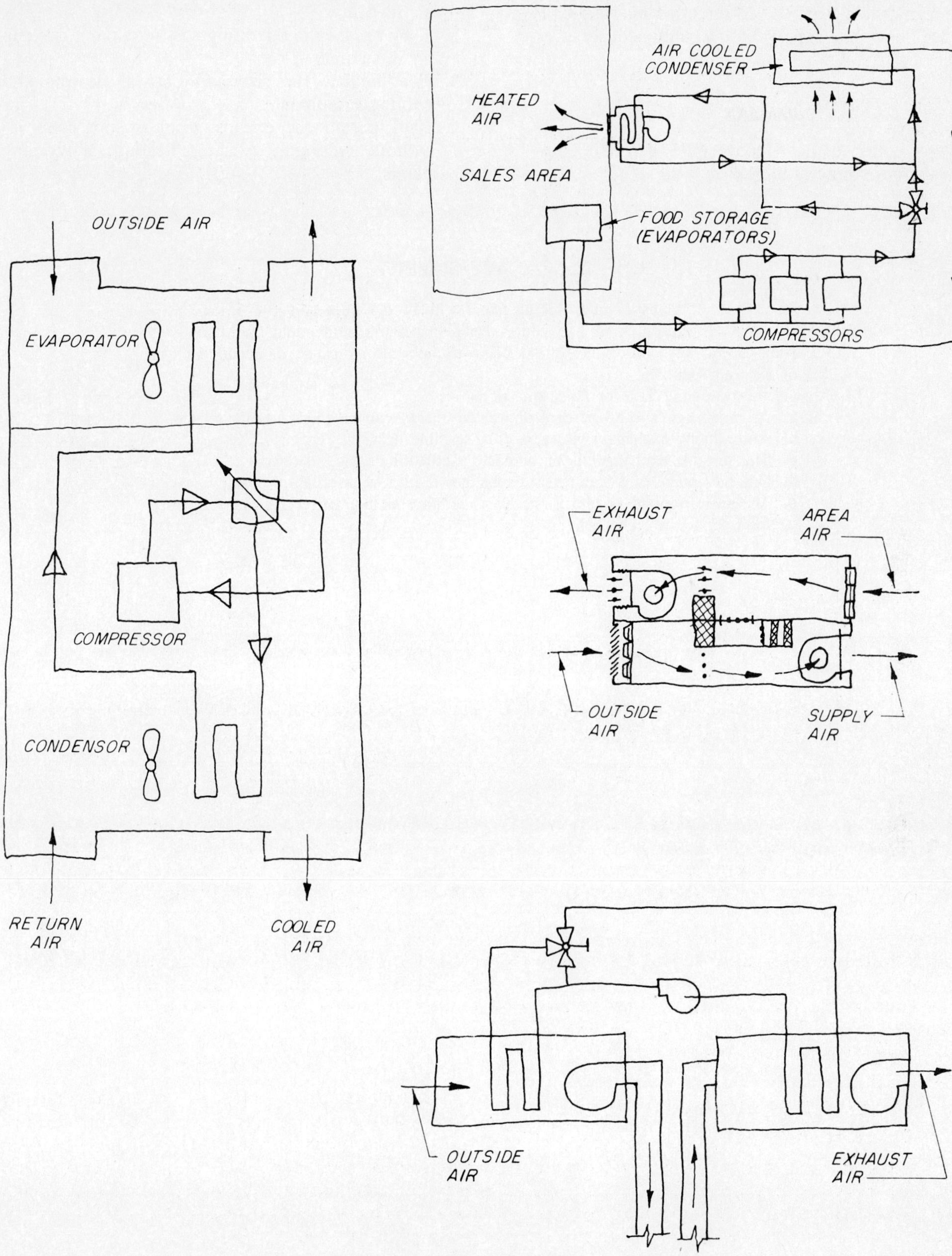

Fig. 22-5. Assignment sketches for this unit.

All of these systems can provide either heating or cooling, can switch from one to the other automatically as needed, and can supply both simultaneously if so desired.

SUMMARY

The installation of heat recovery devices offers several advantages. Some of these are:

1. They reduce the operating cost of heating makeup air in winter.
2. They reduce the operating cost of cooling makeup air in summer.
3. They reduce the size and cost of heating and cooling equipment.
4. They permit the expansion of existing facilities without enlarging existing heating or cooling plants.

UNIT 22 — ASSIGNMENT

1. Prepare a sheet of 11-by-17-inch tracing paper with border lines and title block.
2. Disregarding the title block, divide your sheet of paper into four equal parts.
3. Referring to Fig. 22–5, draw the four sketches to the scale indicated, centering each in one of the four sections.
4. Answer the following True or False questions.

	True	False
a. As much as 50 to 80 percent of wasted energy can be salvaged and reused.	____	____
b. An enthalpy exchanger recovers only sensible heat.	____	____
c. The sensible exchanger is particularly useful for drying systems.	____	____
d. One advantage of a static heat exchanger is its low installation cost.	____	____
e. The all-refrigerant system is the least complex method of recovering heat from food service refrigeration.	____	____

UNIT 23

Control Diagrams for HVAC Equipment

OBJECTIVES: To teach the basic function of HVAC controls

To learn how to draw HVAC-control wiring diagrams for use in conjunction with working drawings

BASIC HVAC CONTROLS

For our purpose, the term "basic HVAC controls" refers to a device that starts, stops, regulates, and/or protects the HVAC system and its related components. Typical devices are thermostats and humidistats.

In general, thermostats and humidistats act similarly to electrical single-pole switches in that they turn the proper component *on* or *off*. When a heating system, for example, is properly controlled by a thermostat, the system becomes energized when the room temperature becomes too low, since the thermostat is designed to sense the room temperature and then to signal the heating unit for the proper response. When the room temperature reaches the level needed to satisfy the thermostat setting, the thermostat signals the heating system to shut off the heat.

A humidistat works in a similar manner in that it is designed to maintain a certain level of humidity. It operates on the change in tension of hair springs, caused by variation in humidity. When the humidity is below the desired percentage, the humidistat activates a humidifier (usually installed within the ductwork) which ejects water vapor into the conditioned air. Fig. 23-1 shows a cutaway view of a humidifier installed on an air plenum.

The sequence of operation for the humidifier shown in Fig. 23-1 is explained in the following paragraphs.

Fill Cycle

The solid-state filling cycle is independent of the humidifying operation and consists of the following components:

A—Solid-state circuit board
B—Water-level probes
G—Fill valve
J—Control transformer

During the fill cycle, the control transformer converts the line-voltage power supply to 24 volts of control power. The fill-valve operation is controlled by a flush cycle, the stainless-steel water-level probes, and the solid-state circuit board. A constant water level is maintained except during the flush cycle.

Humidification Cycle

Almost instantaneous water evaporation is accomplished by the following components:

D—Heating element
I—Control relay
J—Control transformer

Whenever the humidistat senses a need for more humidity, it energizes the control relay and starts the heating element and the system blower (if not already in operation). When the humidistat is satisfied, it deenergizes the control relay, shutting off the heating element and the system blower (providing the blower operation is not required by other controls). The power and water supply to the humidifier must be left on at all times. During summer or other periods where humidification is not required, the humidistat control is turned to *off* and the heating element will not operate. However, the solid-state fill controls will continue to keep the res-

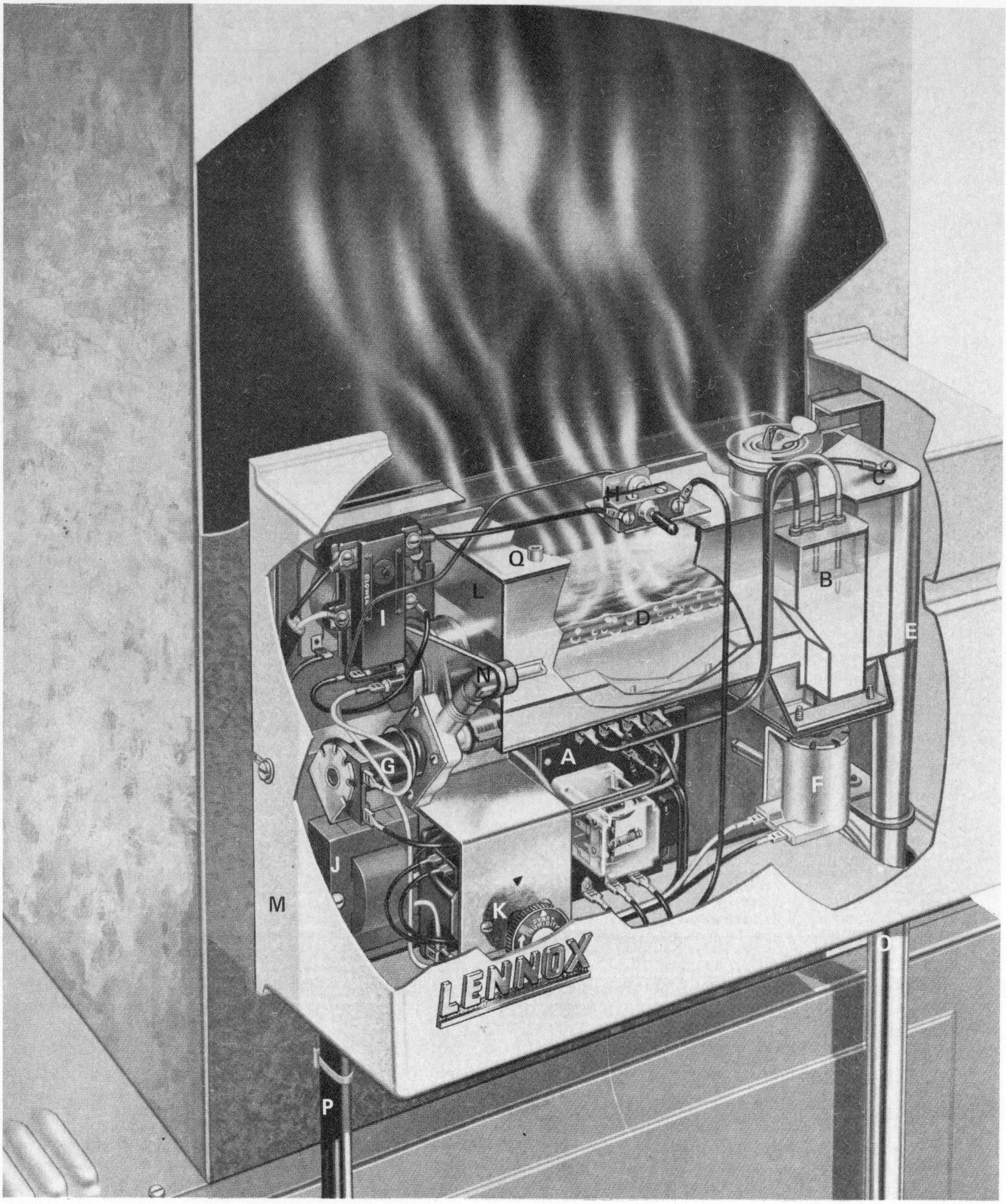

Fig. 23-1. Cutaway of a humidifier installed on an air plenum.

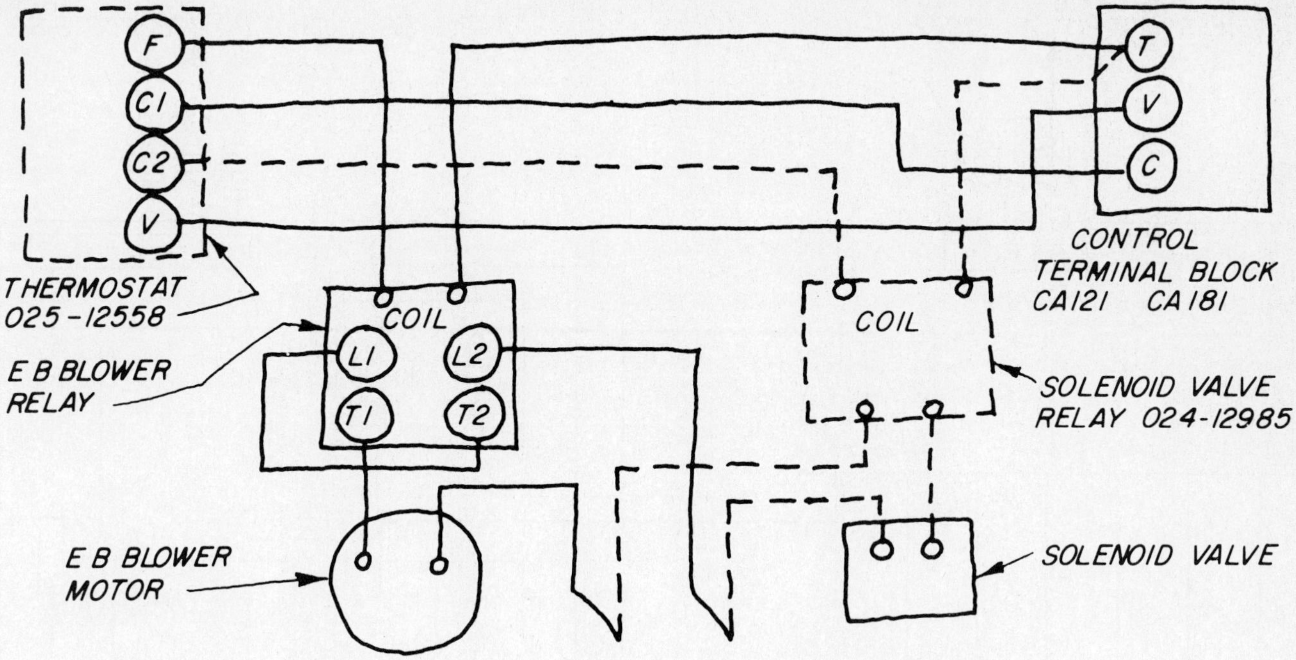

Fig. 23-2. Wiring diagram of a blower motor in a fan-coil unit (engineer's sketch).

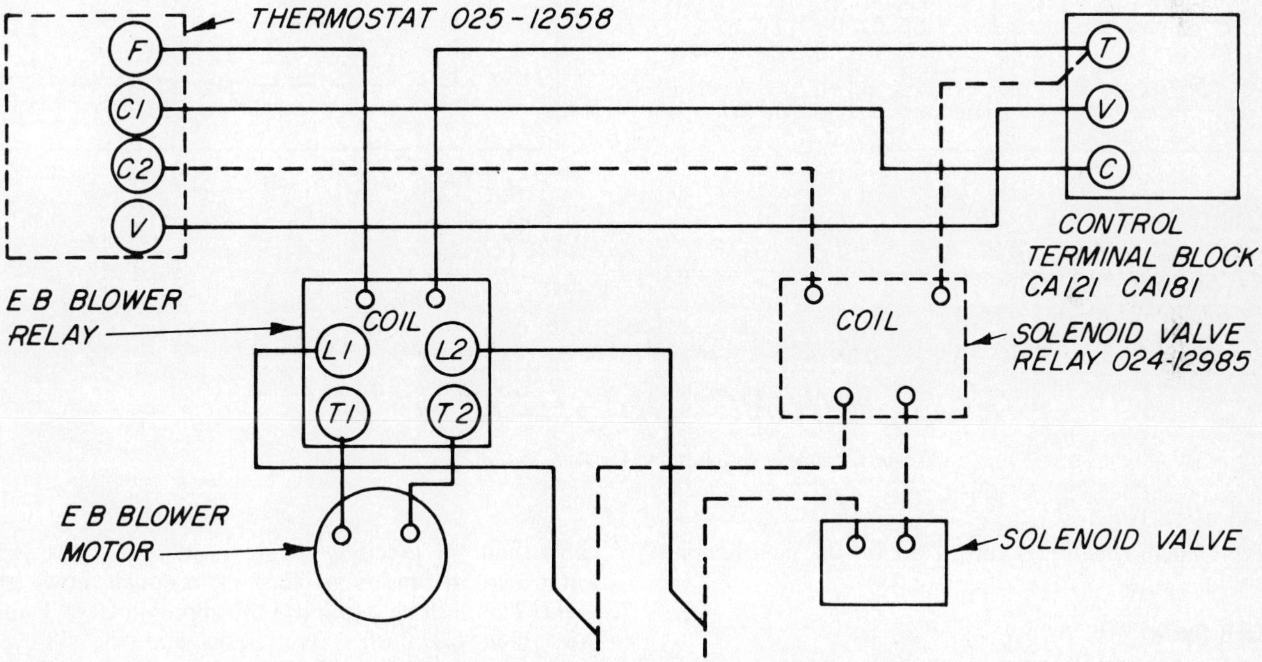

Fig. 23-3. Finished drawing of Fig. 23-2.

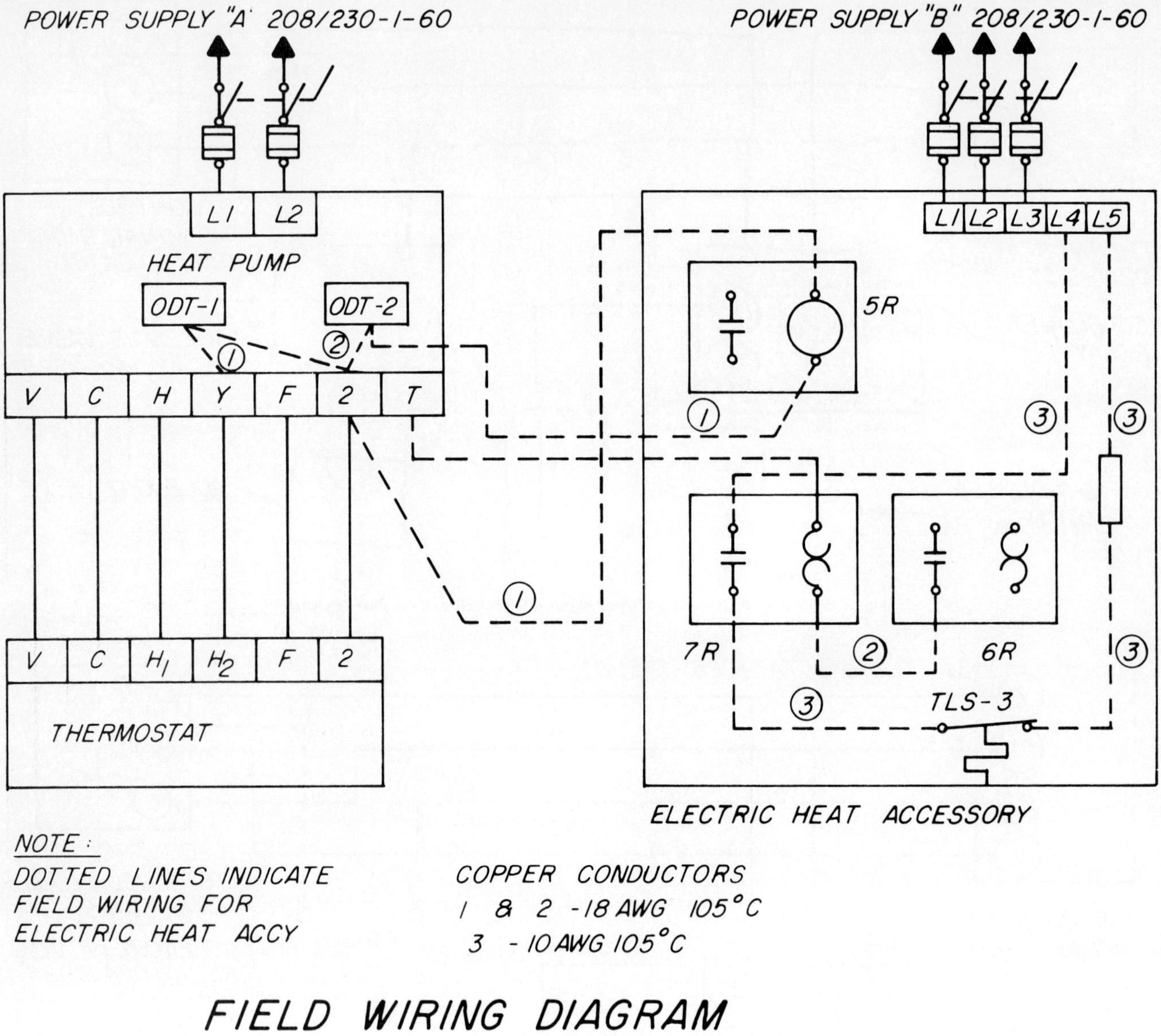

Fig. 23-4. Field wiring diagram for a heat pump with auxiliary electric heat.

ervoir filled, ensuring a clean start for the next heating season.

Flush Cycle

The timed flush cycle removes residuals from the reservoir and consists of the following components:

F—Flush valve
G—Fill valve
J—Control transformer
K—Clock timer
O—Drain

During the flush cycle, the clock timer opens the flush valve; the solid-state fill controls open the fill valve; and six precisely located water-inlet jets swirl water into the reservoir, flushing residuals down the drain. The flushing continues for approximately 3 minutes; then the flush valve closes and the fill-cycle controls shut off the fill valve when the reservoir is refilled. At the factory the humidifier is adjusted to give one flush cycle for every 2 hours of heating-element operation. However, field adjustment of the timer can give a choice of flush-cycle frequency from 2 to 6 hours. A flush cycle every 2 hours of *on* time is recommended for most jobs. Where soft water is prevalent or where a water softener is in the plumbing system, the adjustable timed flush cycles can be decreased to 6 hours of on time.

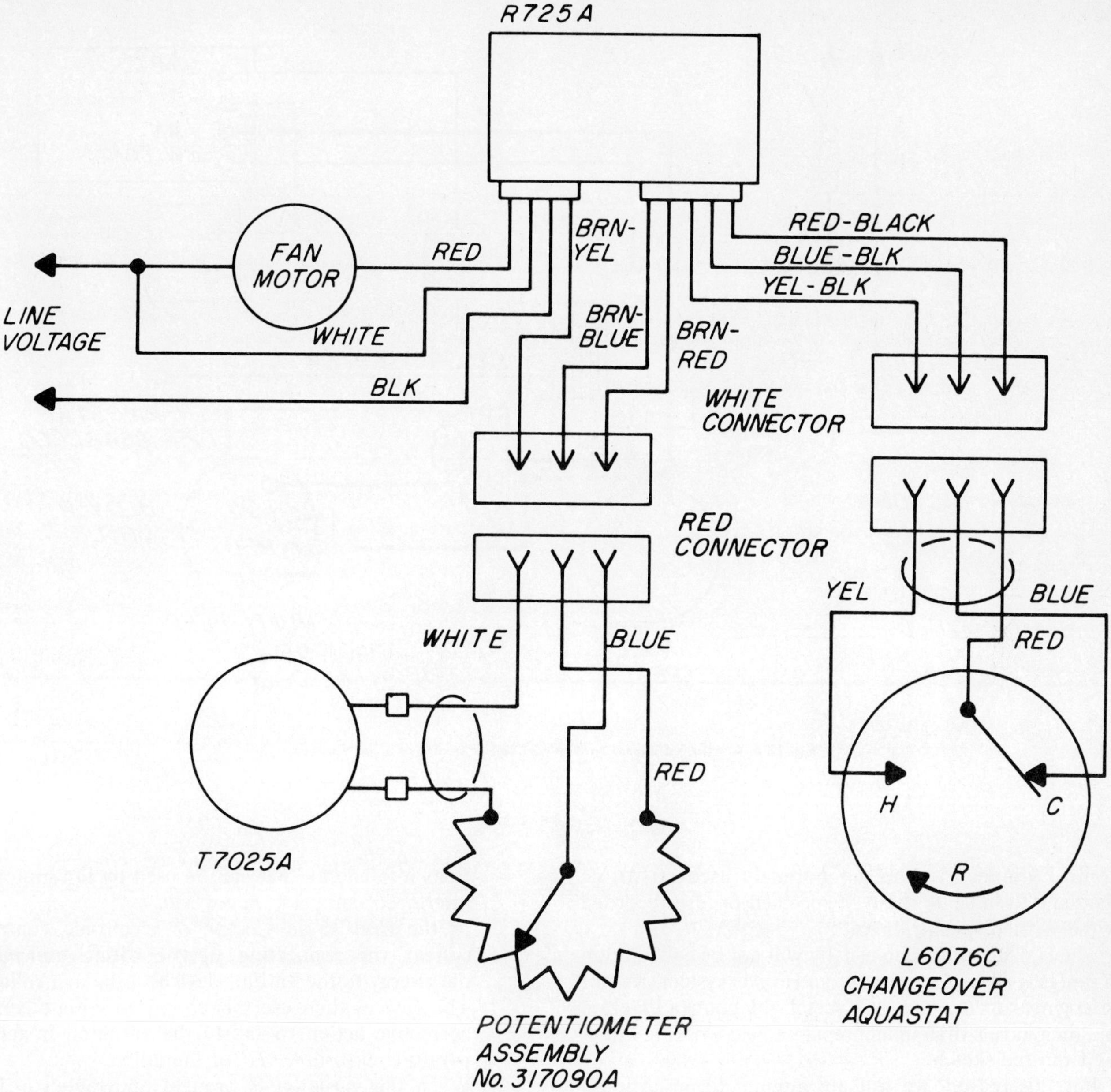

Fig. 23-5. Wiring diagram of an HVAC control system containing a changeover aquastat.

The timer has a manual override control knob which, when turned to the proper position, will give a *flush* whenever desired. Turning this knob also gives a positive "check-out" of all the humidifier operation functions.

Overflow Protection

C—Overflow safety probe
D—Overflow drain

The transparent reservoir has a separate overflow drain compartment. In the event of flush or fill-valve failure, water will run out this independent overflow drain compartment. The overflow safety probe is in the solid-state circuit and will shut off the fill valve in the event the drain line becomes plugged.

The manually reset capillary-actuated limit control (H) senses the heating-element temperature and shuts off the element if a temperature of 260°F is reached.

CONTROL TYPES

Controls for HVAC equipment are either electric, electronic, pneumatic, or combinations of the three.

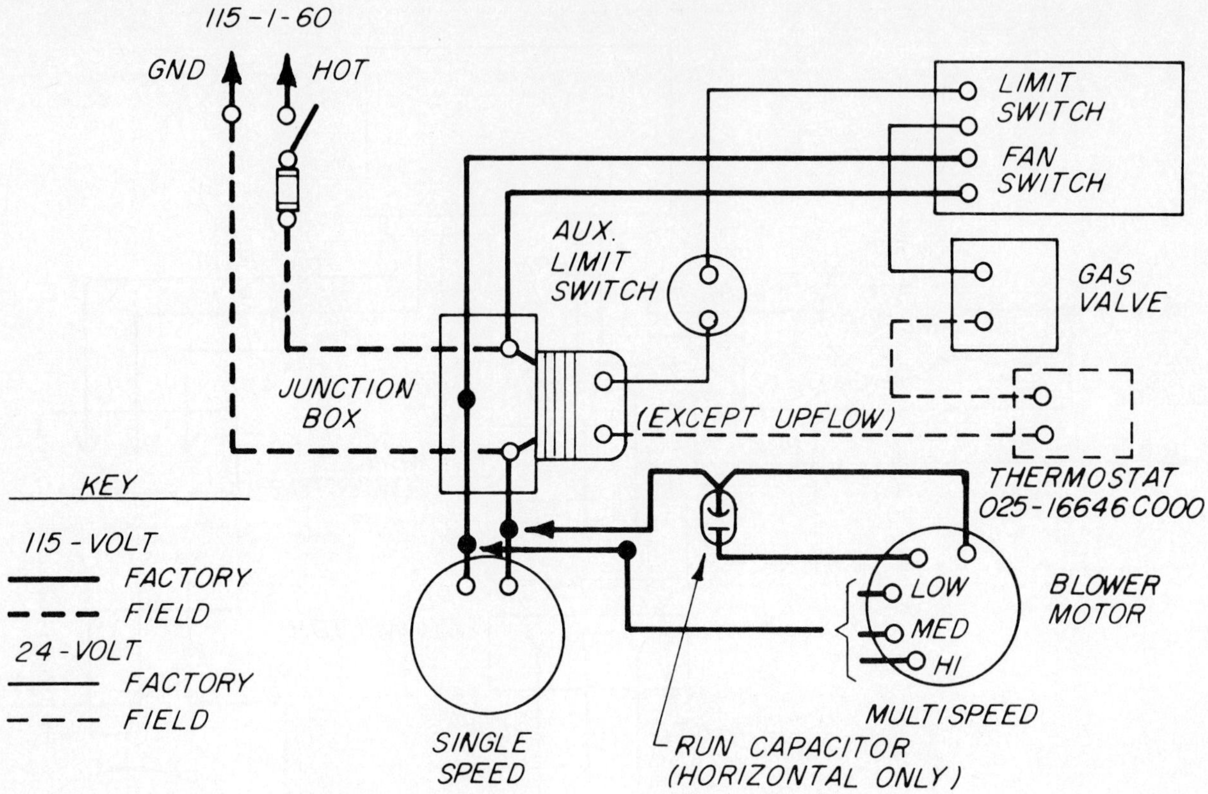

Fig. 23-6. A wiring diagram for low-voltage controls for a gas furnace.

Control graphic symbols are normally used on wiring diagrams to indicate the type specified or else recorded in the written specifications.

The HVAC draftsman usually will not be called upon to design a control system for an HVAC system. Rather an engineer or designer will sketch the control diagram and then have a draftsman prepare a neat working drawing from the sketch.

For this reason, we will not attempt to describe all the combinations of control systems in use, as this alone would fill volumes. However, the HVAC draftsman should have a basic understanding of the various controls used in conjunction with HVAC equipment in order to approach his work more intelligently and more informed.

BASIC FUNCTIONS OF HVAC CONTROL SYSTEM

A fully automatic HVAC control system performs the following basic functions:

1. Sensing elements measure variations in humidity, pressure, and temperature.
2. The control mechanism converts these variations into energy that can be used by fan motors, motorized dampers, valves, etc.
3. In the case of an electric or electronic control system, the connecting electric wiring transmits the energy to the various devices to be controlled.
4. The devices then use this energy to initiate some corrective action to satisfy the variation in temperature, pressure, and/or humidity.
5. When the variation is brought under control by the sensing elements, the sensing element signals the control mechanism or connecting means to stop the corrective action, since the desired conditions have been satisfied.

TYPES OF CONTROL CIRCUITS

Fig. 23-2 shows an engineer's sketch of a wiring diagram of a blower motor in a fan-coil unit. Fig. 23-3 shows this same wiring diagram after it has been prepared by an HVAC draftsman.

Fig. 23-4 shows another field wiring diagram, this one for a heat pump with auxiliary resistance electric heat. The student is not expected to fully understand the function of these wiring diagrams; they are merely

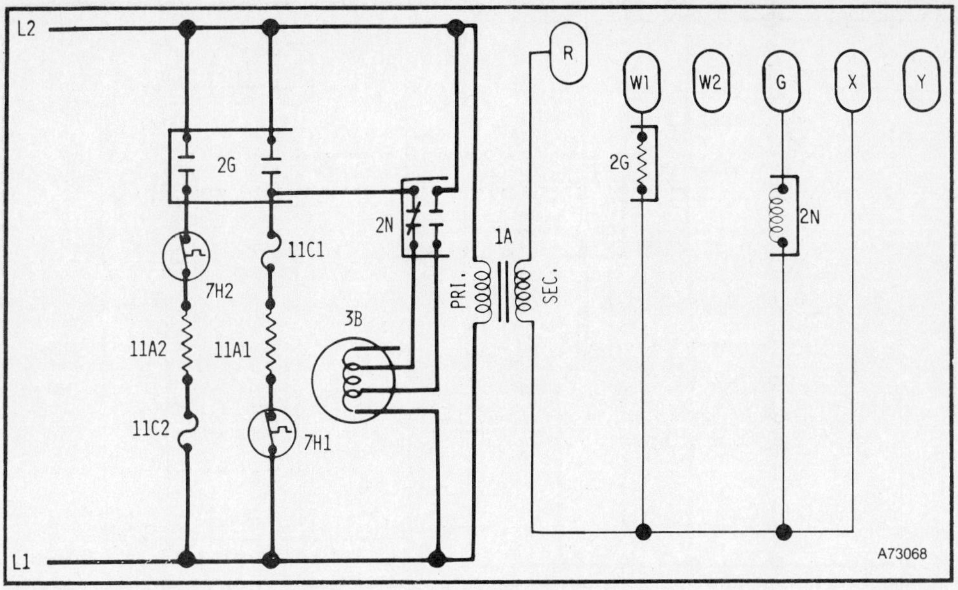

Fig. 23-7. Line-to-line wiring diagram for 8- to 10- kilowatt electric furnace.

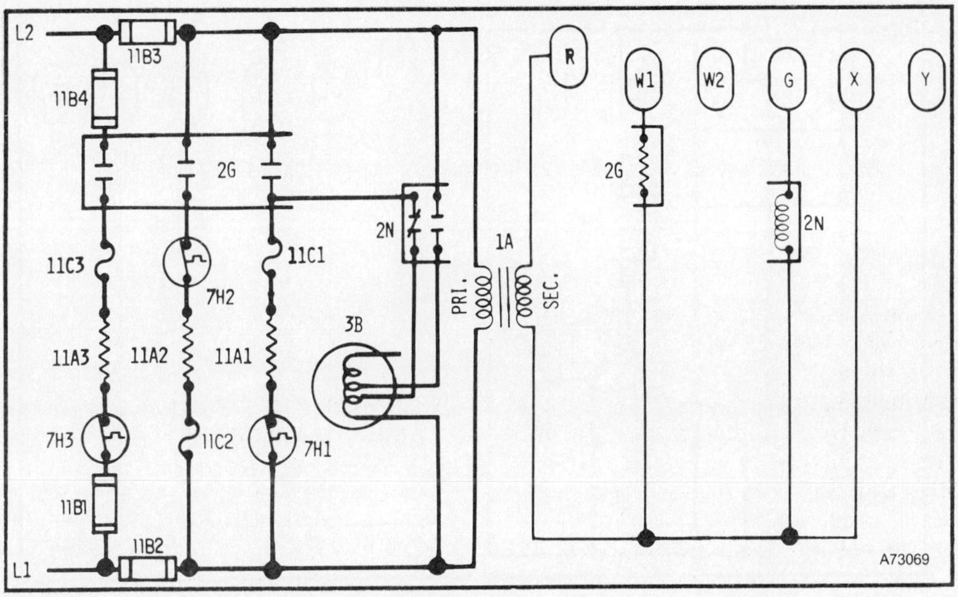

Fig. 23-8. Line-to-line wiring diagram for 15-kilowatt electric furnace.

used to show how such diagrams appear on working drawings.

Figs. 23-5 through 23-12 represent other control wiring diagrams actually used on HVAC working drawings. When reviewing these drawings, try to visualize how each symbol was drawn and what templates were used, if any.

SUMMARY

Controls for HVAC equipment are either electric, electronic, pneumatic, or combinations of the three.

Controls are used to sense and correct variations in temperature, pressure, or humidity.

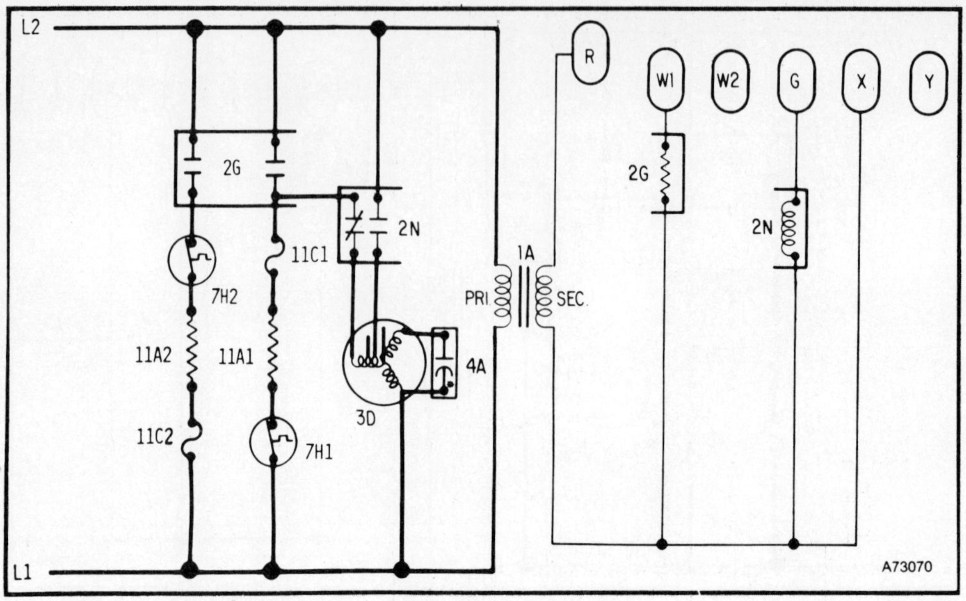

Fig. 23-9. Line-to-line wiring diagram for 7- and 10-kilowatt electric furnace.

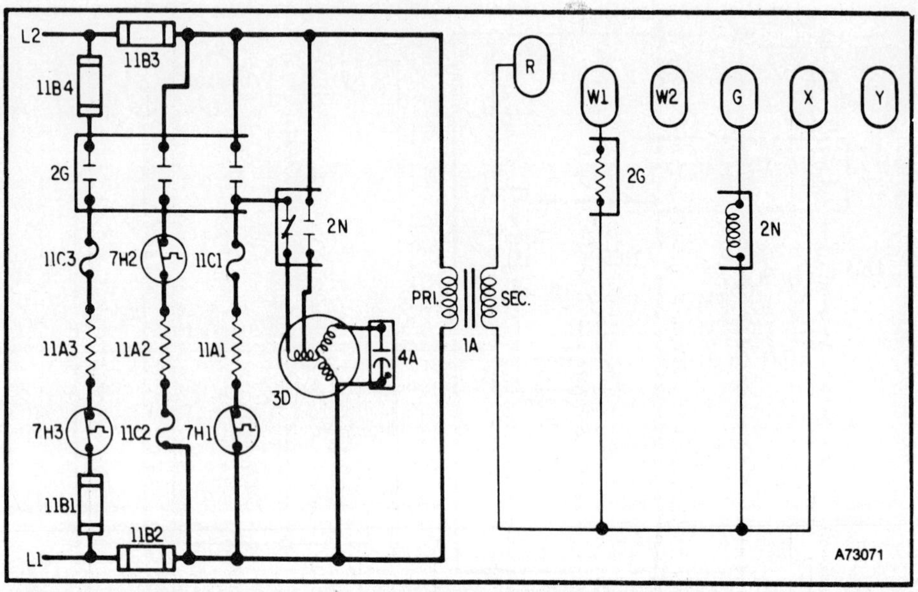

Fig. 23-10. Line-to-line wiring diagram for 15-kilowatt electric furnace.

UNIT 23 — ASSIGNMENT

1. Prepare a sheet of 11-by-17-inch tracing paper with title block and border lines; turn to Fig. 23–13.
2. In the center of your paper, draw the wiring diagram shown in the figure. This is just a freehand sketch; your drawing should be done with a straightedge, triangles, and templates. Be sure to include all lettering.
3. List the components of the following cycles:

a. Fill cycle ______________________

____

____

b. Humidification cycle ______________________

____

____

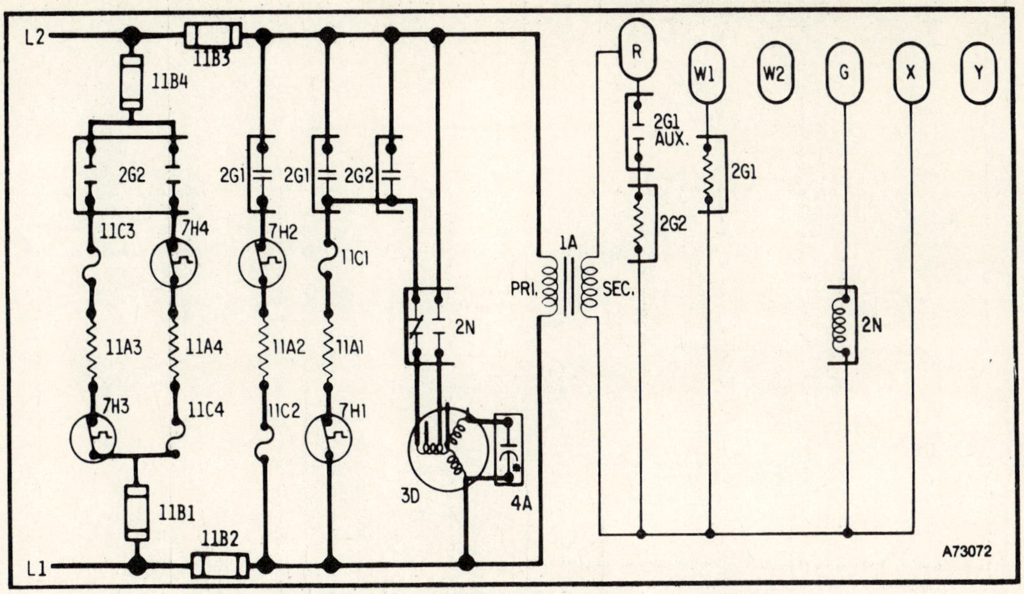

Fig. 23-11. Line-to-line wiring diagram for 20-kilowatt electric furnace.

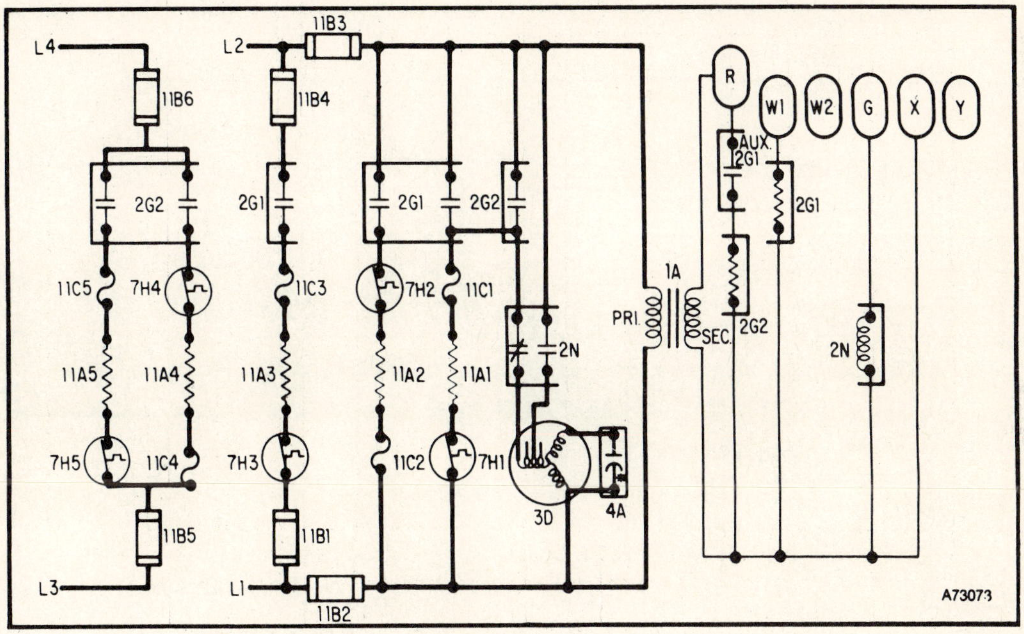

Fig. 23-12. Line-to-line wiring diagram for 25-kilowatt electric furnace.

c. Flush cycle ______________________________

d. Overflow protection ______________________________

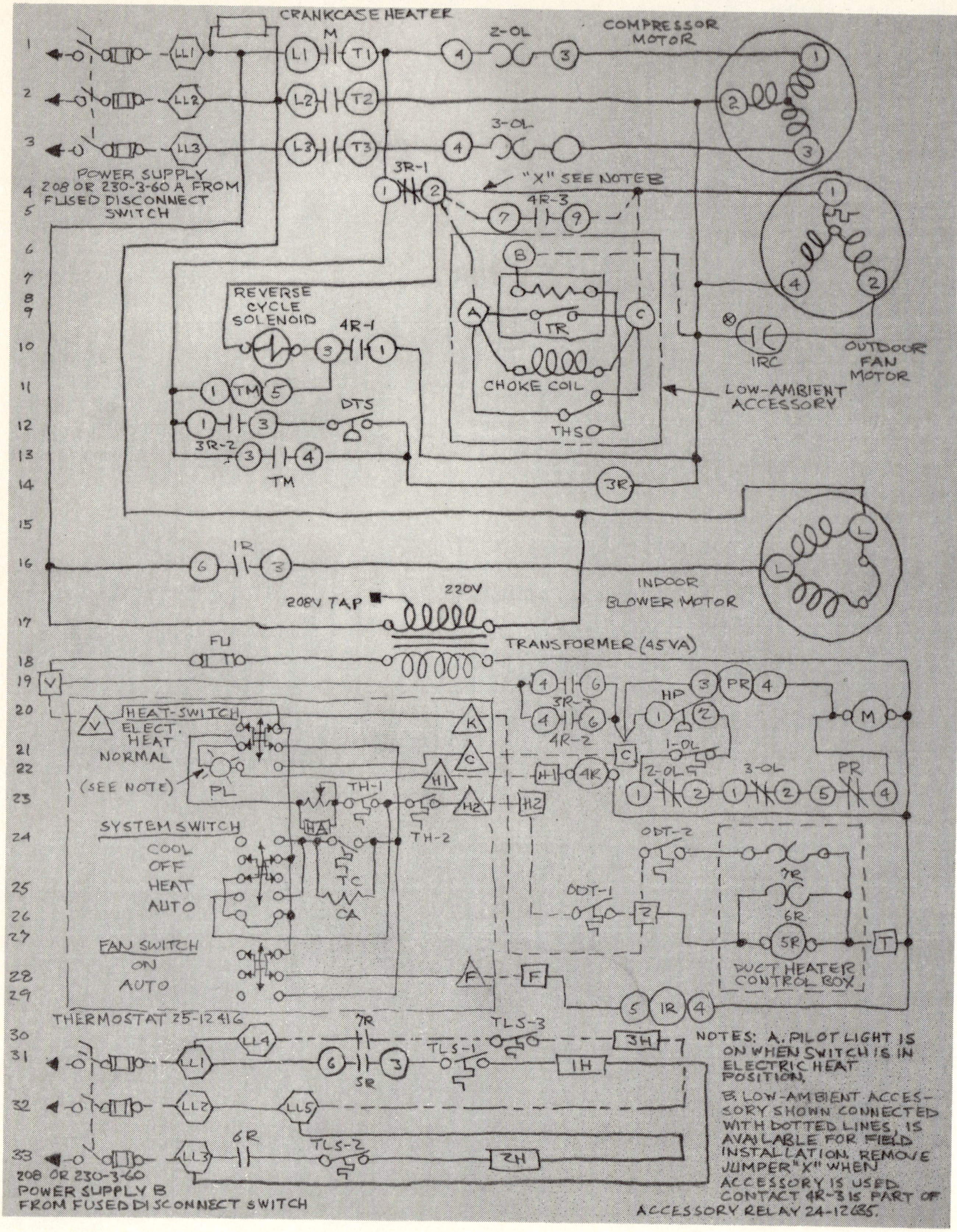

Fig. 23-13. Engineer's sketch for use in the assignment in this unit.

SECTION V

Appendices

APPENDIX A

Heating and Ventilating Specifications

The specifications for a building or project are the written description of what is required by the owner, architect, or engineer. Together with the drawings, the specifications form the basis of the contract requirements for the construction. The following sample illustrates the general wording and contents of a typical heating and ventilating specification.

DIVISION 15 — MECHANICAL

HEATING AND VENTILATING

1. GENERAL:
   (A) The "Instructions to Bidders," "General Conditions" of the architectural specifications govern work under this Section.
   (B) It is understood and agreed that the Contractor has, by careful examination, satisfied himself as to the nature and location of the work under this Section and all conditions which must be met in order to carry out the work under this Section of the Specifications.
   (C) The Drawings are diagrammatic and indicate generally the locations of materials and equipment. These drawings shall be followed as closely as possible. The architectural, structural, and drawings of other trades shall be checked for dimensions and clearance before installation of any work under this Section. The heating and ventilating contractor shall cooperate with the other trades involved in carrying out the work in order to eliminate interference between the trades.
   (D) The Drawings and Specifications are complementary each to the other, and work required by either shall be included in the Contract as if called for by both.
   (E) It shall be the responsibility of the Heating and Ventilating Contractor to obtain drawings of equipment and materials which are to be furnished by the Owner under separate contract, to which this Contractor is to connect, and for which it is necessary that the Heating and Ventilating Contractor coordinate the work under this Section.
   (F) All services that are interrupted or disconnected shall be rerouted and reconnected in order to provide a complete installation.
   (G) All work shall be performed by mechanics skilled in the particular class or phase of work involved.
   (H) All equipment shall be installed in strict accordance with the respective manufacturer's instructions or recommendations.
   (I) Where appropriate, all equipment shall be UL approved.
   (J) The Heating and Ventilating Contractor shall present five (5) copies of shop drawings or brochures for all fixtures, equipment, and fabricated items to the Engineer for the Engineer's approval. The Heating and Ventilating Contractor shall not proceed with ordering, purchasing, fabricating, or installing any equipment prior to the Engineer's approval of the shop drawings and brochures. Checking is only for general conformance with the design concept of the project and for general compliance with the information given in the contract documents. Any action shown is subject to the requirements of the Plans and Specifications. The contractor is responsible for dimensions that shall be confirmed and correlated at the job site, fabrication processes and techniques of construction, coordination of his work with that of all other trades, and the satisfactory performance of his work.
   (K) All materials and equipment shall be new and undamaged and shall be fully protected throughout the construction period in order that all equipment and materials shall be in perfect condition at the time of acceptance of the building by the Owner. It shall be the responsibility of the Heating and Ventilating Contractor to replace any damaged equipment or materials he is furnishing.
   (L) The naming of a certain brand or manufacturer in the Specifications is to establish a quality standard for the article desired. This Contractor is not restricted to the use of the specific brand or manufacturer named. However, where a substitution is requested, a substitution will be permitted only with the written approval of the Engineer. The Heating and Ventilating Contractor shall assume all responsibility for additional expenses as required in any way to meet changes from the original materials or equipment specified. If notice of substitution is not furnished to the Engineer within fifteen days after the General Contract is awarded, then equipment named in the Specifications and materials named in the Specifications are to be used.
   (M) EXCAVATION AND BACKFILLING:
      (1) The Heating and Ventilating Contractor shall do necessary excavation, shoring, and backfilling to complete the work under this Section of the Specifications under the supervision of the Gen-

eral Contractor. No foundation or structural member shall be undermined or weakened by cutting, unless provisions are made to strengthen the member so weakened as necessary.

(2) Excavation shall be cut to provide firm support for all underground conduit, pipe, etc.

(3) Backfill shall be provided as specified under *Backfilling*.

(N) CODES AND STANDARDS: All materials and workmanship shall comply with all applicable codes, specifications, industry standards, and utility company regulations.

(1) In cases of differences between the building codes, Specifications, state laws, industry standards, and utility company regulations in the Contract Documents, the most stringent shall govern. The Contractor shall promptly notify the Architect in writing of any such existing difference.

(2) *Noncompliance:* Should the Contractor perform any work that does not comply with the requirements of the applicable building codes, state laws, industry standards, and utility company regulations, he shall bear the cost arising from the correction of the deficiency.

(3) Applicable codes and all standards shall include all state laws, utility company regulations, and applicable requirements of the following nationally accepted codes and standards:

a. National Building Code
b. National Electrical Code
c. Industry codes, Standards, and Specifications:
1. AMCA—Air Moving and Conditioning Association
2. ASHRAE—American Society of Heating, Refrigeration, and Air Conditioning Engineers
3. SMACNA—Sheet Metal and Air Condition Contractors' National Association

(O) This Contractor shall do all cutting and patching of work as necessary to complete the work under this Section. All finished work in conjunction with this work shall be repaired to perfectly match adjoining finished work. This work shall conform to each respective specification section for the particular phase involved. The Heating and Ventilating Contractor shall employ tradesmen skilled in the particular trade involved in carrying out this work. This work shall proceed only with the approval of the Architect. This Contractor shall perform the work under this Section of the Contract in such a way that the amount of cutting and patching of other work shall be kept to an absolute minimum.

(P) This Contractor shall obtain all permits and arrange for all inspections necessary for the installation of this work. All fees in this relation shall be paid for by the Heating and Ventilating Contractor. This Contractor shall provide the Owner with certificates of inspection from all authorities having jurisdiction. This Contractor shall be responsible for notifying the authorities having jurisdiction at each inspection stage, and no work shall progress until the inspection has been completed and the work approved.

(Q) COORDINATION:

(1) It is called to the Heating and Ventilating Contractor's attention that the ductwork and piping shown on the base bid plan have extremely close clearance.

(2) It shall be the responsibility of the Heating and Ventilating Contractor to coordinate the work under this Section with the Plumbing, and electrical Contractor and other subcontractors.

(3) This Contractor shall check approved equipment drawings of other trades so that all roughing-in work shall be of proper size and in the proper location, for all specific equipment used.

(4) The Heating and Ventilating Contractor shall furnish necessary information to the related trades and shall properly coordinate this information.

(5) The Heating and Ventilating Contractor shall cooperate with the other trades involved in order to eliminate any interference between the trades. This Contractor shall make minor field adjustments to accomplish this.

(6) This Contractor shall cooperate with the other subcontractors in order to establish the responsibilities of each so that work can be completed without delay or interference by this Contractor.

2. WORK INCLUDED:

(A) The work in this Section shall include furnishing all labor, equipment, materials, supplies, and components for complete heating and ventilating systems as indicated on the Drawings and Specifications.

3. ELECTRICAL WORK:

(A) The Heating and Ventilating Contractor is to furnish and attach all necessary components for the heating and control system. The control wiring is to be performed by a qualified electricial contractor employed by the Heating and Ventilating Contractor.

(B) The Heating and Ventilating Contractor is to furnish all motor starters and control components that are specified in the mechanical section of these specifications or shown in schedules on the drawings.

(C) The Electrical Contractor shall furnish all other starters and disconnect switches. The Heating and Ventilating Contractor shall be responsible for the electrical connections and for the electrical wiring between the elements of the pneumatic control system.

(D) The Electrical Contractor shall furnish all conduit, fittings, and materials for the electrical work to connect to the heating and ventilating equipment.

(E) The electrical work required in the Heating and Ventilating Section shall conform to all requirements of the Electrical Section of these Specifications.

(F) All equipment shall be suitable for 208-volt, three-phase, 60-hertz electrical characteristics except motors under ½ hp, which are to be 120-volt, single-phase, 60-hertz. Motors of other characteristics are to be used only where specifically indicated.

4. EQUIPMENT:

(A) VENTILATING UNITS:

(1) Ventilating units shall be the draw-through type as scheduled on the drawings.

(2) Units shall be installed in accordance with the manufacturer's recommendations and be factory assembled and tested.

(3) Casings shall be constructed of steel, reinforced and braced for maximum rigidity. Casings shall be of sectionalized construction and have removable panels for access to all internal parts.

(4) Fans shall be DIDW Class 1 centrifugal and shall be statically and dynamically balanced at the factory. Fan shafts shall run in grease-lubricated ball bearings, the grease line for internal bearings being brought out to the exterior of the casing.

(5) The interior of the unit casing shall be insulated with 1-inch blanket fiber glass insulation, coated to prevent erosion.

(6) Casings and all accessories, with the exception of the coils described in the following paragraphs, shall be galvanized steel or shall be given a protective baked-enamel finish over a suitable rust inhibitor.

(7) Coils shall have seamless copper tubes and aluminum fins. Fins shall be bonded to tubes by mechanical expansion of the tubes. No soldering or tinning shall be used in the bonding process. Coils shall have a galvanized steel casing no lighter than 16 gauge, shall be mounted so that they are accessible for service, and shall be removable without dismantling the entire unit. Steam coils shall be a nonfreeze, distributing type. Hot-water coils shall be type WS, and the steam coils shall be type NS steam distributing.

(8) Heating coils for hot-water service shall be pitched in the unit casing for proper drainage, and tested at 250 psig air pressure under water.

(9) Each unit shall be provided with an adjustable-speed V-belt drive, having a variable-pitch motor sheath to provide approximately ten-percent variation in speed above and below the factory setting. Drive shall be protected by a suitable guard with openings at fan and motor shafts for use of a tachometer.

(10) Motor shall be general-purpose, squirrel-cage type with same type bearings as air-handling unit. Nameplate rating of motor shall not be exceeded by fan BHP requirements.

(11) See details on drawings for mounting and isolation of units.

(12) Filter boxes shall be furnished with throwaway filters, as scheduled. Filter area shall be such that filter velocity is in accordance with the filter manufacturer's recommendation. Filter boxes shall have access doors on both sides. Filters shall fit snugly to prevent air bypass. Both sides of the filter box shall be flanged for fastener holes. Filter shall be not less than 2 inches thick.

(B) Convertor:

(1) Convertor shall be designed for heating hot water with steam. The unit shall be constructed in accordance with ASME Code and labeled for 150-psi working pressure in shell and tubes.

(2) Materials of construction shall be as follows:
a. Water chamber—cast iron.
b. Shell—carbon steel.
c. Tube sheet—rolled steel.
d. Support and Cradles—steel or cast iron.
e. Tubes—¾ inch O.D. 18-gauge seamless drawn copper.
f. Tube Spaces—soft bronze or copper alloy.

(3) Capacity given in schedule on drawings shall be based on a fouling factor of 0.0005.

(4) A valved ________ No. 60 vacuum breaker shall be provided on shell. At each convertor in water piping, a ________ Series 240 Relief valve, ASME rated and sized for full capacity of convertor, shall be installed and set to relieve at pressure shown on Drawings. Valve shall be located so it cannot be valved off from convertor. Pipe discharge shall be full size to the floor drain.

(5) The Contractor shall provide a stand for support of convertor, constructed of welded structural steel or steel pipe.

(6) Convertor shall be as manufactured by ________ Company, Type SU1.

(C) Pumps:

(1) Base-mounted pumps shall be ________ "Universal" of vertical split-case design, equipped with mechanical seals for 225°F operating temperature and built for not less than 125-psi working pressure. Motor shall be 1750 RPM, drip proof and specifically designed for quiet operation. The coupler between motor and pump shall be ________ flexible type. Spring-type couplers will not be acceptable. Pump and motor shall have oil-lubricated bronze sleeve bearings. Pump shall have removable bearings frame to permit disassembly of pump without disconnection of piping, hydraulically balanced bronze impeller, and solid cast-iron volute. Pump and motor shall be mounted on a steel base plate.

(2) Base-mounted pumps shall have pump connectors as manufactured by ________ and shall be stainless steel, Type SPCF. They shall be sized the same as the connecting pipe.

(D) Cabinet Unit Heaters:

(1) Cabinet unit heaters shall be self-contained and factory assembled, and shall consist generally of a filter, heating coil, fan with driving motor, and casing. Heater element shall consist of nonferrous fuses and fins.

(2) Cabinet shall be of steel and be wall mounted or ceiling mounted as indicated on the Drawings.

(3) Front of cabinet shall be removable and not lighter than 16-gauge steel. The back shall be not less than 18-gauge steel. Cabinet shall be provided with steel supports for fan, heating coil, and filter, and shall be steel reinforced members as required to provide a rigid exposed casing and silent operation.

(4) Enclosures shall be complete with inlet and outlet grilles and with doors for access to valves. The element shall be easily accessible for repair after installation of unit. Enclosures shall be galvanized, bonderized, or painted with rust-inhibiting paint at factory, and finished in baked enamel in standard color selected.

(5) Fans shall be centrifugal type with one or more wheels mounted on a single shaft. Fan wheels, shaft, bearings, and fan housing shall be mounted as an integral assembly on a heavy steel mounting plate and securely fastened to enclosures. Bearings of fan and motor shall be self-aligning, permanently lubricated ball type, or sleeve type with ample provision for lubrication and oil reservoirs. Bearings shall be effectively sealed against loss of oil and entrance of dirt.

(6) Filters shall be of the permanent type, mounted in tight-fitting slide-out frames arranged to permit filter renewal or cleaning without removal of front panel.

(7) Heaters shall be tested and rated in accordance with standard test codes adopted jointly by the Unit Heater Division, Air Moving and Conditioning Association, Inc., and the American Society of Heating, Refrigerating, & Air Conditioning Engineers.

(8) Motors shall be resilient mounted on a cushion base, shall be for constant or multispeed operation, and shall have overload protection.

(9) See drawings for capacity, type, and arrangement.

(E) Fin-Tube Radiation: Fin-tube heating elements and enclosures, together with required mounting components and accessories, shall be furnished and installed as indicated on plans. Material shall be as manufactured by ______ Radiator Company or approved equal.

(1) *Nonferrous Heating Elements:* Nonferrous heating elements shall consist of full-hard aluminum plate fins, permanently bonded to copper seamless-drawn tube and guaranteed for working pressure at 300°F not less than 150 psi for 1-inch tube and 200 psi for 1¼-inch tube. Fins shall be embedded in copper tube at least .007 inch.

(2) *Enclosures and Accessories:* Enclosures and accessories shall be of style and dimensions indicated on plans and shall be fabricated from electrozinc-coated, rust-resistant, bonderized steel. Enclosures shall be 14 gauge. Enclosure louvers shall be "pencil-proof" type. On wall-to-wall application, enclosures shall be furnished in one piece up to a maximum of 10-feet enclosure length for rooms or spaces measuring a maximum of 10 feet 10 inches in wall length, and a 6-inch end trim shall be used at each end. Enclosures shall be furnished in two or more lengths for wall lengths exceeding 10 feet 10 inches.

Corners and end enclosures shall have same method of joining.

End trims, furnished with roll-flanged edges, shall be used between ends of enclosures and walls on wall-to-wall applications. End trims shall be 6-inches maximum length and shall be attached without visible fasteners. Corners and end enclosures shall be furnished where indicated, shall be the same gauge as enclosures, and shall fit flush with enclosures.

(3) *Enclosure-Supports:* Type-A Dura-Mount Back: Enclosures shall be supported at top by means of a 20-gauge roll-formed continuous mounting channel fabricated from electrozinc-coated, rust-resistant, bonderized steel. There shall be a minimum of eight bends forming strengthening ribs running the length of the channel. The top projection of channel shall position the enclosure ⅞ inch from wall to prevent "gouging" of walls when enclosure is installed or removed. A sponge-rubber dirt sealer shall be provided to the offset section of channel permanently anchored to wall, allowing enclosure installation or removal without disturbing the sealer. Dirt sealers attached to rear flange of enclosure are not permitted.

(4) *Enclosure Brackets and Element Hangers*:

a. Enclosure brackets and element hangers shall be installed not farther than 4 feet. Brackets shall be die-formed from 3/16-inch-thick stock, 1½ inches wide, and shall be lanced to support and position lower flange of enclosure. Enclosures shall be firmly attached to brackets by setscrews, operated from under the enclosure. Devices that do not provide positive fastening of enclosures are not acceptable. Brackets shall be inserted in prepunched slots in mounting channel to ensure correct alignment and shall be fastened securely to wall at bottom.

b. Sliding saddles shall support heating element and provide positive positioning of element in enclosure to ensure maximum heating efficiency while preventing any possibility of fin impingement on brackets or enclosure joints during expansion or contraction. Element supports shall be double-saddle design fabricated from 16-gauge electrozinc-coated, rust-resistant, bonderized steel. Saddle shall slide freely on saddle support arm bolted to support bracket. Support arm shall allow 1½-inch height adjustment for pitch. The element support saddle shall allow 1⅝-inch lateral movement for expansion and contraction of heating element. Rod or wire hangers are not acceptable.

(5) *Access Doors:* Access doors shall be provided where noted on plans. Doors shall be at least 8 by 8 inches, shall be located in 12-inch-long access panels, and shall have the same gauge as enclosures.

(6) *Enclosure Dampers:* Enclosure dampers shall be provided where indicated. Damper blades shall be fabricated from 18-gauge electrozinc-coated, rust-resistant, bonderized steel, flanged for added rigidity, and shall be permanently attached to enclosures. Threaded damper screw and trunnion shall provide positive operation of blade in any position between open and closed. Damper shall be operated by knob.

(F) FANS: Fans shall be provided as shown and scheduled on drawings. Care shall be taken when mounting in-line fans between joists to provide access to all parts of the fan. (No fans shall be mounted on roof.)

(1) *In-Line Fans:* All these fans shall be equal to ______ "Square Line Centrifugal Fan" with backward-inclined centrifugal wheel, belt-driven Model SQ. B. See Schedule on drawings.

(2) *Fume Centrifugal Fan:* (Welding Fumes) This fan shall be equal to "ILG" Type PE-Belted with cast-iron wheel. See drawings for capacity and mounting.

(3) *Propeller Fans:* See drawings for location and capacity.

(4) Note: See drawings for mounting and isolation.

(G) UNDERGROUND PIPE CONDUIT: (See Alternate A)

(1) *Scope:* The underground conduit shall include reinforced concrete foundation slabs, unit cast-iron pipe supports, unit sleeve-alignment guides, unit pipe anchors and tile conduit envelope, and shall be insulated as specified in these Specifications.

(2) The tile shall be "Therm-O-Tile," or approved equal, and shall be vitrified and equal to or better than "extra-strength" clay pipe of the same diameter as per ASTM Specification C-278. All joints in the conduit envelope shall be sealed with ______ cement mortar mixed in the proportion of one part ______ cement to two parts clean, sharp sand, and coated after cement has set with "Therm-O-Mastic" compound. Additional weatherproofing consisting of one layer of 30-pound felt shall be applied completely over the tile and lapped over the concrete foundation slab. Laps shall be at least 3 inches and shall be sealed with "Therm-O-Mastic" or hot asphalt.

(3) *Tile Conduit Foundation:* The foundation slab shall be at least 4-inch-thick concrete as specified under Section 3, reinforced with 6 by 6 by No. 10 welded wire mesh, and shall have on emergency drain throughout its run. The foundation slab shall extend through building or manhole walls and shall rest on the masonry of these walls. The emergency drain shall be continued to building or manhole sumps, to sewers, to dry wells, or to outfalls. Where con-

duit passes under roadway, it shall be reinforced with an envelope of concrete, 4-inches thick.

(4) *Pipe Supports, Anchors, and Guides:* These accessories shall be as regularly furnished by the manufacturer of the "Therm-O-Tile" conduit and especially designed to fit within the conduit envelope. Location shall be as shown on the Plot Plan. There shall be two guides on each side of each drop.

(5) Conduit shall be sized per the manufacturer's recommendation for the pipe sizes enclosed.

(6) *Shop Drawings:* The successful contractor for the conduit work will be required to furnish, for approval, scale drawings showing cross sections through each separate conduit run and giving all required dimensions of the conduit and centers on which each pipe line will be located within it.

(7) *Wall Openings:* The space around the conduit where it enters walls shall be closed with brick and mortar and completely covered with weatherproof mastic. The space on the inside of the conduit on the inside of building or manhole wall shall be filled with brick and mortar.

(8) *Backfilling:* Backfilling shall be done carefully, and the surface restored to its original condition. Backfilling and tamping shall proceed simultaneously on both sides of the conduit (lengthwise of the trench) until the arch is covered to a depth of at least 18 inches. Rock or stone that might damage the tile shall not be used in the backfill. Backfilling shall comply with Section 2 of the Specifications.

(9) Expansion joints shall be furnished or installed where indicated on Drawings. Model and size shall be as shown on Drawings. Expansion joints shall be as manufactured by ______ Corporation.

(10) Conduit and accessories shall be installed in accordance with the manufacturer's recommendations.

(H) Steam Pressure Reducing Valve Assembly:

(1) *SPRV:* Valve shall be as manufactured by ______ Specialty Manufacturing Corp. Valve body shall be iron and shall have flanged connections. Valve shall be single seated and shall have a stainless-steel valve and seat and deep gland packed stuffing box for minimum friction on valve stem. Valve shall have maximum working pressure of 250 psig, an initial pressure of 75 psig, and a reduced pressure of 10 psig. All parts shall be renewable and interchangeable, and valve seat and disc shall be such that they can be changed without removing valve body from the line. Valve shall be ______ No. 7000 3-inch size, capacity 8600 pounds condensate per hour.

(2) Components of pressure-reducing valves shall be installed as recommended by the manufacturer. See diagram on Drawing M-5.

(3) Safety valve on reduced-pressure side of pressure-reducing station shall be as manufactured by ______ Company, ASME tested and rated, having cast-iron body and 250-pound flanged inlet connection. Valve shall be Catalog No. 630 and shall have capacity indicated on Drawings based on a 3-percent accumulation. Pipe discharge of safety valve shall be full size to exterior of building.

(I) Steam Specialties:

(1) *High-Pressure Steam:* Traps shall be ______ side inlet inverted bucket traps. Traps shall be built for up to 125-psig working steam pressure.

(2) *Low-Pressure Steam:* Traps shall be ______ B series, float and thermostatic type, built for up to 15-psig working steam pressure. Traps shall be sized at ¼-pound of differential.

(3) Traps shall have cast-iron or semisteel body with stainless-steel trim.

(4) Flash tank shall be as detailed on the Drawings.

(J) Heating Water Specialties:

(1) Hot-water expansion tank shall be ______ built for 125-pounds water working pressure, and shall be equipped with proper size "Airtrol" tank fitting and air charger fitting. Tank shall be ______, ASME of size as indicated on drawings.

(2) Air separaters shall be ______ "Rolairtrol." They shall be installed as instructed by manufacturer and shall afford adequate clearance to remove built-in strainer.

(3) Water makeup pressure-reducing valves shall be ______ Type 12, with screwed ends.

(4) Air vents installed on heating coils and at high points in water piping and on all upfeed radiation units shall consist of an air chamber and manual air vent of a type suitable for the particular application where used. See detail on drawings.

(5) *Balancing Fittings:* The Heating and Ventilating Contractor shall furnish and install ______ Series 700 flow meter fittings as a permanent part of the water piping systems for use with ______ No. SD-400-4 flow meter in determining and balancing water flow in all systems. These fittings shall be at points indicated on Drawings and shall be located so as to provide 15 diameters upstream and 5 diameters downstream of uninterrupted straight pipe. The systems shall be balanced as closely as practicable by use of the aforementioned flow meter.

(6) Water relief valves shall be ______ Series 230 or 240 as indicated on Drawings. Valves shall be set to relieve at the pressures indicated on Drawings. Pipe discharge shall be full size to floor drain.

(7) *Thermometers:* Thermometers shall be ______, Type 105 straight form or Type 115-90 degree back angle form, as applicable, industrial type, with standard separable socket, 9-inch scale, 30°F to 240°F range, cast aluminum case, with red-reading mercury tube.

(8) *Pressure Gauges:* Pressure gauges for pipe lines shall be ______ drawn case gauges, No. 1000 for pressure, No. 1002 for vacuum, and No. 1004 for compound. Pressure gauges shall have brass bourdon tube soldered to socket and tip, brass movement, and white-coated metal dial 3½ inches in diameter, graduated to meet system design requirements, normal operating pressure being indicated middial. ______ No. 1106B brass pulsation dampener shall be provided on gauges near pumps and ______ No. 1092 T-handled cock in ¼-inch line to all pressure gauges. No. 1100 pigtail siphon shall be provided on all steam gauges.

(K) Condensate Pump:

(1) One duplex condensation pump shall be furnished and installed where shown on the Plans. The

pump shall be manufactured by the ______ Co. The pump shall be driven by open, drip-protected ball-bearing motors, rated for single-phase, 60-hertz, 120-volt ac operation. Each pumping unit shall have capacity at 205°F as indicated. The Heating and Ventilating Contractor shall furnish combination starters with disconnect switch and motor overload protection.

(2) The equipment is to include one cast-iron receiving tank, two pumping units, and accessories listed below:

a. The pumping units shall be bronze fitted throughout, shall be of the centrifugal type, and shall have rotating parts that have been dynamically balanced. The receiving tank, pumping units, and motors shall be assembled by the manufacturer to form an integral unit. A strainer with movable screen is to be installed in the receiving tank.

b. Automatic controls shall consist of a float switch, combination starters, and alternator. Controls shall provide automatic alternation between one pump motor and the other.

(3) If requested, the installing contractor shall secure from the pump manufacturer a factory test report which is to be submitted to the specifying engineer for approval. This report shall show the actual condensate capacity for the pumping units and the power input to the units, all as determined by tests of the actual equipment furnished. The test report is to be certified by the manufacturer as to its correctness and all particulars.

(L) DIFFUSERS, REGISTERS, AND GRILLES:

(1) *Diffusers:* Diffuser shall be ______ type SFSV with deflectrol, No. 7 finish. Dropped Collar shall be provided where called for on drawings.

(2) *Exhaust Registers:* Exhaust registers shall be ______ type GMRV or GFRV, No. 4 finish. Grille in Dark Room shall be light tight.

(3) *Supply Registers:* Supply registers shall be ______ type GMAV with No. 4 finish.

(4) *Exhaust Grilles:* Exhaust grilles shall be ______ type EMR or GFR with No. 7 finish.

(5) *Door Grilles:* Shall be provided by General Contractor. (See Drawing for location only.)

(M) MECHANICAL EXPANSION JOINTS: Mechanical expansion joints shall be provided as called for on drawings. Size, capacity, and type are given on drawings. Expansion joints and guides shall be equal to ______.

(N) FLEXIBLE CONNECTIONS: Bronze unbraided flexible metal nipples shall be provided to connect to steel heating piping below floor and shall extend from ¾-inch inlet to fin tube radiation above floor. See detail on drawings. Nipples shall be furnished in 18-inch length, ¾-inch hex male end, and suitable for temperatures up to 350°F.

(O) UNIT HEATERS: Horizontal hot-water–type heaters shall be provided by ______ or equal. See drawing for Model number, size, and capacity.

(P) FAN-COIL UNITS: Two units of the following type and capacity shall be provided: Model OH 600, 4.2 GPM, 42.2 mbh, 200°F EWT, 180°F LWT, P.D. - 12 feet, 3-speed, and a return-air plenum with permanent filter and a fan control switch. Motor shall have split capacitor, 1/20 hp, 120 volt/1 hp.

(Q) AUTOMATIC TEMPERATURE CONTROLS:

(1) *General:* The System shall be a complete system of automatic temperature regulation of the pneumatic type with electric accessories and components as indicated. Component parts of the system shall be manufactured by ______; the base bid shall be ______ and can be considered as an alternate at the option of the Contractor. The entire system shall be installed by the Control Manufacturer. All control items except room thermostats shall be properly identified with engraved plastic nameplates permanently attached. Room thermostat locations shall be coordinated to align vertically or horizontally with adjacent light switches or control instruments. Room thermostats shall be 5 feet 6 inches (nominal) above the floor. Room thermostat covers shall be open.

(2) MATERIALS: (Thermostats)

a. Firestats shall be UL approved, manual-reset type T-7602 with an adjustable temperature setting. Range hood shall be set at 250°F; all others at 125°F.

b. Freezestats shall be T-7606 with 20-foot temperature-sensitive element, located downstream from the coil. If any portion of the element senses a temperature below its setting, the contacts shall break.

c. Unit heater thermostats for space mounting shall be T-7162, line-voltage type with SP-ST switching action rated 6 amperes for full load and rated 36 amperes at 120 volts for lock rotor.

d. Surface-mounted aquastats shall be type T-7912 with adjustable set point and 10° differential. Contacts shall be rated 10 amperes at 120 volts.

e. Thermostatic sensors shall be T-5210 style "B" bulbs with 5½-inch and 4-foot capillaries or 8-foot and 17-foot averaging element. Sensors shall be designed to measure a temperature and to convert the measurement to an air-pressure signal that is transmitted to a receiver, a 3- to 15-psi signal. The sensor shall have pneumatic feedback.

f. Fluidic controlling receivers shall be T-9000. The instrument shall accept a 3- to 15-psi pneumatic signal from one or two temperature-relative humidity or pressure transmitters, and have pneumatic feedback and enclosed fluidic circuitry.

g. Receiver controller shall be T-5312 and shall accept a 3- to 15-psi signal from a remote transmitter.

h. Pneumatic thermometer shall be T-5500 to provide visual indication of the temperature. The thermometer shall accept a 3- to 15-psi signal from a remote transmitter. The T-5500 shall be for flush-mounted applications.

i. Day-night thermostats shall be T-4502 equipped with two (2) separate bimetallic elements for day and night operation. The thermostats shall be equipped with an indexing switch for changing the thermostat to day or night operation as desired, or from a control location by changing the air pressure from 15 to 20 pounds.

j. Remote capillary pneumatic thermostats shall be T-8000 for direct or reverse acting. The T-8000 shall be equipped with the proper capillary, *B* bulb for sensing outside air and water temperature and also of the averaging type for all other applications.

k. Remote capillary pneumatic thermostats for two-position applications shall be T-8000.

(3) VALVES:

a. Valves shall be sized by the control manufacturer and shall have threaded connections, except valves over 2 inches shall have flanged connections. Valve packing shall be U-cap silicone except where indicated. Maximum allowable pressure drop shall be 5-feet water column for water valves and 60-percent steam pressure.

b. Valve operators for valve ½ inch to 2 inches shall be V-3000 piston operated. The diaphragm shall be manufactured of Butyl rubber enclosed in a heavy die-cast aluminum housing.

c. Valves for steam or water service shall be V-3752 normally open type. The valve shall be equipped with a V-3000 piston operator. The V-3752 bodies shall be of high-grade cast red brass in sizes ½ inch through 2 inches and shall have a back-setting feature that permits changing the stem packing without interrupting service to the system. The modulating plug shall have a replaceable composition disc, especially compounded for steam or hot water, and shall provide an equal percentage relationship between valve lift and flow at a constant pressure drop.

d. Connector valves shall be V-3800 for steam or water with restrictor mounting space. The valve shall be equipped with a heavy-duty moulded rubber diaphragm of the oval piston type. The valve body shall be of cast red brass in sizes ½ inch through 2 and ¾ inch.

e. Valve for steam or water service shall be V-3970 normally closed type. The valve shall be equipped with a V-3000 piston operator. The V-3870 bodies shall be of high-grade cast red brass in sizes ½ inch through 2 inches and shall have a back-setting feature that permits changing the stem packing without interrupting service to the system. The modulating plug shall have a replaceable compulsion disc especially compounded for steam or hot water, and shall provide an equal percentage relationship between valve lift and flow at a constant pressure drop.

f. Three-way mixing valves shall be V-4322 and ½ inch through 2 inches in size. The body shall be three-way with screwed ends and made of high-grade cast red brass. The body shall be suitable for pressure to 150 psi.

(4) DAMPERS AND DAMPER MOTORS:

a. Automatic control dampers shall be interlocking and airtight. They shall be of opposed-blade construction for modulating service and of parallel-blade construction for two-position service. Dampers shall be of the multilouver construction with brass bearings, channel iron frame, and maximum width of 6 feet.

b. Control dampers shall be D-1200 or D-1300 and manufactured specifically to control the airflow in heating, ventilating, and air-conditioning systems.

c. Frames shall be made of No. 13 galvanized sheet steel, formed into channels and riveted. In addition to the rigid frame construction, corner brackets shall be used to maintain perfect alignment of the damper.

d. Blades shall consist of two formed No. 22 galvanized sheets, spot-welded together for strength to withstand high velocities and static pressures. Square blade pins shall be furnished to ensure nonslip pivoting of the blades when a damper is used as a single module or when it is interconnected with others.

e. Bushings shall be made of oil-impregnated sintered bronze and shall provide constant lubrication.

f. Synthetic elastomer seals shall be provided on the blade edges and on the top, bottom, and sides of the frame, and shall be capable of withstanding air temperatures from —20 to 200°F or from —65 to 400°F, as required for the application. The material and extruded form of the blade edge seals shall create a positive seal when the blades are closed. The seals shall be replaceable if they become damaged. Leakage shall be less than 0.5 percent when closing against 4-inch w.g. static pressure, based on conventional velocity of 2000 FPM.

g. Damper motors shall be provided for all automatic dampers and shall be of sufficient capacity to operate the connected damper. Where required, damper motors shall be equipped with positive positioners.

h. Automatic control dampers are specified to be provided as an integral part of the air units.

(5) PANELS:

a. Control cabinets shall be furnished where specified. In general, it is the intention of this specification that control cabinets be furnished for each air-handling unit, major equipment components, and elsewhere as specified. Control cabinets shall be fabricated of extruded aluminum or steel. The cabinets shall have a face panel for flush-mounting gauges and a subpanel for mounting of controllers, relays, etc. Those controls which require manual positioning or visual indication shall be flush mounted and identified with engraved nameplates on the face panel. The controls that must be accessible for maintenance and calibration only are to be mounted on the subpanel inside cabinet. Each item shall be identified by engraved nameplates.

(6) CONTROL PIPING:

a. Control piping shall be hard-drawn copper tubing where exposed and may be either hard- or soft-drawn copper tubing where it is to be concealed. Either solder or compression fitting shall be used. Tubing shall be run in a neat and workmanlike manner and shall be fastened securely to the building structure. All tubing in finished rooms shall be concealed. Where exposed in unfinished rooms, tubing shall be run either parallel to or at right angles to the building structure. In lieu of copper tubing, plastic tubing may be used in end compartment of unit and in control panel.

b. *Plastic Tubing:* In lieu of copper tubing, high-density virgin polyethylene may be substituted, subject to the following requirements. The tubing shall be rated 600-psi burst pressure at 72°F and 300-psi burst pressure at 140°F. Each tube shall be individually numbered at intervals not exceeding 1 inch. All tubing except tubing in control panels and junction boxes shall be installed in EMT conduit in sizes of ½ inch through 2 inches. Standard electrical fittings shall be used. All bends for conduit 1 inch and larger shall be standard purchased bends. There shall not be more than three 90-degree bends between pull boxes. Termination shall be made in standard electric junction boxes or enclosed control panels. Where connections are made from plastic to copper, protective grommets shall be used to prevent electrolysis. The conduit installation shall conform to the same standard as set forth in the copper piping requirements.

(7) INSTALLATION:

a. EXPOSED:

1. Single polyethylene tubing and soft copper aluminum tubing may be run exposed for a length of 18 inches or less. For lengths that exceed 18 inches, the lines shall be run within enclosed trough or conduit, and this tube carrier system shall be installed in a workmanlike manner, parallel to building lines, adequately supported, etc. All connections, except for terminal connections to valves, damper operators, etc., shall be made inside troughs, junction boxes, or control cabinets.
2. Factory-manufactured bundles of polyethylene tubing, with protective outer sheath and hard copper or aluminum tubing, may be installed without an additional trough or conduit envelope, provided that the tube system is installed in same workmanlike manner as specified above for trough and conduit systems.

b. CONCEALED-ACCESSIBLE:

1. Single polyethylene tubing and soft copper or aluminum tubing, either individual or bundled, shall be installed in a workmanlike manner, securely fastened to fixed members of the building structure at sufficient points to avoid excessive freedom of movement. Field-fabricated bundles shall be tied together with a sufficient number of nylon ties to present a neat, uniform appearance.

c. *Concealed-Inaccessible*: Single polyethylene tubes shall be run within enclosed trough or conduit. Factory-manufactured bundles of polyethylene tubing, with protective outer sheath and soft copper or aluminum tubing, may be installed without an additional trough or conduit envelope. Fitting connections to polyethylene tubing shall not be made within the inaccessible area.

d. *Piping Test:* The piping system shall be tested and made tight under pressure of 30 psi. Leakage will not exceed 10 psi in 12 hours.

(8) Miscellaneous relays, pressure switches, disconnect switches, PE and EP relays, time clocks and other items shall be provided as required for the sequence of control indicated. Time clock shall be seven-day type. The PE relays shall be located within 5 feet of the motor control device.

(9) *Air Compressor:* Air compressor shall be of the electric type complete with tank, gauges, combination pressure-reducing valves, low-pressure relief valve, filter assembly, and necessary accessories. The unit shall be of ample capacity to automatically maintain the desired air pressure with an idle period equal to at least twice that of the operating period. Compressor shall be single-stage, high-pressure (60- to 75-psi) type fitted with galvanized ASME reservoir. Motor shall be provided with built-in overload protection. Compressor and motor shall be sized so that their capacity is sufficient for future addition to building.

(10) WORK BY OTHERS:

a. Dampers and valves will be installed by the Mechanical Contractor.

b. Temperature-control wiring shall be the responsibility of the Temperature Control Contractor. This responsibility shall consist of wiring the following items:

1. P.E. switches
2. E.P. switches
3. Firestats
4. Freezestats
5. Air compressor
6. All interlocking wiring required for the system to function properly.

c. The electrical contractor shall be responsible for all power wiring to the equipment.

(11) SEQUENCE OF OPERATION:

a. *Air-Handling Unit Control:* When the unit fan is started, EP-1 is energized to position the outdoor air damper open through D-1 and through PE-1 on units 7 and 8 which energize their respective exhaust fans. When the unit fan is stopped, PE-1 is de-energized and the outdoor air damper is positioned fully closed. Firestat T-3, located in the filter discharge, stops the fan if the discharge temperature rises above 125°F. T-1, located in the discharge duct, controls V-1 modulater to bypass an increasing amount of water around the coil. T-2, with its element located on the face of the heating coil, prevents the discharge temperature from falling below 35°F by overriding T-1 and modulating V-1. Five (5) units have hot-water coils and require three-way valves; three (3) units are steam coils and require steam valves.

b. *Radiation Control:* During the day operation, T-1 controls V-1 to maintain the desired space temperature. During the night operation, T-1 controls V-1 to maintain a reduced night temperature of 60°F.

c. *Exhaust Fan Control:* T-1, located in the intake of each exhaust fan, stops the unit fan should the intake temperature rise above 125°F.

d. *Pump Bypass Control:* T-1, with its sensor located in the supply and return lines, controls V-1 to maintain the desired differential pump pressure.

e. *Convertor Control:* T-1, with its sensor TT-2 located in the convertor supply, is reset inversely with changes in the outside air temperature by TT-1, to control V-1 and maintain the desired convertor discharge temperature inversely with changes in the outside air temperature. T-2 located in the outside air, stops the pump should the outside air temperature rise above 70°F. When the pump is stopped, EP-1 is de-energized to position V-1, fully closed.

f. *Domestic Hot-Water Control:* T-1, located in the discharge of the convertor, controls V-1 to maintain the desired domestic hot-water temperature. T-2, located in the tank, starts the circulating pump if the temperature rises above a predetermined setting. Aquastat T-3 starts the recirculating pump if the temperature rises above the predetermined setting.

g. *Air Supply:* During the day, seven-day time clock C-1 is energized. It, in turn, energizes EP-1 which positions VA-1 to supply 15 pounds of air to day-night thermostats for day operation. During the night operation, C-1 is de-energized. It, in turn, de-energizes EP-1 which positions VA-1 to supply 20 pounds of air to day-night thermostats for night operation.

h. *Valve and Mechanical Rooms Exhaust-Fan Control:* T-1 cycles the fan to maintain a desired temperature of 85°F. A motorized damper, located at the outside air intake louver, interlocks with fan to open when fan is energized and to close when fan is de-energized. A low-limit thermostat, located downstream of the damper in the outside air duct, shuts off fan if temperature drops below 35°F.

i. *Fan-Coil, Unit-Heater, and Cabinet Unit-Heater Control:* T-1 cycles fans to maintain the space temperature. Aquastat T-2 stops the fans if the supply-water temperature falls below 90°F.

(12) *Service and Guarantee:* The entire control system shall be serviced and maintained in first-class condition by the control manufacturer for a period of one year after acceptance at no extra cost to the Owner. At the end of the one-year guarantee period, the Control Contractor must be capable of furnishing emergency service within a normal requested time.

5. DUCTWORK:

(A) Duct thickness, duct breaking, duct joints (both longitudinal and transverse), duct hangers, and all general ductwork shall be in accordance with the recommendations of "Duct Manual and Sheet Metal Construction for Ventilating and Air-Conditoning Systems," as prepared by the Sheet Metal and Air-Conditioning Contractors' National Association.

(B) Ductwork shall be galvanized steel, manufactured in gauges recommended in the above manual.

(C) Duct fittings shall be equivalent to Air Distribution Institute Standard fittings as a minimum requirement.

(D) Where mains split, splitter dampers must be furnished.

(E) All duct joints are to be airtight at ½-inch water pressure.

(F) *Turning Vanes:* ________ or ________ air-turns are to be used in all elbows except round pipes.

(G) EXTERNAL DUCT INSULATION*:*

(1) *Exposed Ductwork:* Supply ducts and outside air ducts shall be insulated on the outside with 1-inch-thick J-M No. 814, nonflexible, SPIN-GLAS fiber glass duct insulation. Insulation shall have factory-applied facing—FSK (Foil Skrim Kraft). Insulation shall have an average thermal conductivity not to exceed .23 Btu-inch per square foot per °F per hour at a mean temperature of 75°F. All insulation shall be applied with edges tightly butted. Insulation shall be impaled on pins welded to the duct and secured with speed clips. Pins shall be clipped off close to speed clips. Spacing of pins shall be as required to hold insulation firmly against duct surface, but not less than one pin per square foot. All joints and speed clips shall be sealed with ________ 207 glass fabric set in ________ 30-35, on a 3-inch-wide strip of same facing adhered with ________ 85-20 adhesive.

(2) *Concealed Ductwork:* Supply ducts and outside air ducts shall be insulated on the outside with 1-inch-thick J-M flexible Microlite or approved equal. Insulation shall be cut slightly longer than the perimeter of the duct to ensure full thickness at corners. The insulation shall have an average thermal conductivity not to exceed 0.25 Btu-inch per square foot per °F per hour at a mean temperature of 75°F. All insulation shall be applied with edges tightly butted. Insulation shall be secured with ________ 85-20 adhesive. Adhesive shall be applied so that insulation conforms to duct surface uniformly and firmly. All joints shall be taped and sealed with 3-inch-wide strips of the facing applied with 85-20 adhesive.

(H) Where ducts change shape, enlarge, or reduce, transition is to be made with a maximum angle of 15 degrees, except where it is specifically shown otherwise.

(I) *Volume-Control Duct Dampers and Duct Damper Hardware:* Volume-control duct dampers shall be placed as indicated on drawings and constructed as shown in the SMACNA Duct Manual, Plate No. 28 and multiple-plate volume dampers SMACNA Manual, Plate No. 29. Vent-lok control hardware shall be appropriate for specific use.

(J) FIRE DAMPERS*:*

(1) Fire dampers shall be installed at locations indicated on the Plans, in full conformance with NFPA Bulletin No. 90-A, and in complete accordance with city, state, and local codes.

(2) All fire dampers shall bear label of UL and be listed under the continuing inspection service of UL, where applicable, and shall have been successfully tested for 1½ hours, up to 1800°F. In mounting conditions not in conformance with UL testing, units shall be built in full conformance with standards of the American Insurance Association and the NFPA Bulletin No. 90-A.

(3) Fire dampers shall be as manufactured by "Fire-Seal" damper. Dampers shall be provided with interlocking blades to form a solid coating of steel when closed. When open, the blades are to be completely concealed in the head of the frame, allowing 100-percent-free undisturbed airflow with minimum turbulence. Entire assembly shall be galvanized for corrosion resistance and shall conform to ASTM Specification A-90-63-T.

(K) *Control Dampers:* Control dampers shall be furnished to the Mechanical Contractor by the Control Manufacturer and shall be installed in accordance with the Control Manufacturer's instructions in the duct system.

(L) *Canvas Connections:* Canvas connections shall be Vent-Fabric "Vent-Fab," 20-ounce waterproof and fireproof canvas approved by Underwriters' Laboratories. Each air-handling unit return and supply shall be connected with canvas connection.

(M) *Sawdust Ductwork:* Sawdust ductwork shall be round, 16 gauge with fittings as shown. Ductwork shall be equal to ______. All joints shall be welded air-tight. Welding frame fume exhaust shall be of same material.

(N) Flexible metal exhaust duct shall be equal to ______ or ______.

(O) This Contractor shall provide all louvers equal to Air Balance, 4 inches deep, rain and storm check, and of all-aluminum construction. Installation in wall shall be done by the General Contractor.

(P) *Paint Spray Booth Vent:* Paint spray booth vent shall be fabricated of No. 20 gauge galvanized sheet steel equal to vent manufactured by ______. Vent shall be furnished with a roof flange and weather canopy with rain guard, as manufactured by ______. See drawings for support detail.

6. WIRING DIAGRAMS:

(A) This Contractor shall furnish to the Architect for the Architect's approval five (5) copies of wiring diagrams (i.e., all individual wires diagrammed) showing the complete control wiring system diagram for the heating and ventilating system. The Contractor shall furnish a framed, glass-protected copy of this wiring diagram for the complete system *as installed.* This Contractor shall place this wiring diagram in the Equipment Room.

7. TESTS, CLEANING, AND GUARANTEE:

(A) This Contractor shall provide all pumps, gauges, and other instruments necessary to perform tests as required.

(B) This Contractor shall hydrostatically test the piping of the steam system and heating-water system, to a pressure of 150 percent of the system working pressures. The pressure test shall be for at least 8 hours, at which time pressure shall remain constant without additional pumping. After satisfactorily completing tests and before permanently connecting equipment, this Contractor shall blow and flush piping thoroughly so that interiors of all piping shall be free of foreign matter. All traps, strainers, etc., shall be cleaned at the time of flushing.

(C) The Contractor shall adjust and regulate the completed system under actual heating and ventilating conditions to produce a satisfactory system. All automatic temperature controls shall be adjusted for satisfactory operation during the first heating and ventilating seasons. This Contractor shall make all necessary adjustments during the first heating and ventilating seasons without additional cost to the Owner. (This does not mean that the Heating and Ventilating Contractor is responsible for any negligence of operation by the Owner.)

(A) This Contractor shall furnish complete instructions covering the operation of the heating, ventilating, and control systems.

(B) A framed, glass-protected copy of operating instructions is to be placed in the Mechanical Equipment Room by this Contractor.

9. PERFORMANCE TESTS:

(A) This Contractor shall provide all necessary instruments to perform tests as required.

(B) The Heating and Ventilating Contractor shall conduct the following tests upon completion of installation of the system under the direction of the Architect.

(1) *Air Distribution System:* Performance test after proper balancing of system, showing airflow measurements through each supply and return.

(2) *Ventilation System:* Performance test after proper balancing, showing airflow measurements through each exhaust grille.

(3) *Heating System:* Operating test of the entire system during cold weather with fiinal adjustment to outdoor design conditions as necessary and to the heating system. In the event the weather prevents testing before acceptance of the building, the building will be accepted subject to successful completion of the above tests.

(4) *Air-Handling Equipment:* Airflow tabulation and listing of outlet and inlet temperatures and heating medium inlet-outlet temperatures.

(C) Performance tests shall include the complete heating and ventilating systems and all their parts, including thermostatic and electrical controls, in order to determine that the systems are in compliance with the Contract. Tests shall show that the heating and ventilating systems are acceptable before the installation is approved for acceptance by the Owner. The Contractor shall furnish the Owner with four (4) copies of the finds of the approved texts, including tabulation of all readings for the job and computations. Final payment to the Contractor, less an amount to cover the cost of certain tests, shall be made when weather conditions have caused postponement of certain tests.

(D) The heating units shall be checked for performance to determine that the Specifications are met in every respect.

(E) This Contractor shall guarantee all materials and equipment installed by him, and all workmanship for a period of one year after final acceptance of the heating work against all defects occurring during that period.

(F) The heating and ventilating system shall be tested under operating conditions for a period of five 8-hour days or as necessary to demonstrate that the requirements of the Contract are fulfilled.

(G) This Contractor shall adjust and regulate the completed system under actual heating conditions to produce a satisfactory system. All automatic temperature controls shall be adjusted for satisfactory operation in the first heating season. This Contractor shall make all necessary adjustments in the first heating season without additional cost to Owner. (This does not mean that the Heating Contractor is responsible for any negligence in operation by Owner.)

10. MARKING OF CONTROLS, VALVES, AND ELECTRIC CONTROLS:

(A) All valves, electric controls, and electric starters and switches are to be marked with permanent metal tags. The tags shall be black enamel finish with engraved, white enameled letters that are at least ¼ inch high.

(B) The location at ceiling of all damper electric controls and balancing valves and duct balancing dampers shall be marked above the ceiling with tag, as described above and screwed to the metal ceiling runners at each location.

11. TEMPORARY HEATING:

(A) The General Contractor shall be responsible for furnishing temporary heating throughout the period of construction of the building.

12. PAINTING:

(A) Factory-painted equipment and materials shall receive primer coat and two coats of enamel, factory applied.

(B) Factory-painted equipment shall be touched up as necessary and where factory-painted color does not match the Architect's color scheme in finished areas.

(C) Factory-primed equipment and materials shall be painted with two coats of enamel in accordance with the Painting Section of these Specifications. This paint shall be applied by the Painting Contractor who shall use experienced painters for this purpose in accordance with the Painting Section of these Specifications.

(D) Unpainted equipment and materials shall receive one coat of primer and two coats of enamel in accordance with the requirements of the Painting Section of these Specifications for exposed metal. This paint shall be applied by the Painting Contractor who shall use experienced painters for this purpose in accordance with the Painting Section of these Specifications.

(E) The inside of all ductwork visible in finished building shall be painted dull black, the paint to be applied in accordance with Painting Section of the Specifications.

(F) All colors shall be as approved by the Architect.

13. TEMPORARY POWER:

(A) The General Contractor shall provide, as required, 208/120-volt, single-phase, 60-hertz temporary power for use of the Heating and Ventilating Contractor.

14. LUBRICATION PRIOR TO START-UP:

(A) Prior to start-up, the Heating and Ventilating Contractor shall fully lubricate all equipment under this Section in accordance with the respective manufacturers' recommendations.

Supply and Return Air Outlets

Manufacturers of diffusers, registers, and grilles usually provide tables that show the proper size according to CFM requirements. Typical examples are shown herein. All Tables courtesy York Air Conditioning & Refrigeration, Division of Borg-Warner.

Table B-1. Side Wall Registers

CFM	Size	Throw-Feet
60	10 x 4	10
80	10 x 6	12
100	14 x 6	14
120	14 x 6	16
140	16 x 6	16
160	16 x 6	18
180	20 x 6	14
200	20 x 6	16
220	20 x 6	18
240	20 x 6	20
260	24 x 6	20
280	30 x 6	18
300	30 x 6	20

Table B-2. Round Ceiling Diffusers

CFM	Diameter (in)
100	6
120	8
140	8
160	8
180	8
200	10
220	10
240	10

Table B-3. Floor Outlets

CFM	Size	Spread (ft)
60	2¼ x 10	7
80	2¼ x 12	11
100	2¼ x 14	11
120	4 x 10	11
140	4 x 10	13
160	4 x 12	13

Table B-4. Return Air Grilles

CFM	Free Area Sq in	Side Wall Return Grilles	Floor Grilles
60- 140	40	10 x 6	4 x 14
140- 170	48	12 x 6	4 x 18 or 6 x 10
170- 190	55	10 x 8	4 x 18 or 6 x 12
190- 235	67	12 x 8	6 x 14
235- 260	74	18 x 6	6 x 16 or 8 x 14
260- 370	106	12 x 12	8 x 20
370- 560	162	18 x 12	8 x 30
560- 760	218	24 x 12	10 x 30 or 12 x 24
760- 870	252	18 x 18	12 x 30
870- 960	276	30 x 12	12 x 30
960-1170	340	24 x 18	
1170-1470	423	30 x 18	18 x 30 14 x 30
1470-1580	455	24 x 24	20 x 30
1580-1770	510	36 x 18	22 x 30
1770-1990	572	30 x 24	24 x 30
1990-2400	690	36 x 24	24 x 36
2400-3020	870	36 x 30	30 x 36

Table B-5. Recommended and Maximum Velocities For Residences Duct

	Recommended Velocities (FPM)	Maximum Velocities (FPM)
Filters (total face area)	250	300
Main Ducts (net free area)	700-900	800-1200
Branch Ducts (net free area)	600	700-1000
Branch Risers (net free area)	500	650-800

Table B-6. Recommended Return Intake Face Velocities

Intake Location	Velocity Over Gross Area (FPM)
Above occupied zone	800-Up
Within occupied zone, not near seats	600-800
Within occupied zone, near seats	400-600
Door or wall louvers	200-300
Undercutting of doors (through undercut area)	200-300

Table B-7. Rectangular Main Ducts 6-in Size

		Recommended (800 FPM)		Maximum (1000 FPM)	
6-In Size	Equiv Dia in Inches	Approx CFM	Approx Inches-Static Pressure Drop Per 100 Equivalent Feet	Approx CFM	Approx Inches-Static Pressure Drop Per 100 Equivalent Feet
4 x 6	5.3	120	.25	150	.35
5 x 6	6.0	160	.20	200	.30
6 x 6	6.8	200	.17	250	.26
8 x 6	7.5	250	.15	310	.23
10 x 6	8.4	310	.13	380	.20
12 x 6	9.1	360	.12	450	.18
14 x 6	9.8	420	.11	520	.16
16 x 6	10.4	460	.105	600	.15
18 x 6	11.0	530	.094	650	.14
20 x 6	11.5	590	.089	710	.135
22 x 6	12.0	630	.084	790	.13
24 x 6	12.4	680	.080	840	.125
26 x 6	12.8	710	.076	900	.120
28 x 6	13.2	750	.073	940	.115
30 x 6	13.6	800	.071	1000	.110
32 x 6	14.0	850	.069	1090	.105
34 x 6	14.4	900	.067	1170	.100
36 x 6	14.7	940	.065	1200	.096

Table B-8. Rectangular Main Ducts 8-in Size

8-In Size	Equiv Dia in Inches	Recommended (800 FPM)		Maximum (1000 FPM)	
		Approx CFM	Approx Inches-Static Pressure Drop Per 100 Equivalent Feet	Approx CFM	Approx Inches-Static Pressure Drop Per 100 Equivalent Feet
4 x 8	6.1	160	.19	200	.29
5 x 8	6.9	210	.165	260	.25
6 x 8	7.5	250	.15	300	.23
8 x 8	8.8	340	.13	420	.19
10 x 8	9.8	420	.11	520	.17
12 x 8	10.7	500	.095	630	.145
14 x 8	11.5	580	.089	710	.135
16 x 8	12.2	650	.083	810	.125
18 x 8	12.9	715	.078	900	.118
20 x 8	13.5	790	.073	990	.110
22 x 8	14.1	870	.069	1100	.105
24 x 8	14.6	930	.066	1190	.100
26 x 8	15.2	1000	.063	1290	.095
28 x 8	15.6	1080	.060	1375	.090
30 x 8	16.1	1150	.058	1450	.086
32 x 8	16.5	1200	.056	1500	.085
34 x 8	17.0	1250	.055	1590	.083
36 x 8	17.4	1320	.054	1620	.080

Table B-9. Rectangular Main Ducts 10-in Size

10-In Size	Equiv Dia in Inches	Recommended (800 FPM)		Maximum (1000 FPM)	
		Approx CFM	Approx Inches-Static Pressure Drop Per 100 Equivalent Feet	Approx CFM	Approx Inches-Static Pressure Drop Per 100 Equivalent Feet
4 x 10	6.8	200	.170	260	.260
6 x 10	8.4	310	.130	380	.200
8 x 10	9.8	420	.110	520	.170
10 x 10	11.0	520	.093	670	.140
12 x 10	12.1	630	.081	800	.130
14 x 10	13.0	730	.075	920	.120
16 x 10	13.8	820	.071	1020	.110
18 x 10	14.8	940	.064	1200	.098
20 x 20	15.5	1050	.061	1325	.092
22 x 10	16.0	1150	.058	1400	.088
24 x 10	16.7	1250	.055	1500	.084
26 x 10	17.3	1350	.053	1650	.080
28 x 10	17.8	1425	.052	1725	.078
30 x 10	18.4	1500	.049	1800	.074
32 x 10	19.0	1575	.048	1950	.071
34 x 10	19.5	1650	.046	2050	.070
36 x 10	20.0	1725	.044	2150	.068
40 x 10	21.0	1900	.042	2300	.064
44 x 10	21.8	2100	.040	2600	.061
48 x 10	22.8	2250	.038	2900	.058
54 x 10	23.8	2400	.036	3100	.055
60 x 10	25.0	2800	.034	3400	.051

Table B-10. Rectangular Main Ducts 12-in Size

		Recommended (800 FPM)		Maximum (1000 FPM)	
12-In Size	Equiv Dia in Inches	Approx CFM	Approx Inches-Static Pressure Drop Per 100 Equivalent Feet	Approx CFM	Approx Inches-Static Pressure Drop Per 100 Equivalent Feet
4 x 12	7.4	240	.150	300	.230
6 x 12	9.1	370	.120	450	.180
8 x 12	10.7	500	.095	630	.145
10 x 12	12.1	630	.081	800	.130
12 x 12	13.0	720	.075	930	.120
14 x 12	14.3	880	.068	1100	.100
16 x 12	15.2	1000	.062	1250	.095
18 x 12	16.1	1150	.058	1450	.088
20 x 12	17.0	1250	.055	1600	.082
22 x 12	17.8	1350	.052	1750	.078
24 x 12	18.4	1450	.049	1850	.073
26 x 12	19.2	1600	.048	2000	.070
28 x 12	19.8	1700	.046	2150	.068
30 x 12	20.3	1800	.044	2250	.066
32 x 12	21.0	1900	.042	2400	.064
34 x 12	21.5	2000	.041	2500	.061
36 x 12	22.2	2100	.039	2700	.058
40 x 12	23.2	2300	.037	2900	.056
44 x 12	24.0	2500	.036	3150	.054
48 x 12	25.3	2800	.033	3500	.050
54 x 12	26.5	3100	.032	3850	.047
60 x 12	27.8	3400	.030	4150	.045

Table B-11. Round Branch Ducts

	Recommended (600 FPM)		Maximum (900 FPM)	
Dia Size in Inches	Approx CFM	Approx Inches-Static Pressure Drop Per 100 Equivalent Feet	Approx CFM	Approx Inches-Static Pressure Drop Per 100 Equivalent Feet
4	52	.20	80	.40
5	82	.15	125	.31
6	118	.12	180	.25
7	160	.095	240	.20
8	210	.080	320	.17
9	270	.070	400	.15
10	323	.060	490	.15
12	470	.050	700	.105

Table B-12. Branch Risers

Size	Equiv Dia in Inches	Recommended (500 FPM)		Maximum (700 FPM)	
		Approx CFM	Approx Inches-Static Pressure Drop Per 100 Equivalent Feet	Approx CFM	Approx Inches-Static Pressure Drop Per 100 Equivalent Feet
2¼ x 12	5.3	76	.098	108	.18
3¼ x 8	5.5	83	.090	118	.17
3¼ x 10	6.0	98	.082	138	.15
3¼ x 12	6.5	114	.077	158	.14
3¼ x 14	7.2	140	.065	200	.12

Notes
1. Do not interpolate static pressure drop between different sizes of ducts.
2. It is permissable to interpolate between recommended and maximum columns for static pressure drop for the same size duct if CFM falls between the CFM's shown.

Table B-13. Rectangular Return Ducts 6-in Size

6-In Size	Equiv Dia in Inches	Recommended (500 FPM)		Maximum (700 FPM)	
		Approx CFM	Approx Inches-Static Pressure Drop Per 100 Equivalent Feet	Approx CFM	Approx Inches-Static Pressure Drop Per 100 Equivalent Feet
4 x 6	5.3	75	.095	110	.180
5 x 6	6.0	95	.081	140	.150
6 x 6	6.8	130	.070	180	.130
8 x 6	7.5	150	.062	210	.120
10 x 6	8.4	190	.053	260	.100
12 x 6	9.1	220	.048	320	.090
14 x 6	9.8	275	.043	375	.083
16 x 6	10.4	310	.040	410	.080
18 x 6	11.0	330	.038	450	.073
20 x 6	11.5	370	.036	510	.068
22 x 6	12.0	390	.035	550	.065
24 x 6	12.4	410	.033	590	.062
26 x 6	12.8	450	.032	620	.060
28 x 6	13.2	475	.031	650	.058
30 x 6	13.6	500	.030	700	.055
32 x 6	14.0	520	.029	750	.053
34 x 6	14.4	550	.028	780	.051
36 x 6	14.7	580	.027	800	.050

Table B-14. Rectangular Return Ducts 8-in Size

8-In Size	Equiv Dia in Inches	Recommended (500 FPM)		Maximum (700 FPM)	
		Approx CFM	Approx Inches-Static Pressure Drop Per 100 Equivalent Feet	Approx CFM	Approx Inches-Static Pressure Drop Per 100 Equivalent Feet
4 x 8	6.1	100	.080	140	.150
5 x 8	6.9	130	.070	180	.130
6 x 8	7.5	150	.062	210	.120
8 x 8	8.8	210	.050	290	.095
10 x 8	9.8	275	.043	360	.084
12 x 8	10.7	320	.040	440	.074
14 x 8	11.5	360	.037	500	.070
16 x 8	12.2	400	.035	570	.064
18 x 8	12.9	450	.032	630	.060
20 x 8	13.5	490	.031	690	.056
22 x 8	14.1	550	.029	760	.052
24 x 8	14.6	575	.027	810	.051
26 x 8	15.2	610	.027	870	.049
28 x 8	15.6	680	.025	910	.048
30 x 8	16.1	710	.024	1000	.045
32 x 8	16.5	750	.023	1050	.043
34 x 8	17.0	780	.022	1100	.042
36 x 8	17.4	810	.022	1150	.041

Table B-15. Rectangular Return Ducts 10-in Size

10-In Size	Equiv Dia in Inches	Recommended (500 FPM)		Maximum (700 FPM)	
		Approx CFM	Approx Inches-Static Pressure Drop Per 100 Equivalent Feet	Approx CFM	Approx Inches-Static Pressure Drop Per 100 Equivalent Feet
4 x 10	6.8	135	.070	175	.130
6 x 10	8.4	190	.053	260	.100
8 x 10	9.8	275	.043	360	.084
10 x 10	11.0	330	.038	460	.072
12 x 10	12.1	400	.034	570	.063
14 x 10	13.0	470	.032	640	.058
16 x 10	13.8	510	.029	720	.053
18 x 10	14.8	600	.027	830	.050
20 x 10	15.5	650	.025	920	.047
22 x 10	16.0	700	.024	980	.045
24 x 10	16.7	750	.023	1050	.043
26 x 10	17.3	810	.022	1150	.041
28 x 10	17.8	880	.021	1200	.039
30 x 10	18.4	920	.021	1300	.038
32 x 10	19.0	975	.020	1400	.037
34 x 10	19.5	1025	.019	1450	.036
36 x 10	20.0	1100	.018	1500	.035
40 x 10	21.0	1200	.017	1700	.033
44 x 10	21.8	1300	.017	1800	.032
48 x 10	22.8	1400	.016	1950	.030
54 x 10	23.8	1500	.015	2150	.028
60 x 10	25.0	1750	.014	2300	.026

Table B-16. Rectangular Return Ducts 12-in Size

12-In Size	Equiv Dia in Inches	Recommended (500 FPM)		Maximum (700 FPM)	
		Approx CFM	Approx Inches-Static Pressure Drop Per 100 Equivalent Feet	Approx CFM	Approx Inches-Static Pressure Drop Per 100 Equivalent Feet
4 x 12	7.4	145	.064	210	.120
6 x 12	9.1	220	.048	320	.090
8 x 12	10.7	320	.040	440	.074
10 x 12	12.1	400	.034	570	.063
12 x 12	13.0	460	.031	630	.058
14 x 12	14.3	530	.028	780	.051
16 x 12	15.2	620	.026	880	.048
18 x 12	16.1	710	.024	1000	.044
20 x 12	17.0	790	.023	1100	.041
22 x 12	17.8	880	.022	1200	.040
24 x 12	18.4	910	.021	1300	.038
26 x 12	19.2	990	.020	1400	.036
28 x 12	19.8	1050	.019	1450	.035
30 x 12	20.3	1125	.018	1550	.034
32 x 12	21.3	1200	.018	1650	.033
34 x 12	21.5	1275	.017	1750	.032
36 x 12	22.2	1325	.017	1850	.031
40 x 12	23.2	1450	.016	2050	.029
44 x 12	24.0	1500	.015	2150	.028
48 x 12	25.3	1700	.014	2450	.026
54 x 12	26.5	1900	.013	2650	.024
60 x 12	27.8	2100	.012	2900	.023

Table B-17. Round Return Branch Ducts

Diameter in Inches	Recommended (500 FPM)		Maximum (700 RPM)	
	Approx CFM	Approx Inches-Static Pressure Drop Per 100 Equivalent Feet	Approx CFM	Approx Inches-Static Pressure Drop Per 100 Equivalent Feet
4	42	.140	60	.250
5	68	.105	95	.190
6	98	.081	140	.160
7	135	.069	185	.130
8	175	.058	250	.110
9	220	.050	310	.092
10	275	.045	380	.080
12	390	.035	550	.065

Table B-18. Return Air Grille and Register Sizing

Size or Equivalent	Free Area (Sq Ft)	Air Capacities CFM	
		Recommended Velocity (500 FPM)	Maximum Velocity (700 FPM)
10 x 6	.275	137	192
12 x 6	.336	168	235
10 x 8	.382	191	267
12 x 8	.467	233	327
18 x 6	.517	258	362
12 x 12	.730	365	511
18 x 12	1.120	560	784
24 x 12	1.517	758	1062
18 x 18	1.743	871	1220
30 x 12	1.911	955	1338
24 x 18	2.336	1168	1635
30 x 18	2.942	1471	2059
24 x 24	3.156	1578	2209
36 x 18	3.549	1774	2484
30 x 24	3.974	1987	2782
36 x 24	4.792	2396	3354

APPENDIX C

Duct Sizing Chart

Example of Use:

A small utility room requires 35 CFM of air at a velocity of 400 fpm. What size duct must be used to deliver this volume of air?

1. Find 35 CFM on the CFM scale at the left of the chart.
2. Move horizontally to the right, to the 400 fpm line.
3. At the intersection of 35 CFM and 400 fpm, locate the duct diameter line.
4. Read 4-inch duct diameter.
5. Move downward from the intersection to the friction loss scale and read .09-inch water friction loss.

Chart C-1. Duct Sizing Chart

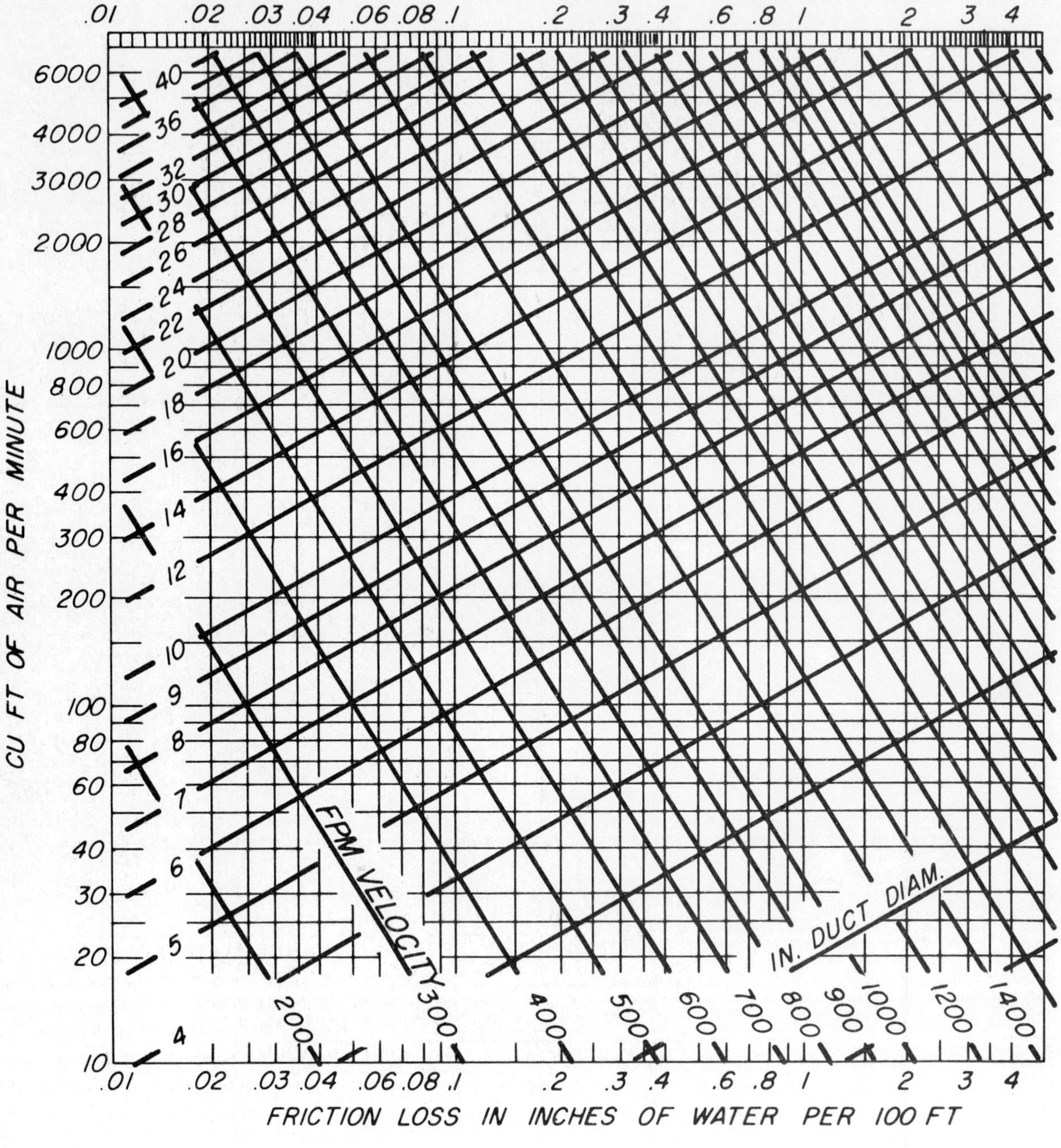

Chart C-2. Equivalent Rectangular Duct Chart

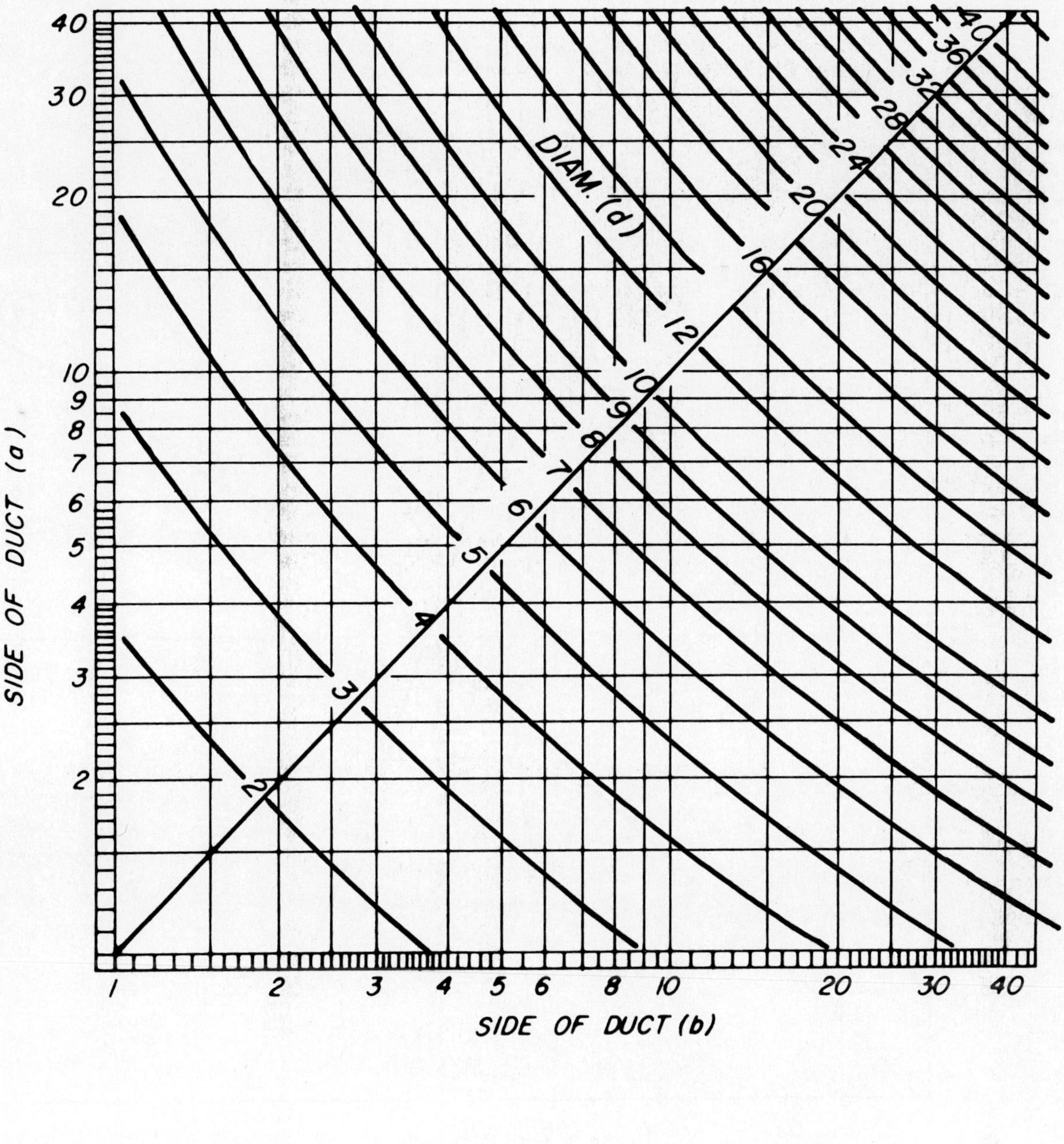

Chart C-3. Use of Duct Sizing Chart

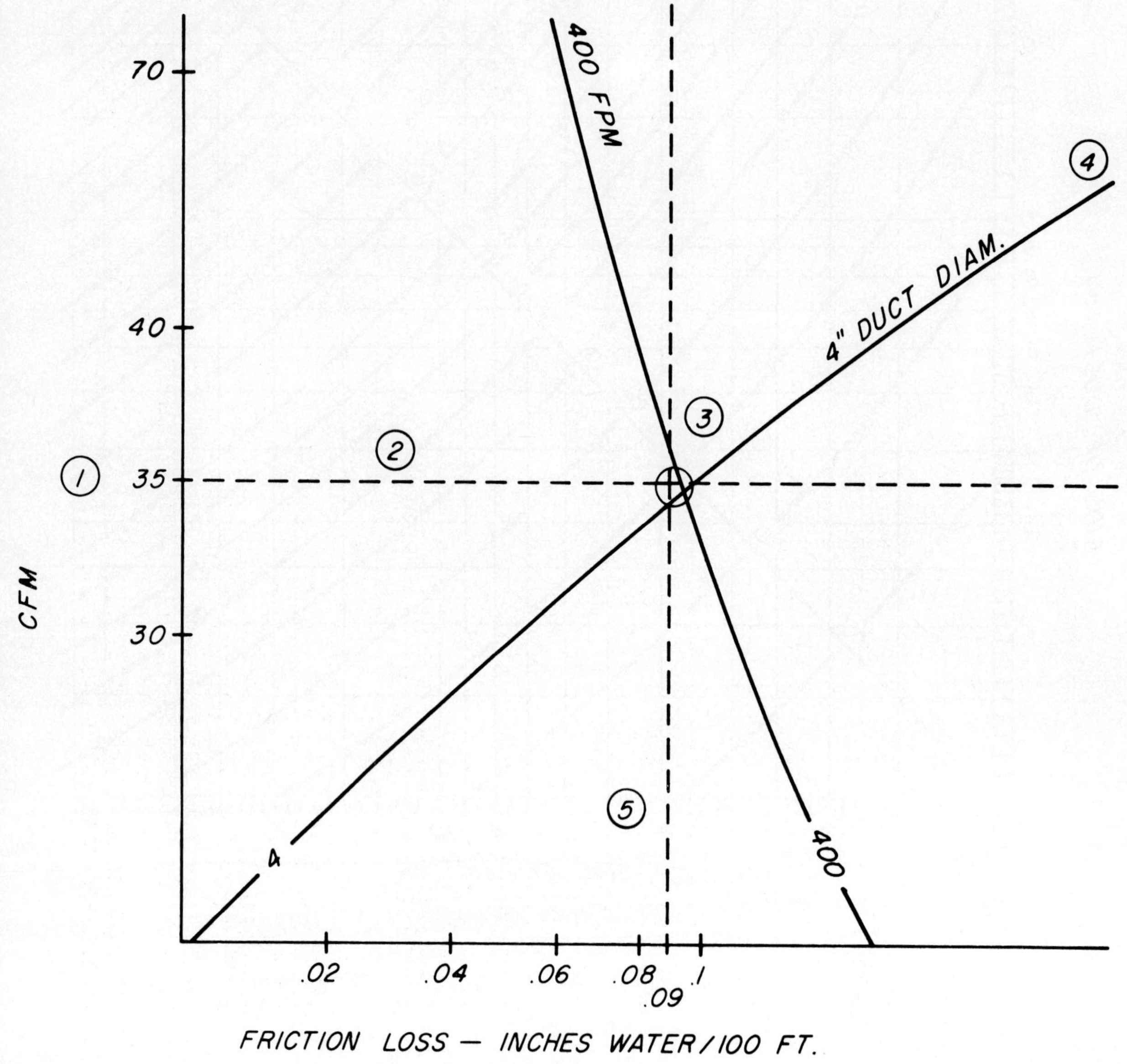

Chart C-4. Friction Loss Chart

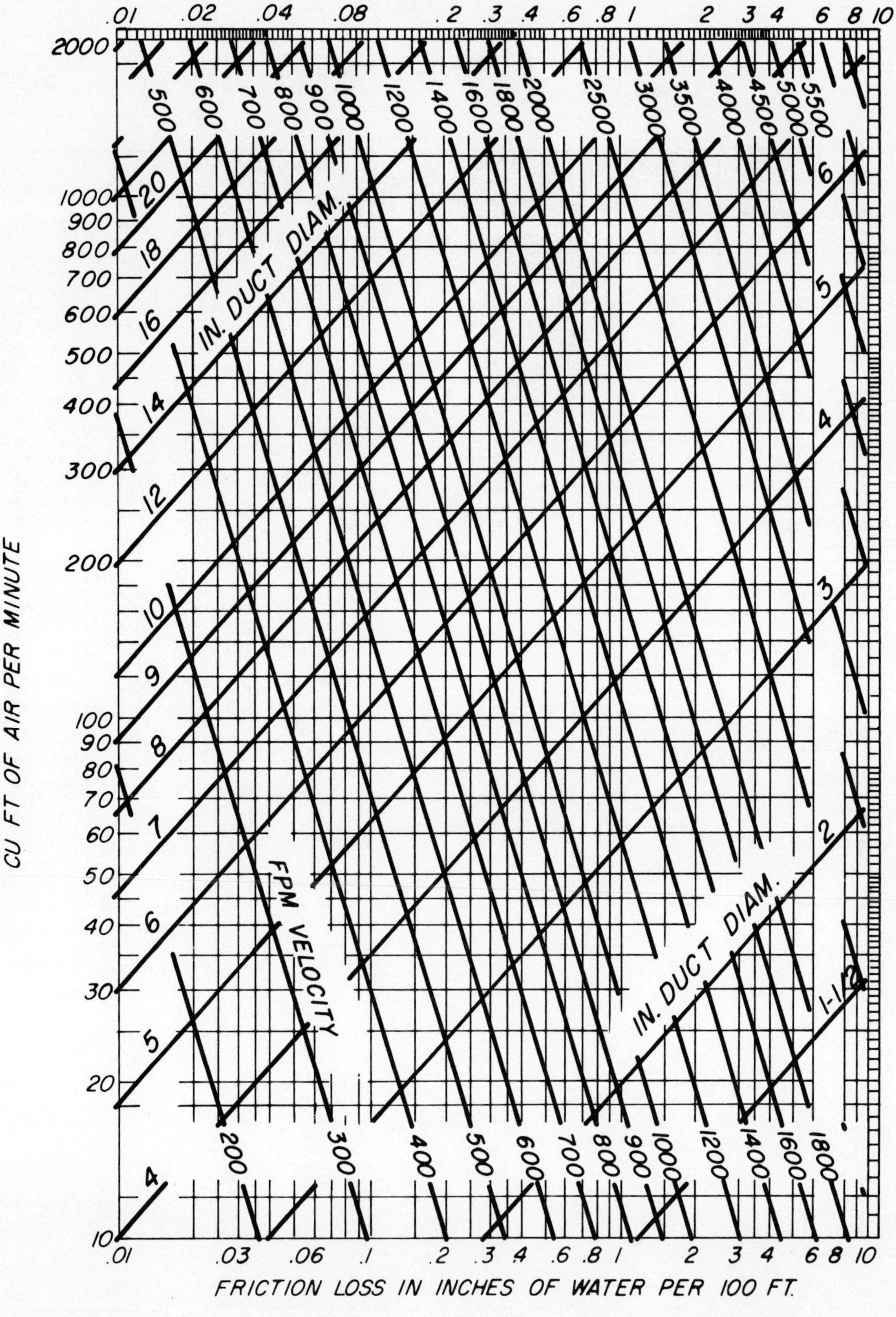

Table C-1. Comparative Analysis of Outlets

	FLOOR	BASEBOARD	LOW SIDEWALL	HIGH SIDEWALL	CEILING
COOLING PERFORMANCE	Excellent	Excellent if used with perimeter systems	Excellent if designed to discharge upward	Good	Good
HEATING PERFORMANCE	Excellent	Excellent if used with perimeter systems	Excellent if used with perimeter systems	Fair—should not be used to heat slab houses in Northern climates	Good—should not be used to heat slab houses in Northern climates
INTERFERENCE WITH DECOR	Easily concealed because it fits flush with the floor and can be painted to match	Not quite so easy to conceal because it projects from the baseboard	Hard to conceal because it is usually in a flat wall	Impossible to conceal because it is above furniture and in a flat wall	Impossible to conceal but special decorative types are available
INTERFERENCE WITH FURNITURE PLACEMENT	No interference—located at outside wall under a window	No interference—located at outside wall under a window	Can interfere because air discharge is not vertical	No interference	No interference
INTERFERENCE WITH FULL-LENGTH DRAPES	No interference—located 6 or 7 inches from the wall	When drapes are closed, they will clover the outlet	When located under a window, drapes will close over it	No interference	No interference
INTERFERENCE WITH WALL-TO-WALL CARPETING	Carpeting must be cut	Carpeting must be notched	No interference	No interference	No interference
OUTER COST	Low	Medium	Low to medium, depending on the type selected	Low	Low to high—wide variety of types are available
INSTALLATION COST	Low because the sill need not be cut	Low when fed from below—sill need not be cut	Medium—requires wall stack and cutting of plates	Low on furred ceiling system: high when using under-floor system	High because attic ducts require insulation

Courtesy Carrier Air Conditioning Co.

APPENDIX D

Floor Plans

The two floor plans at the back of this book are typical of the type of plan views a draftsman will be using. Thus, they are used in assignment problems in this book. A floor plan for a residential structure (Fig. D-1) and a business building (Fig. D-2) were chosen to permit a variety of HVAC examples to be used in the exercises.

Answer Sheet

Unit 1
1. False
2. True
3. True
4. True
5. False
6. True
7. True
8. True
9. False
10. True

Unit 4, Assignment 4
A. 6′-4″
B. 12′-8″
C. 2′-6″
E. 0′-6″
F. 1′-0″
G. 2′-0″
H. 0′-4″
I. 1′-9″

Unit 6, Assignment 5
a. False
b. True
c. False
d. True
e. False

Unit 7, Assignment 4
a. False
b. False
c. True
d. False
e. False

Unit 8, Assignment 5
a. supply-air outlet
b. grille
c. fixed-louver
d. ceiling diffuser
e. adjustable-louver

Unit 9, Assignment 4
a. False
b. False
c. True
d. True
e. False

Unit 10, Assignment 5
a. isometric projection
b. isometric planes
c. isometric axes
d. isometric lines
e. circle

Unit 11, Assignment 4
a. False
b. True
c. True
d. False
e. True

Unit 12, Assignment 4
a. sleeve
b. chase
c. visualization
d. cross section
e. recessed

Unit 13, Assignment 3
1. c
2. f
3. b
4. e
5. h
6. a
7. g
8. d

Unit 14, Assignment 3
a. False
b. True
c. True
d. False
e. True

Unit 15, Assignment 5
a. False
b. True
c. True
d. False
e. True

Unit 16, Assignment 4
a. True
b. False
c. True
d. False
e. True

Unit 17, Assignment 6
a. True
b. True
c. False
d. True
e. False

Unit 18, Assignment 4
a. False
b. True

Unit 19, Assignment 3
a. fans
b. centrifugal
c. vaneaxial, tubeaxial
d. propeller
e. direct connection

Unit 20, Assignment 3
a. liquify
b. double-pipe
c. shell-and-coil
d. air
e. propeller-type

Unit 21, Assignment 6
a. True
b. False
c. True
d. False
e. True

Unit 22, Assignment 4
a. True
b. False
c. True
d. True
e. False

Unit 23, Assignment 3
a. solid-state circuit board
water-level probes
fill valve
control transformer
b. heating element
control relay
control transformer
c. flush valve
fill valve
control transformer
clock timer
drain
d. overflow safety probe,
overflow drain

Index

Index

L

M

N

O

P

R